2006

INTERNATIONAL RESIDENTIAL CODE®

STUDY COMPANION

2006 International Residential Code
Study Companion

ISBN-13: 978-1-58001-519-0
ISBN-10: 1-58001-519-0

Cover Design:	Linda Cohoat
Publications Manager:	Mary Lou Luif
Project Editor:	Roger Mensink
Illustrator/Interior Design:	Mike Tamai
Typesetting:	Yolanda Nickoley
Manager of Development:	Doug Thornburg

INTERNATIONAL
CODE COUNCIL®

First Printing:

Printed in the United States of America

TABLE OF CONTENTS

INTRODUCTION

This study companion provides practical learning assignments for independent study of the provisions of the 2006 *International Residential Code (*IRC®). The independent study format affords a method for the student to complete the program in an unregulated time period. Progressing through the workbook, the learner can measure his or her level of knowledge by using the exercises and quizzes provided for each study session.

The workbook is also valuable for instructor-led programs. In jurisdictional training sessions, community college classes, vocational training programs and other structured educational offerings, the study guide and the IMC can be the basis for classroom instruction.

All study sessions begin with a general learning objective specific to the session, the specific code sections or chapter under consideration and a list of questions summarizing the key points of study. Each session addresses selected topics from the IRC and includes code text, a commentary on the code provisions, illustrations representing the provisions under discussion and multiple choice questions that can be used to evaluate the student's knowledge. Before beginning the quizzes, the student should thoroughly review the IRC, focusing on the key points identified at the beginning of each study session.

The workbook is structured so that after every question the student has an opportunity to record his or her response and the corresponding code reference. The correct answers are found in the back of the workbook in the answer key.

This study companion was developed by Douglas W. Thornburg, AIA, C.B.O., Technical Director of Product Development for the International Code Council. In addition to authoring numerous educational texts and resource materials, he instructs seminars nationally on both the IBC and IRC. Mr. Thornburg has over 25 years experience in the application and enforcement of building codes.

The information presented in this publication is believed to be accurate; however, it is provided for informational purposes only and is intended for use only as a guide. As there is a limited discussion of selected code provisions, the code itself should always be referenced for more complete information. In addition, the commentary set forth may not necessarily represent the views of any enforcing agency, as such agencies have the sole authority to render interpretations of the IRC.

Questions or comments concerning this study companion are encouraged. Please direct your comments to ICC at *studycompanion@iccsafe.org*.

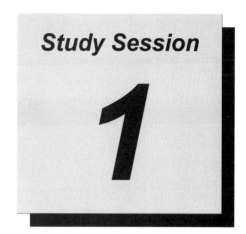

2006 IRC Chapters 1 and 43
Administration and Referenced Standards

OBJECTIVE: To obtain an understanding of the administrative provisions of the *International Residential Code*® (IRC®), including the scope and purpose of the code, duties of the building official, issuance of permits, submission of construction documents, inspection procedures, certificate of occupancy and referenced standards.

REFERENCE: Chapters 1 and 43, 2006 *International Residential Code*

KEY POINTS:
- What is the purpose and scope of the *International Residential Code*?
- When materials, methods of construction or other requirements are specified differently in separate provisions, which requirement shall govern?
- When there is a conflict between a general requirement and a specific requirement, which provision shall be applicable?
- When do the provisions of the appendix apply?
- How are existing buildings to be addressed?
- What term is applied to the building department? How are deputies appointed?
- What are the powers and duties of the building official in regard to the application and interpretation of the code? In regard to right of entry?
- What degree of liability does the building official have in regard to performance of their duties?
- Under what conditions may the building official grant modifications to the code?
- How may alternative materials, designs and methods of construction be approved?
- When is a permit required? What types of work are exempted from permits?
- Is work exempted from a permit required to comply with the provisions of the code?
- What is the process outlined for obtaining a permit?
- What information is required on the permit application?
- What conditions or circumstances would bring the validity of a permit into question?
- When does a permit expire? What must occur when a permit expires prior to completion of a building?
- When are construction documents required? When must plans be prepared by a registered design professional?

- How long must approved construction documents be retained by the building department?
- How are temporary buildings addressed? To what levels of conformance must they comply?
- What fees are set forth in the code? How are fees to be determined?
- What types of inspections are specifically required by the code? When are inspections required?
- When is a certificate of occupancy required? For what reasons is revocation permitted?
- When may a temporary certificate of occupancy be issued?
- What is the purpose of a board of appeals? Who shall serve on the board?
- What limitations are placed on the authority of the board of appeals?
- When should a stop-work order be issued?
- What are referenced standards? How are they applied to the IRC®?

Code Text: *Detached one- and two-family dwellings and multiple single-family dwellings (townhouses) not more than three stories above grade plane in height with a separate means of egress and their accessory structures shall comply with the* International Residential Code.

Discussion and Commentary: Many residential structures are exempt from the requirements of the *International Building Code®* and are regulated instead by a separate and distinct document, the *International Residential Code. The IRC contains prescriptive requirements for the construction of detached single-family dwellings, detached duplexes, townhouses and all structures accessory to such buildings. Limited to three stories in height with individual egress facilities, these residential buildings are fully regulated by the IRC for building, plumbing, mechanical, electrical and energy provisions.*

Single-family dwelling

Two-family dwelling

Townhouses

Accessory structures

In addition to the general requirements, a townhouse is also limited by definition in the *International Residential Code*. To fall under the scope of the IRC, each townhouse must be a single unit from the foundation to the roof, with at least two sides open to the exterior.

Code Text: *The provisions of the* International Residential Code for One- and Two-Family Dwellings *shall apply to the construction, alteration, movement, enlargement, replacement, repair, equipment, use and occupancy, location, removal and demolition of detached one- and two-family dwellings and townhouses not more than three stories above-grade in height with a separate means of egress and their accessory structures.*

Discussion and Commentary: The *International Residential Code* is intended to regulate the broad spectrum of construction activities associated with residential buildings and structures. The provisions address building planning and construction, as well as energy efficiency, mechanical, plumbing and electrical aspects.

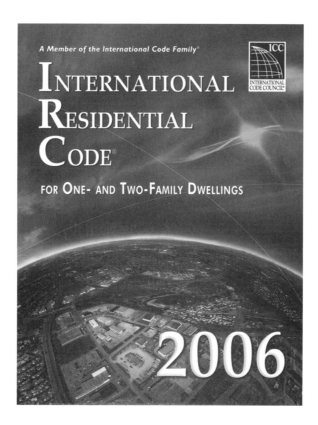

Accessory buildings—such as private garages, carports, sheds, playhouses and similar structures incidental to the major dwelling or townhouse—are also regulated by the IRC, provided each accessory structure is limited to 3,000 square feet in floor area and two stories in height.

Code Text: *Provisions in the appendices shall not apply unless specifically referenced in the adopting ordinance.*

Discussion and Commentary: The appendix chapters of the IRC address subjects that are inappropriate as a mandatory portion of the code. Rather, the appendices are optional, with each jurisdiction adopting all, some or none of the appendix chapters, depending on its needs for enforcement in any given area. There are various reasons why certain issues are placed in the appendix. Often, the provisions are limited in application or interest. Some appendix chapters are merely extensions of requirements set forth in the body of the code. Others address issues that are often thought of as outside of the scope of a traditional building code. For whatever reason, no appendix chapter is applicable unless specifally adopted.

IRC Appendix Chapters

Appendix A	Sizing and Capacities of Gas Piping
Appendix B	Sizing of Venting Systems Serving Appliances Equipped with Draft Hoods, Category I Appliances, and Appliances Listed for Use and Type B Vents
Appendix C	Exit Terminals of Mechanical Draft and Direct-Vent Venting Systems
Appendix D	Recommended Procedure for Safety Inspection of an Existing Appliance Installation
Appendix E	Manufactured Housing Used as Dwellings
Appendix F	Radon Control Methods
Appendix G	Swimming Pools, Spas and Hot Tubs
Appendix H	Patio Covers
Appendix I	Private Sewage Disposal
Appendix J	Existing Buildings and Structures
Appendix K	Sound Transmission
Appendix L	Permit Fees
Appendix M	Home Day Care – R-3 Occupancy
Appendix N	Venting Methods
Appendix O	Gray Water Recycling Systems
Appendix P	Sprinkling
Appendix Q	*International Residential Code* Electrical Provisions/National Electrical Code Cross Reference

---

Appendix chapters not adopted as a portion of a jurisdiction's building code may still be of value in application of the code. Provisions in the appendices might provide some degree of assistance in evaluating proposed alternative designs, methods or materials of construction.

Code Text: *The department of building safety is hereby created and the official in charge thereof shall be known as the building official. The building official shall be appointed by the chief appointing authority of the jurisdiction. In accordance with the prescribed procedures of this jurisdiction and witht he concurrenece of the appointing authority, the building official shall have the authority to appoint a deputy building offical, the related technical officers, inspectors, plan examiners and other employees. Such employees shall have powers as delegated by the building official.*

Discussion and Commentary: The building official is an appointed officer of the jurisdiction and charged with the administrative responsibilities of the department of building safety. It is not uncommon for the jurisdiction to use a different position title to identify the building official, such as Chief Building Inspector, Superintendent of Central Inspection, or Director of Code Enforcement. Regardless of the jurisdictional title, the code recognizes the individual in charge as the building official.

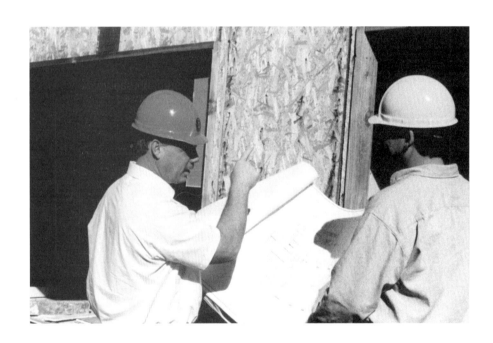

Inspectors, plan reviewers and other technical staff members are typically given some degree of authority to act for the building official in the decision-making process, including the making of appropriate interpretations on various provisions of the code.

Code Text: *The building official is hereby authorized and directed to enforce the provisions of the IRC. The building official shall have the authority to render interpretations of the IRC and to adopt policies and procedures in order to clarify the application of its provisions. Such interpretations, policies and procedures shall be in conformance with the intent and purpose of the IRC. Such policies and procedures shall not have the effect of waiving requirements specifically provided for in the IRC.*

Discussion and Commentary: The building official must be knowledgeable to the extent that he or she can rule on those issues that are not directly addressed or are unclear in the code. The basis for such a determination is the intent and purpose of the *International Residential Code*, which often takes some research to discover.

City of (Jurisdiction)

Department of Building Safety

Photo

_____
Name of individual

_____
Job function

The individual identified on the badge is a duly authorized employee of (the Jurisdiction) and is a designated representative of the Department of Building Safety.

Valid
through   _____   _____
                Date   Building Official

Proper Identification Mandated by Section R104.5
when Inspecting Structures or Premises

Although the IRC gives broad authority to the building official in interpreting the code, this authority also comes with great responsibility. The building official must restrict all decisions to the intent and purpose of the code; the waiving of any requirements is strictly prohibited.

Code Text: *The building official, member of the board of appeals or employee charged with the enforcement of the IRC, while acting for the jurisdiction in good faith and without malice in the discharge of the duties required by the IRC or other pertinent law or ordinance, shall not thereby be rendered liable personally and is hereby relieved from personal liability for any damage accruing to persons or property as a result of any act or by reason of an act or omission in the discharge of official duties.*

Discussion and Commentary: The protection afforded by the code regarding employee liability is limited to only those acts that occur in good faith. Absolute immunity from all tort liability is not provided where an employee acts maliciously. An employee is not relieved from personal liability where malice can be shown, and it is probable that the jurisdiction will not provide for the employee's defense.

Public officials should familiarize themselves with the laws of their respective states regarding their exposure to tort liability.

Topic: Alternative Materials

Category: Administration

Reference: IRC R104.11

Subject: Duties and Powers of Building Official

Code Text: *The provisions of the IRC are not intended to prevent the installation of any material or to prohibit any design or method of construction not specifically prescribed by the IRC, provided than any such alternative has been approved.*

Discussion and Commentary: The building official is granted broad authority in the acceptance of alternative materials, designs and methods of construction. Arguably the most important provision of the IRC, the intent is to implement the adoption of new technologies. Furthermore, it gives the code even more of the character of a performance code. The provisions encourage state-of-the-art concepts in construction, design and materials as long as they meet the performance level intended by the IRC. In a portion of the provision not quoted above, reference is also made to the use of any performance-based provisions of the other *International Codes*® as acceptable alternates.

Building official may approve alternative materials, design and methods of construction, if:

- Proposed alternative is satisfactory

- Proposed alternative complies with intent of code

- Material, method or work is equivalent in:

  (1) Quality

  (2) Strength

  (3) Effectiveness

  (4) Fire Resistance

  (5) Durability

  (6) Safety

Advisable for building official to:

- Require sufficient evidence or proof

- Record any action granting approval

- Enter information into files

The building official should ensure that any necessary substantiating data or other evidence be submitted to show that the alternative is in fact equivalent in performance. Moreover, where tests are performed, reports of such tests must be retained by the building official.

Topic: Permits Required and Exempted   **Category:** Administration

Reference: IRC R105.1, R105.2   **Subject:** Permits

Code Text: *Any owner or authorized agent who intends to construct, enlarge, alter, repair, move, demolish, or change the occupancy of a building or structure . . . shall first make application to the building official and obtain the required permit. See multiple exceptions where a permit is not required. Exemption from permit requirements of the IRC shall not be deemed to grant authorization for any work to be done in any manner in violation of the provisions of the IRC or any other laws or ordinances of the jurisdiction.*

Discussion and Commentary: Except in those few instances specifically listed, such as minor repairs, maintenance, finish work and limited equipment replacement, all construction-related work requires a permit and is subject to subsequent inspections.

Work exempt from building permit:

- One-story detached accessory buildings limited to 120 square feet
- Fences not over 6 feet high
- Retaining walls not over 4 feet high
- Water tanks supported directly on grade, if the capacity does not exceed 2 to 1
- Sidewalks and driveways
- Painting, papering, tiling carpeting, cabinets, counter tops and similar finish work
- Prefabricated swimming pools less than 24 inches deep
- Swings and other playground equipment
- Window awnings supported by an exterior wall that do not project more than 54 inches from the exterior wall

Whether or not a building permit is required by the code, all work must be done in accordance with the code requirements. It is important that the owner be responsible for proper and safe construction.

Code Text: *Construction documents, special inspection and structural observation programs, and other data shall be submitted in one or more sets with each application for a permit. See exception that authorizes building official to waive the submission of construction documents. Manufacturer's installation instructions, as required by the IRC, shall be available on the job site at the time of inspection. The construction documents submitted with the application for permit shall be accompanied by a site plan showing the size and location of new construction and existing structures on the site and distances from lot lines.*

Discussion and Commentary: After the construction documents have been submitted, they are to be reviewed by the building department for compliance with the adopted codes and other jurisdictional requirements. Once the construction documents have been approved, a permit is to be issued.

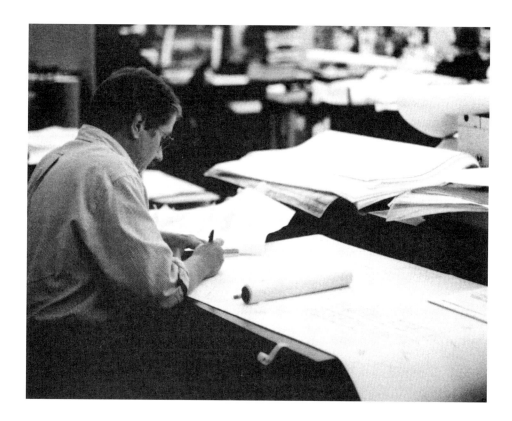

As a part of the necessary recordkeeping process, the retention of at least one set of approved construction documents is required. Such documents must be retained for a minimum of 180 days after the completion date of the work, or as otherwise mandated by state or local laws.

Code Text: *A permit shall not be valid until the fees prescribed by law have been paid. On buildings, structures, electrical, gas, mechanical and plumbing systems or alterations requiring a permit, a fee for each permit shall be paid as required, in accordance with the schedule as established by the applicable governing authority. Building permit valuation shall include total value of the work for which a permit is being issued, such as electrical, gas, mechanical, plumbing equipment and other permanent systems, including materials and labor.*

Discussion and Commentary: Fees are typically established at a level that will provide enough funds to adequately pay for the costs of operating the various building department functions, including administration, plan review and inspection. It is common that the fee schedule be based on the projected construction cost (valuation) of the work. Although there are several different methods for determining the appropriate valuation, it is important that a realistic and consistent approach be taken so that permit fees are applied fairly and accurately.

APPENDIX L

PERMIT FEES

TOTAL VALUATION	FEE
$1 to $500	$24
$501 to $2,000	$24 for the first $500; plus $3 for each additional $100 or fraction thereof, to and including $2,000
$2,001 to $40,000	$69 for the first $2,000; plus $11 for each additional $1,000 or fraction thereof, to and including $40,000
$40,001 to $100,000	$487 for the first $40,000; plus $9 for each additional $1,000 or fraction thereof, to and including $100,000
$100,001 to $500,000	$1,027 for the first $100,000; plus $7 for each additional $1,000 or fraction thereof, to and including $500,000
$500,001 to $1,000,000	$3,827 for the first $500,000; plus $5 for each additional $1,000 or fraction thereof, to and including $1,000,000
$1,000,001 to $5,000,000	$6,327 for the first $1,000,000; plus $3 for each additional $1,000 or fraction thereof, to and including $5,000,000
$5,000,001 and over	$18,327 for the first $5,000,000; plus $1 for each additional $1,000 or fraction thereof

---

A fee schedule is set forth in Appendix L of the IRC. Like the other appendix provisions, the fee schedule is only applicable where specifically adopted by the jurisdiction.

Code Text: *For onsite construction, from time to time the building official, upon notification from the permit holder or his agent, shall make or cause to be made any necessary inspections and shall either approve that portion of the construction as completed or shall notify the permit holder or his or her agent wherein the same fails to comply with the IRC.*

Discussion and Commentary: The inspection function is possibly the most critical activity in the entire code enforcement process. At the various stages of construction, an inspector often provides the final check of the building for safety-related compliance. If necessary, the building official may accept inspection reports from approved agencies verifying compliance with the applicable provisions of the code. It is important that such agencies be highly qualified and reliable.

Required inspections (where applicable):

- Foundation
- Plumbing, mechanical, gas and electrical
- Floodplain
- Frame and Masonry
- Fire-resistance-rated construction
- Others as required by the building official
- Final

The permit holder or authorized agent must notify the building official that the work is ready for inspection. Furthermore, the responsibility to make the work accessible and available for inspection rests with the person requesting the inspection.

Code Text: *Work shall not be done beyond the point indicated in each successive inspection without first obtaining the approval of the building official. The building official, upon notification, shall make the requested inspections and shall either indicate the portion of the construction that is satisfactory as completed, or shall notify the permit holder or an agent of the permit holder wherein the same fails to comply with the IRC. Any portions that do not comply shall be corrected and such portion shall not be covered or concealed until authorized by the building official.*

Discussion and Commentary: Each successive inspection must be approved prior to further work. This practice helps to control the concealment of any work that must be inspected, which would result in the removal of materials that block access.

The building official must establish a consistent and convenient procedure for notification of the appropriate persons when an inspection approval is not granted. The notification method should encourage an efficient and effective resolution to the issues under consideration.

Topic: Use and Occupancy

Reference: IRC R110.1

Category: Administration

Subject: Certificate of Occupancy

Code Text: *No building or structure shall be used or occupied, and no change in the existing occupancy classification of a building or structure or portion thereof shall be made until the building official has issued a certificate of occupancy therefor as provided herein.* See exceptions for 1) work exempt from permits, and 2) accessory structures. *Issuance of a certificate of occupancy shall not be construed as an approval of a violation of the provisions of the IRC or of other ordinances of the jurisdiction.*

Discussion and Commentary: The certificate of occupancy is the tool by which the building official regulates and controls the uses and occupancies of the various buildings and structures within the jurisdiction. The code makes it unlawful to use or occupy a building unless a certificate of occupancy has been issued for that specific use. A temporary certificate of occupancy may be issued prior to completion of all of the work, but only in those portions of the building that can be safely occupied.

> ## Certificate of Occupancy
> (Address of Structure)
>
> This (applicable portion of structure) has been inspected for compliance with the laws and ordinances of (jurisdiction) and is hereby issued a Certificate of Occupancy
>
> Building permit number _____
>
> Applicable edition of code _____
>
> Sprinkler system provided/ required _____
>
> Special conditions _____
>
> _____
>
> _____
>
> _____
>
> Building Official _____
>
> Name and address of owner _____
>
> _____

The building official is permitted to suspend or revoke a certificate of occupancy based on any of the following reasons: 1) when the certificate is issued in error, 2) when incorrect information is supplied or 3) when the building is in violation of the code.

Code Text: *In order to hear and decide appeals of orders, decisions or determinations made by the building official relative to the application and interpretation of the IRC, there shall be and is hereby created a board of appeals. An application for appeal shall be based on a claim that the true intent of the code or the rules legally adopted thereunder have been incorrectly interpreted, the provisions of the code do not fully apply, or an equally good or better form of construction is proposed. The board shall have no authority to waive requirements of the IRC.*

Discussion and Commentary: Any aggrieved party with a material interest in the decision of the building official may appeal such a decision before a board of appeals. This provides a forum other than the court system in which the building official's action can be reviewed.

Members of the board of appeals shall be qualified by experience and training to rule on issues pertaining to building construction. Appendix B of the *International Building Code* provides greater detail on administrative procedures and the necessary qualifications for board service.

Code Text: *This chapter lists the standards that are referenced in various sections of the IRC. The standards are listed herein by the promulgating agency of the standard, the standard identification, the effective date and title, and the section or sections of the IRC that reference the standard. The application of the referenced standard shall be as specified in Section R102.4.*

Discussion and Commentary: Limited in their scope, standards define more precisely the general provisions in the code. The *International Residential Code* references several hundred standards, each addressing a specific aspect of building design or construction. In general terms, the standards referenced by the IRC are primarily materials, testing and installation standards.

The codes and standards referenced in the IRC are an extension of the code, but only to the degree prescribed by the IRC. Where there is a conflict between the provisions in the IRC and any referenced code or standard, the provisions of the IRC shall apply.

Quiz

Study Session 1
IRC Chapters 1 and 43

1. The *International Residential Code* is applicable to single-family dwellings a maximum of _____ stories in above-grade height.

    a. one   b. two

    c. three   d. four

    Reference _R101-2_

2. If there is a conflict in the code between a general requirement and a specific requirement, the _____ requirement shall apply.

    a. general   b. specific

    c. least restrictive   d. most restrictive

    Reference _R102.1_

3. Provisions of the appendices do not apply unless _____.

    a. specified in the code

    b. applicable to unique conditions

    c. specifically adopted

    d. relevant to fire or life safety

    Reference _R102.5_

4. The _____ is used in the code as the term to describe the individual in charge of the department of building safety.

    (a.) building official   b. code official

    c. code administrator   d. chief building inspector

Reference __R103.1__

5. The building official has the authority to _____ the provisions of the code.

    a. ignore   b. waive

    c. modify   (d.) interpret

Reference __R104.1__

6. Used materials may be utilized under which one of the following conditions?

    a. they meet the requirements for new materials

    (b) when approved by the building official

    c. used materials may never be used in new construction

    d. a representative sampling is tested for compliance

Reference __R104.9.1__

7. The building official has the authority to grant modifications to the code _____.

    (a.) for only those issues not affecting life safety or fire safety

    (b.) or individual cases where the strict letter of the code is impractical

    c. where the intent and purpose of the code cannot be met

    d. related only to administrative functions

Reference __R.104.10__

8. The building official shall not grant modifications to any provision related to areas prone to flooding without the _____.

    (a.) granting of a variance by the board of appeals

    b. approval of the jurisdiction's governing body

    c. presentation of an approved design certificate

    d. submission by a registered design professional

Reference __R104.10.1__

9. Tests performed by _____ may be required by the building official where there is insufficient evidence of code compliance.

    a. the owner   b. the contractor

    (c.) an approved agency   d. a design professional

Reference R104·11·1

10. A permit is not required for the construction of a one-story detached accessory structure when it has a maximum floor area of _____ square feet.

    a. 100   (b.) 120

    c. 150   d. 200

Reference R 105·2 #1

11. Prefabricated swimming pools are subject to a building permit where they have a minimum depth of _____.

    a. 12 inches   (b.) 24 inches

    c. 30 inches   d. 36 inches

Reference R105·2 #7

12. Unless an extension is authorized, a permit becomes invalid when work does not commence within _____ after permit issuance.

    a. 90 days   (b.) 180 days

    c. one year   d. two years

Reference R·105·5

13. The building permit, or a copy of the permit, shall be kept _____ until completion of the project.

    (a.) at the job site   b. by the permit applicant

    c. by the contractor   d. by the owner

Reference R105·7

14. When a building permit is issued, the construction documents shall be approved _____.

    (a) in writing or by stamp

    b. and two sets returned to the applicant

    c. and stamped as "Accepted as Reviewed"

    d. pending payment of the plan review fee

    Reference _R106.3.1_

15. Unless otherwise mandated by state or local laws, the approved construction documents shall be retained by the building official for a minimum of _____ from the date of completion of the permitted work.

    a. 90 days   (b.) 180 days

    c. one year   d. two years

    Reference _R106.5_

16. Where applicable, which one of the following inspections is not specifically identified by the *International Residential Code* as a required inspection?

    a. foundation inspection

    b. frame inspection

    c. fire-resistance-construction inspection

    (d.) energy efficiency inspection

    Reference _R109.1_

17. Whose duty is it to provide access to work in need of inspection?

    (a.) the permit holder or agent

    b. the owner or owner's agent

    c. the contractor

    (d.) the person requesting the inspection

    Reference _R109.3_

18. The certificate of occupancy shall contain all of the following information except:

    a.   the name and address of the owner

    b   the name of the building official

    c   the edition of the code under which the permit was issued

    (d)   the maximum occupant load

Reference _R110.3_

19. A temporary certificate of occupancy is valid for what period of time?

    a.   30 days

    b.   60 days

    c.   180 days

    (d)   a period set by the building official

Reference _R110.4_

20. The board of appeals is not authorized to rule on an appeal based on a claim that _____.

    a.   the provisions of the code do not fully apply

    (b.)   a code requirement should be waived

    c.   the rules have been incorrectly interpreted

    d.   a better form of construction is provided

Reference _R112.2_

21. The membership of the board of appeals _____.

    a.   shall include a jurisdictional member

    b.   must consist of at least five members

    (c)   shall be knowledgeable of building construction

    d.   must include an engineer or an architect

Reference _R112.1_

22. The notice of a code violation shall be served to _____.

    a. the permit holder

    b. the person responsible for the work

    c. the owner or owner's agent

    d. the general contractor

Reference _R113.2_

23. When a stop work order is issued, it shall be given to any of the following individuals except for the _____.

    a. owner   b. owner's agent

    c. permit applicant   d. person doing the work

Reference _R114.1_

24. Which of the following standards is applicable for factory-built fireplaces?

    a. ASTM A 951–02   b. CPSC 16 CFR Part 1404

    c. NFPA 259-04   d. UL 127–99

Reference _R1004.1_

25. ACI 318-05 is a reference standard addressing _____.

    a. structural concrete   b. wood construction

    c. structural steel buildings   d. gypsum board

Reference _R402.2_

26. All of the following issues are specifically identified for achieving the purpose of the *International Residential Code*, except for _____.

    a. affordability   b. structural strength

    c. energy conservation   d. usability and accessibility

Reference _R101.3_

27. A permit is not required for the installation of a window awning provided the awning projects a maximum of _____ inches from the exterior wall.

    a. 30   b. 36

    c. 48   (d.) 54

Reference _R105.2_

28. Where a self-contained refrigeration system contains a maximum of _____ pound(s) of refrigerant, a permit is not required.

    a. 1   b. 2

    c. 5   (d.) 10

Reference _R105.2_

29. It is the duty of the _____ or their agent to notify the building official that work is ready for inspection.

    (a.) permit holder   b. owner

    c. contractor   d. design professional

Reference _R109.3_

30. What is the role of the building official in relationship to the board of appeals?

    a. advisor only   (b.) ex officio member

    c. full voting member   d. procedural reviewer

Reference _R112.1_

31. Where the enforcement of a code provision would violate the conditions of an appliance's listing, the conditions of the listing _____ .

    a. are no longer valid   b. may be disregarded

    (c.) shall apply   d. are optional

Reference _R102.4_

- Exceptions -

32. Inspection reports shall be retained in the official records for what minimum period?

   a. until after the certificate of occupancy is issued

   b. 180 days after the report is issued

   (c.) the time required for the retention of public records

   d. 180 days after the certificate of occupancy is issued

Reference _R104-7_

33. Which of the following reasons is not specically identified by the code as authority to suspend or revoke a permit?

   a. the permit was issued in error

   (b.) required inspections have not been performed

   c. it was issued in violation of a jurisdictional ordinance

   d. inaccurate information was provided at the time of issuance

Reference _R110.5 / R105-6_

34. A permit for a temporary structure is limited as to time of service, but shall not be allowed for more than _____ days.

   a. 30   b. 60

   c. 90   (d.) 180

Reference _R105-5 / R107.1_

35. The issuance of a certificate of occupancy is not required for what maximum size of accessory building?

   a. 120 square feet

   b. 150 square feet

   c. 200 square feet

   (d.) all accessory buildings are exempt

Reference _R110.1_

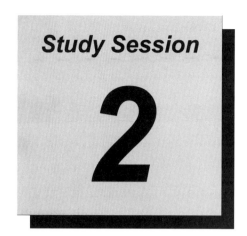

2006 IRC Sections R301 and R302
Building Planning I

OBJECTIVE: To gain an understanding of the design criteria to be used in the application of the prescriptive provisions of the *International Residential Code*, including wind and seismic limitations, snow loads and live loads, as well as the methods for addressing buildings located in very close proximity to lot lines.

REFERENCE: Sections R301 and R302, 2006 *International Residential Code*

KEY POINTS:
- What is the definition of light-framed construction? Under what conditions must such construction be designed in accordance with accepted engineering practice?
- What design standards are identified as acceptable alternatives to the prescriptive structural provisions of the IRC?
- At what wind speed must the structural system be designed based on a source other than the IRC? What design standards are acceptable for residential wind design?
- What is the most common design wind speed in the United States? Where are the hurricane-prone regions?
- How is the appropriate wind exposure category determined? Which category is assumed unless the site meets the definition of another category?
- What are the two methods of establishing wind speed? How do they compare to each other? Which method is used in the IRC?
- Buildings constructed in which seismic design categories are subject to the seismic provisions of the IRC? Which category requires design to the requirements of the IBC?
- In which seismic design categories are irregular buildings required to be designed? How is an irregular building determined?
- What alternate methods are permitted in the determination of a site's seismic design category?
- What is the maximum snow load permitted when using the conventional construction provisions of the IRC?
- What are the limitations on story height for the various types of construction methods?
- What is a dead load? A live load?

KEY POINTS:
(Cont'd)

- What is the design live load to be used for sleeping rooms? For rooms other than sleeping rooms?
- How is an attic defined? How does the minimum live load for attic areas differ based on roof slope?
- What is the maximum deflection permitted for floor structural members? For roof rafters with a ceiling attached directly to the bottom of the rafters?
- What is fire separation distance? How is it measured?
- At what fire separation distance from an interior property line do exterior walls not need a fire-resistance rating? If a rated wall is required, what is the minimum required rating of the wall?
- How far must a projection be located from an interior lot line?
- At what fire separation distance are openings permitted in exterior walls? Where is the amount of openings limited?
- Under what condition may the exterior wall of an accessory structure located on the property line have no fire-resistance rating?
- How are penetrations of exterior walls required to have a fire-resistance rating to be addressed?

Code Text: *The requirements of the IRC are based on platform and balloon-frame construction for light-frame buildings. The requirements for concrete and masonry buildings are based on a balloon framing system. When a building of otherwise conventional construction contains structural elements exceeding the limits of Section R301 or otherwise, not conforming to the IRC, these elements shall be designed in accordance with accepted engineering practice.*

Discussion and Commentary: Light-framed construction is a type of construction whose vertical and horizontal structural elements are primarily formed by a system of repetitive wood or light-gage steel framing members. Platform construction is defined as a method of construction by which floor framing bears on load-bearing walls that are not continuous through the story levels or floor framing. In balloon framing, the wall studs extend beyond the floor line.

Platform framing   **Balloon framing**

There will be times when the methods or materials used in construction do not conform with the prescriptive structural provisions of the IRC. In such situations, it is acceptable to use an engineered solution to satisfy the requirements of the code.

Code Text: *As an alternative to the requirements in Section R301.1 the following standards are permitted subject to the limitations of the IRC and the limitations therein. Where engineered design is used in conjunction with these standards the design shall comply with the* International Building Code. *1) American Forest and Paper Association (AF&PA) Wood Frame Construction Manual (WFCM) and 2) American Iron and Steel Institute (AISI), Standard for Cold-Formed Steel Framing—Prescriptive Method for One-and Two-Family Dwellings (COFS/PM) with supplement.*

Discussion and Commentary: Although the *International Residential Code* is intended to address most structural conditions encountered in residential construction, it may be advantageous to utilize more comprehensive prescriptive design methods that are set forth in the various industry standards.

Where utilizing the design criteria of either the *Wood Frame Construction Manual* or the *Cold-Formed Steel Design Manual*, they must be in accordance with the IBC. The IBC becomes the design basis in those cases where the prescriptive provisions are not used.

Code Text: *Buildings shall be constructed in accordance with the provisions of the IRC as limited by the provisions of Section R301. Additional criteria shall be established by the local jurisdiction and set forth in Table R301.2(1).*

Discussion and Commentary: Table R301.2(1) is designed so that jurisdictions recognize and use certain climatic and geographic design criteria that vary from location to location. Communities are directed to complete the table with a variety of factors. The code references the table in many of its provisions, mandating completion of the table in order to understand the code requirements. Footnotes to the table provide guidance on how and where the determination of the appropriate information is obtained.

TABLE R301.2(1)
CLIMATIC AND GEOGRAPHIC DESIGN CRITERIA

GROUND SNOW LOAD	WIND SPEED[d] (mph)	SEISMIC DESIGN CATEGORY[f]	SUBJECT TO DAMAGE FROM			WINTER DESIGN TEMP[e]	ICE BARRIER UNDERLAYMENT REQUIRED[h]	FLOOD HAZARDS[g]	AIR FREEZING INDEX[i]	MEAN ANNUAL TEMP[j]
			Weathering[a]	Frost line depth[b]	Termite[c]					

For SI: 1 pound per square foot = 0.0479 kPa, 1 mile per hour = 0.447 m/s.

a. Weathering may require a higher strength concrete or grade of masonry than necessary to satisfy the structural requirements of this code. The weathering column shall be filled in with the weathering index (i.e., "negligible," "moderate" or "severe") for concrete as determined from the Weathering Probability Map [Figure R301.2(3)]. The grade of masonry units shall be determined from ASTM C 34, C 55, C 62, C 73, C 90, C 129, C 145, C 216 or C 652.

b. The frost line depth may require deeper footings than indicated in Figure R403.1(1). The jurisdiction shall fill in the frost line depth column with the minimum depth of footing below finish grade.

c. The jurisdiction shall fill in this part of the table to indicate the need for protection depending on whether there has been a history of local subterranean termite damage.

d. The jurisdiction shall fill in this part of the table with the wind speed from the basic wind speed map [Figure R301.2(4)]. Wind exposure category shall be determined on a site-specific basis in accordance with Section R301.2.1.4.

e. The outdoor design dry-bulb temperature shall be selected from the columns of $97^1/_2$-percent values for winter from Appendix D of the International Plumbing Code. Deviations from the Appendix D temperatures shall be permitted to reflect local climates or local weather experience as determined by the building official.

f. The jurisdiction shall fill in this part of the table with the seismic design category determined from Section R301.2.2.1.

g. The jurisdiction shall fill in this part of the table with (a) the date of the jurisdiction's entry into the National Flood Insurance Program (date of adoption of the first code or ordinance for management of flood hazard areas), (b) the date(s) of the currently effective FIRM and FBFM, or other flood hazard map adopted by the community, as may be amended.

h. In accordance with Sections R905.2.7.1, R905.4.3.1, R905.5.3.1, R905.6.3.1, R905.7.3.1 and R905.8.3.1, where there has been a history of local damage from the effects of ice damming, the jurisdiction shall fill in this part of the table with "YES". Otherwise, the jurisdiction shall fill in this part of the table with "NO".

i. The jurisdiction shall fill in this part of the table with the 100-year return period air freezing index (BF-days) from Figure R403.3(2) or from the 100-year (99%) value on the National Climatic Data Center data table "Air Freezing Index- USA Method (Base 32°Fahrenheit)" at www.ncdc.noaa.gov/fpsf.html.

j. The jurisdiction shall fill in this part of the table with the mean annual temperature from the National Climatic Data Center data table "Air Freezing Index-USA Method (Base 32°Fahrenheit)" at www.ncdc.noaa.gov/fpsf.html.

The frost line depth should be established by the local jurisdiction based on past practice or available information. Since frost penetration is dependent on soil conditions and elevations as well as expected winter temperatures, it may vary greatly within a limited geographical area.

Code Text: *Buildings and portions thereof shall be limited by wind speed, as defined in Table R301.2(1), and construction methods in accordance with the IRC. Basic wind speeds shall be determined from Figure R301.2(4). Construction in regions where the basic wind speeds from Figure R301.2(4) equal or exceed 100 miles per hour (45 m/s) in hurricane-prone regions, or 110 miles per hour (49 m/s) elsewhere, shall be designed in accordance with one of five methods. See five possible methods.*

Discussion and Commentary: In most geographical areas, the prescriptive provisions of the IRC are applicable only where the basic wind speed is less than 110 miles per hour. In such cases where the wind speed exceeds this limitation, structures must be designed in accordance with one of five established methods. The code is more stringent in hurricane-prone areas, as a design is required where the wind speed has been established as 100 miles per hour or greater.

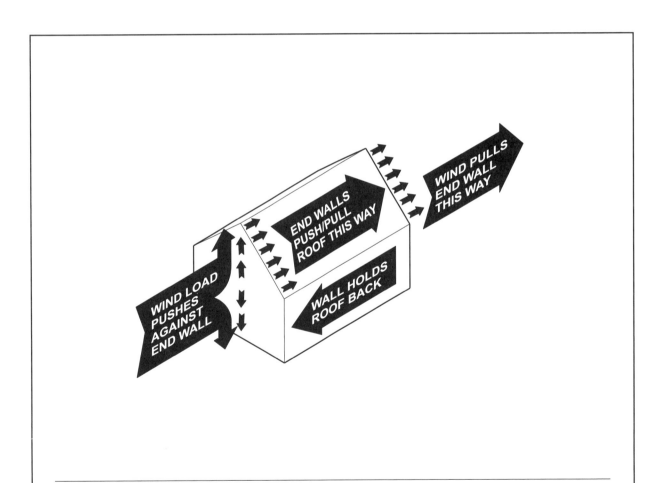

Wind loads are a major consideration in designing a structure's lateral force-resisting system. In addition, other building elements are affected as evidenced by provisions for roof tie-downs, for example. Windows, skylights and exterior doors must withstand the component and cladding pressures.

Topic: Protection of Openings	**Category:** Building Planning
Reference: IRC R301.2.1.2	**Subject:** Design Criteria

Code Text: *Windows in buildings located in windborne debris regions shall have glazed openings protected from windborne debris.* See exception for use of precut wood structural panels with complying attachment hardware.

Discussion and Commentary: Opening protection in windborne debris regions is usually in the form of permanent shutters or laminated glass that meet the requirements of the Large Missile Test of ASTM E 1996 and E 1886. The exception provides a prescriptive approach, which is only permitted in one- and two-story buildings. This method of opening protection uses wood structural panels with a maximum span of 8 feet along with complying fasteners. The builder must precut these panels to fit each glazed opening and provide the necessary attachment hardware.

Fasteners per Table R301.2.1.2

Min $^7/_{16}$ in. precut wood structural panel

8 ft max span

TABLE R301.2.1.2
WINDBORNE DEBRIS PROTECTION FASTENING SCHEDULE
FOR WOOD STRUCTURAL PANELS[a, b, c, d]

FASTENER TYPE	FASTENER SPACING (inches)		
	Panel span ≤ 4 feet	4 feet < panel span ≤ 6 feet	6 feet < panel span ≤ 8 feet
No. 6 Screws	16″	12″	9″
No. 8 Screws	16″	16″	12″

For SI: 1 inch = 25.4 mm, 1 foot = 304.8 mm, 1 pound = 4.448N,
1 mile per hour = 0.447 m/s.

a. This table is based on 130 mph wind speeds and a 33-foot mean roof height.

b. Fasteners shall be installed at opposing ends of the wood structural panel.
   Fasteners shall be located a minimum of 1 inch from the edge of the panel.

c. Fasteners shall be long enough to penetrate through the exterior wall covering and a minimum of $1^1/_4$ inches into wood wall framing and a minimum of $1^1/_4$ inches into concrete block or concrete, and into steel framing a minimum of 3 exposed threads. Fasteners shall be located a minimum of $2^1/_2$ inches from the edge of concrete block or concrete.

d. Where screws are attached to masonry or masonry/stucco, they shall be attached using vibration-resistant anchors having a minimum ultimate withdrawal capacity of 490 pounds.

For SI: 1 inch = 25.4 mm, 1 foot = 304.8 mm

A prescriptive approach for the installation of the wood structural panels as described in Table R301.2.1.2 is permitted, but only for buildings with a mean roof height of 33 feet or less in those areas where the wind speed does not exceed 130 miles per hour.

Topic: Wind Exposure Category
Reference: IRC R301.2.1.4

Category: Building Planning
Subject: Design Criteria

Code Text: *For each wind direction considered, an exposure category that adequately reflects the characteristics of ground surface irregularities shall be determined for the site at which the building or structure is to be constructed. For any given wind direction, the exposure in which a specific building or other structure is sited shall be assessed as being in Exposure Category A, B, C or D.*

Discussion and Commentary: In addition to wind speed, wind loading on structures is a function of the site's exposure category. The exposure category reflects the characteristics of ground surface irregularities and accounts for variations in ground surface roughness that arise from natural topography and vegetation as well as from constructed features.

Exposure B should be used unless the site meets the definition for another category of exposure. Exposure B includes urban and suburban areas, wooded areas and other terrain with numerous closely-spaced obstructions having the size of single-family dwellings or larger.

Topic: Seismic Provisions   **Category**: Building Planning
Reference: IRC R301.2.2, R301.2.2.1   **Subject**: Design Criteria

Code Text: *The seismic provisions of the IRC shall apply to buildings constructed in Seismic Design Categories C, D_0, D_1, and D_2, as determined in accordance with Section R301.2.2.* See exception for detached one- and two-family dwellings in Category C. *Buildings shall be assigned a Seismic Design Category in accordance with Figure R301.2(2).*

Discussion and Commentary: Earthquakes generate internal forces in a structure due to inertia. These forces can cause a building to be distorted and severely damaged. The objective of earthquake resistant construction is to resist these forces and the resulting distortions. Buildings are deemed to be located in one of several seismic design categories, ranging from Category A to Category E. Buildings in Categories A and B as well as detached one- and two-family dwellings in Category C perform satisfactorily when constructed in accordance with the basic prescriptive provisions of the IRC. On the other hand, Category E buildings must typically be designed in accordance with the seismic provisions of the *International Building Code*.

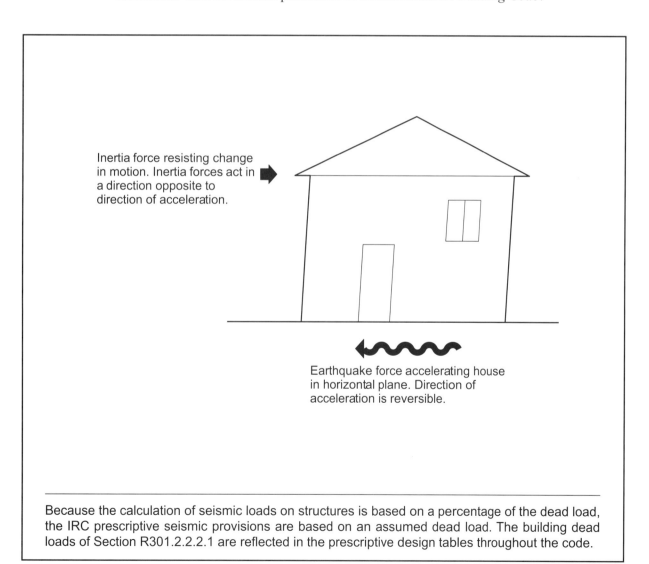

Inertia force resisting change in motion. Inertia forces act in a direction opposite to direction of acceleration.

Earthquake force accelerating house in horizontal plane. Direction of acceleration is reversible.

Because the calculation of seismic loads on structures is based on a percentage of the dead load, the IRC prescriptive seismic provisions are based on an assumed dead load. The building dead loads of Section R301.2.2.2.1 are reflected in the prescriptive design tables throughout the code.

Topic: Irregular Buildings
Reference: IRC R301.2.2.2.2

Category: Building Planning
Subject: Design Criteria

Code Text: *Prescriptive construction as regulated by the IRC shall not be used for irregular structures located in Seismic Design Categories C, D_0, D_1 and D_2. Irregular portions of structures shall be designed in accordance with accepted engineering practice to the extent the irregular features affect the performance of the remaining structural system. A building or a portion of a building shall be considered to be irregular when one or more of seven specific conditions exist. See seven possible conditions.*

Discussion and Commentary: Conventional light-frame construction typically allows cantilevers and offsets within certain limits in order to accommodate common design features and options. These features are not well suited to resisting earthquake loads. This becomes more of a concern in areas of higher seismic hazard.

- Portions of buildings are considered irregular where one or more of seven listed conditions exist:

  - Exterior shear wall lines or braced wall panels not in one plane vertically from foundation to uppermost story where required (exception for limited cantilevers and setbacks).

  - Section of floor or roof not laterally supported by shear walls or braced wall lines on all edges (exception for limited extension beyond shear wall or braced wall line).

  - End of braced wall panel occurs over an opening below and ends more than 1 foot from edge of opening (exception for opening with complying header).

  - Opening in a floor or roof exceeds 12 feet or exceeds 50 percent of least floor or roof dimension.

  - Portions of floor level vertically offset (exceptions for framing directly supported by continuous foundatons or where floor framing lapped or tied together).

  - Shear walls and braced wall lines do not occur in two perpendicular directions.

  - Shear walls or braced wall lines constructed of dissimilar bracing systems on any story above grade.

The IRC, consistent with the IBC, identifies building features that are considered irregular and are not permitted under conventional light-frame construction provisions for the higher seismic design categories (C, D_0, D_1 and D_2).

Topic: Floodplain Construction

Category: Building Planning

Reference: IRC R301.2.4

Subject: Design Criteria

Code Text: *Buildings and structures constructed in whole or in part in flood hazard areas (including A or V Zones) as established in Table R301.2(1) shall be designed and constructed in accordance with Section R324.* See exception for buildings in identified floodways.

Discussion and Commentary: Application of the flood provisions of the code cannot prevent or eliminate all future flood damage. Rather, the provisions represent a reasonable balance of the knowledge and awareness of flood hazards, methods to guide development to less hazard-prone locations, methods of design and construction intended to resist flood damage, and each community's and landowner's reasonable expectations of land use.

The flood hazard areas shown on FEMA maps are determined using the base flood, which is defined as have a 1-percent chance of occurring in any given year. The maps do not show the worst case flood nor the "flood of record," which usually is the most severe in history.

Code Text: *Wood framed construction, cold-formed steel framed construction and masonry and concrete construction in regions with ground snow loads 70 psf (3.35 kPa) or less, shall be in accordance with Chapters 5, 6 and 8. Buildings in regions with ground snow loads greater than 70 psf (3.35 kPa) shall be designed in accordance with accepted engineering practice. The roof shall be designed for the live load indicated in Table R301.6 or the snow load indicated in Table R301.2(1), whichever is greater.*

Discussion and Commentary: The prescriptive tables for floor, wall and roof construction are limited to buildings constructed in areas having a maximum 70 psf snow load. Structures in regions exceeding this limitation require an engineered design for all elements that carry snow loads. The basic 20 psf live load may be reduced based on the tributary area supported by any structural member of the roof, but no reduction is permitted for snow loads.

TABLE R301.6
MINIMUM ROOF LIVE LOADS IN POUNDS-FORCE
PER SQUARE FOOT OF HORIZONTAL PROJECTION

ROOF SLOPE	TRIBUTARY LOADED AREA IN SQUARE FEET FOR ANY STRUCTURAL MEMBER		
	0 to 200	201 to 600	Over 600
Flat or rise less than 4 inches per foot (1:3)	20	16	12
Rise 4 inches per foot (1:3) to less than 12 inches per foot (1:1)	16	14	12
Rise 12 inches per foot (1:1) and greater	12	12	12

For SI: 1 square foot = 0.0929 m², 1 pound per square foot = 0.0479 kPa,
           1 inch per foot = 83.3 mm/m.

For roofs, both the live load and snow load must be considered, with the more restrictive loading condition used. The live load takes into account that individuals and materials may be present on the roof during roofing operations.

Code Text: *Buildings constructed in accordance with these provisions shall be limited to story heights of not more than the following: 1) for wood wall framing, the laterally unsupported bearing wall stud height permitted by Table R602.3(5) plus a height of floor framing not to exceed sixteen inches (see exception for braced walls); 2) for steel wall framing, a stud height of 10 feet, plus a height of floor framing not to exceed 16 inches; 3) for masonry walls, a maximum bearing wall clear height of 12 feet plus a height of floor framing not to exceed 16 inches (see exception for gable end walls); and 4) for insulating concrete form walls, the maximum bearing wall height per story as permitted by Section 611 tables plus a height of floor framing not to exceed 16 inches.*

Discussion and Commentary: This section identifies the maximum heights for use of the IRC's prescriptive provisions, with an allowance for a floor system up to 16 inches in depth.

TABLE R602.3(5)
SIZE, HEIGHT AND SPACING OF WOOD STUDS[a]

STUD SIZE (inches)	BEARING WALLS					NONBEARING WALLS	
	Laterally unsupported stud height[a] (feet)	Maximum spacing when supporting roof and ceiling only (inches)	Maximum spacing when supporting one floor, roof and ceiling (inches)	Maximum spacing when supporting two floors, roof and ceiling (inches)	Maximum spacing when supporting one floor only (inches)	Laterally unsupported stud height[a] (feet)	Maximum spacing (inches)
2 × 3[b]	—	—	—	—	—	10	16
2 × 4	10	24	16	—	24	14	24
3 × 4	10	24	24	16	24	14	24
2 × 5	10	24	24	—	24	16	24
2 × 6	10	24	24	16	24	20	24

For SI: 1 inch = 25.4 mm.

a. Listed heights are distances between points of lateral support placed perpendicular to the plane of the wall. Increases in unsupported height are permitted where justified by analysis.

b. Shall not be used in exterior walls.

These requirements are height-limiting solely due to the prescriptive structural limitations of the code. Such story heights may be increased through the engineered design provisions in the *International Building Code.*

Code Text: *The minimum uniformly distributed live load shall be as provided in Table R301.5.*

Discussion and Commentary: Table R301.5 provides the minimum loads based on the use of a particular area or portion of a structure. These loads must be considered for the design of corresponding structural elements of any residence constructed under the code. For instance, bedrooms (sleeping rooms) require the use of a minimum 30 psf live load, whereas all other rooms must be designed for at least a 40 psf uniform live load. Exterior balconies require a higher design load (60 psf) than decks (40 psf), simply because of the lack of redundancy in structural supports. An exterior balcony is defined as "an exterior floor projecting from and supported by a structure without additional independent supports." A deck is "supported on at least two opposing sides by an adjoining structure and/or posts, piers, or other independent supports."

TABLE R301.5
MINIMUM UNIFORMLY DISTRIBUTED LIVE LOADS
(in pounds per square foot)

USE	LIVE LOAD
Attics with limited storage[b, g, h]	20
Attics without storage[b]	10
Decks[e]	40
Exterior balconies	60
Fire escapes	40
Guardrails and handrails[d]	200[i]
Guardrails in–fill components[f]	50[i]
Passenger vehicle garages[a]	50[a]
Rooms other than sleeping rooms	40
Sleeping rooms	30
Stairs	40[c]

For SI: 1 pound per square foot = 0.0479 kPa, 1 square inch = 645 mm², 1 pound = 4.45 N.

a. Elevated garage floors shall be capable of supporting a 2,000-pound load applied over a 20-square-inch area.

b. Attics without storage are those where the maximum clear height between joist and rafter is less than 42 inches, or where there are not two or more adjacent trusses with the same web configuration capable of containing a rectangle 42 inches high by 2 feet wide, or greater, located within the plane of the truss. For attics without storage, this live load need not be assumed to act concurrently with any other live load requirements.

c. Individual stair treads shall be designed for the uniformly distributed live load or a 300-pound concentrated load acting over an area of 4 square inches, whichever produces the greater stresses.

d. A single concentrated load applied in any direction at any point along the top.

e. See Section R502.2.1 for decks attached to exterior walls.

f. Guard in-fill components (all those except the handrail), balusters and panel fillers shall be designed to withstand a horizontally applied normal load of 50 pounds on an area equal to 1 square foot. This load need not be assumed to act concurrently with any other live load requirement.

g. For attics with limited storage and constructed with trusses, this live load need be applied only to those portions of the bottom chord where there are two or more adjacent trusses with the same web configuration capable of containing a rectangle 42 inches high or greater by 2 feet wide or greater, located within the plane of the truss. The rectangle shall fit between the top of the bottom chord and the bottom of any other truss member, provided that each of the following criteria is met:

    1. The attic area is accessible by a pull-down stairway or framed opening in accordance with Section R807.1; and

    2. The truss has a bottom chord pitch less than 2:12.

h. Attic spaces served by a fixed stair shall be designed to support the minimum live load specified for sleeping rooms.

i. Glazing used in handrail assemblies and guards shall be designed with a safety factor of 4. The safety factor shall be applied to each of the concentrated loads applied to the top of the rail, and to the load on the in-fill components. These loads shall be determined independent of one another, and loads are assumed not to occur with any other live load.

Attic design loads are based on a variety of factors. Where it is assumed that there is insufficient clearance or headroom to accumulate significant storage, only 10 psf is mandated. Where storage is more probable due to a sizable attic height, a 20 psf dead load is to be used.

Topic: Wall Construction and Penetrations   **Category:** Building Planning

Reference: IRC R302.1   **Subject:** Exterior Wall Location

Code Text: *Construction . . . and penetrations of exterior walls of dwellings and accessory buildings shall comply with Table R302.1.* See exception, which allows for nonrated exterior walls of small storage sheds and similar accessory buildings.

Discussion and Commentary: Where an exterior wall has a fire separation distance of less than 5 feet, it is mandated that the wall be of minimum 1-hour fire-resistance-rated construction. The fire-resistive rating must be obtained from both sides of the wall. Because the IRC does not contain specific provisions for fire-resistance ratings of walls and other elements, it will be necessary to utilize the assemblies set forth in the *International Building Code* or another fire-resistance directory. Regardless of their location, exterior walls of storage sheds and similar accessory buildings that do not exceed 120 square feet in floor area are not required to be provided with fire-resistance-rated walls.

TABLE R302.1
EXTERIOR WALLS

EXTERIOR WALL ELEMENT		MINIMUM FIRE-RESISTANCE RATING	MINIMUM FIRE SEPARATION DISTANCE
Walls	(Fire-resistance rated)	1 hour with exposure from both sides	0 feet
	(Not fire-resistance rated)	0 hours	5 feet
Projections	(Fire-resistance rated)	1 hour on the underside	2 feet
	(Not fire-resistance rated)	0 hours	5 feet
Openings	Not allowed	N/A	< 3 feet
	25% Maximum of Wall Area	0 hours	3 feet
	Unlimited	0 hours	5 feet
Penetrations	All	Comply with Section R317.3	< 5 feet
		None required	5 feet

N/A = Not Applicable.

Where a building's exterior wall is located with a fire separation distance of less than 5 feet, the provisions of Section R317.3 regulate the protection of any penetrations that occur within the wall. Several methods are provided for both through penetrations and membrane penetrations.

Code Text: *These provisions* based on fire separation distance *shall not apply to walls, projections, openings or penetrations in walls that are perpendicular to the line used to determine the fire separation distance. FIRE SEPARATION DISTANCE is the distance measured from the building face to one of the following: 1) to the closest interior lot line; or 2) to the centerline of a street, an alley or public way; or 3) to an imaginary line between two buildings on the lot.*

Discussion and Commentary: To reduce the potential for fire spread from one residential structure to another due to their close proximity, the code provides a choice as to the method for addressing radiant heat transfer between structures. If the building is located far enough away from a lot line, the atmospheric separation mitigates the concern. Where the building is located closer to the lot line, a fire-resistive barrier of one hour is mandated to restrict the spread of heat and flame.

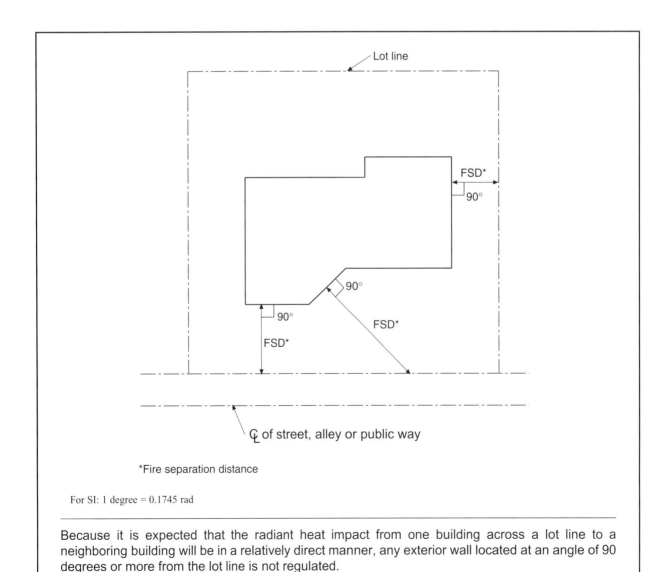

Lot line

FSD*

90°

90°

FSD*

90°

FSD*

℄ of street, alley or public way

*Fire separation distance

For SI: 1 degree = 0.1745 rad

Because it is expected that the radiant heat impact from one building across a lot line to a neighboring building will be in a relatively direct manner, any exterior wall located at an angle of 90 degrees or more from the lot line is not regulated.

Code Text: *. . . projections . . . of exterior walls of dwellings and accessory buildings shall comply with Table R302.1. Projections beyond the exterior wall shall not extend more than 12 inches (305 mm) into the areas where openings are prohibited.* See exceptions for 1) small detached tool shed, storage sheds, playhouse and similar structures, and 2) detached garages.

Discussion and Commentary: Projections, particularly those that are combustible, pose a fire hazard when located in close proximity to a lot line. Therefore, a minimum distance must be maintained between any projection and a neighboring lot. In addition, permitted projections located within 5 feet of an interior lot line must be protected with 1-hour construction on the projection's underside. The code provides for exceptions to the setback requirements for garages and small accessory structures.

For SI: 1 foot = 304.8 mm

The code typically allows for a 12-inch encroachment into areas where openings are prohibited, which would allow for a projection to extend within 2 feet of an interior lot line. However, projections for detached garages and small detached tool sheds, storage sheds, playhouses and similar structures may extend closer.

Topic: Openings

Reference: IRC R302.1

Category: Building Planning

Subject: Exterior Wall Location

Code Text: . . . openings . . . of exterior walls of dwellings and accessory buildings shall comply with Table R302.1. See exception for foundation vents.

Discussion and Commentary: Unprotected openings, typically doors and windows, in exterior walls of adjacent buildings provide greater opportunity for the spread of fire and heat from one structure to another than do the exterior walls. For that reason, there cannot be openings, including those in accessory structures, in exterior walls with a fire separation distance of less than 3 feet. Where less of a potential hazard exists, such as where the fire separation distance is between 3 feet and 5 feet, unprotected openings are permitted, but only in a limited amount. Once a distance of 5 feet or greater is provided, an unlimited amount of unprotected openings is allowed.

Unlimited, unprotected openings permitted

5' or greater

3' to < 5'

Unprotected openings permitted, but limited to 25% of wall area

Less than 3'

No openings permitted

No limitation on openings at 90° to property line

PL

For SI: 1 foot = 304.8 mm, 1 degree = 0.1745 rad

There is no option set forth in the IRC to allow for the protection of openings in lieu of an adequate spatial separation from adjoining structures or lots. The use of fire doors, fire windows or fire shutters is not specifically mentioned as an alternative to an adequate fire separation distance.

Quiz

Study Session 2
IRC Sections R301 and R302

1. What is the weathering potential for all dwelling sites located in the state of Kansas?

    (a.) severe

    b. moderate

    c. negligible

    d. none

    Reference _R301.2(3)_

2. The basic wind speed to be used for most of the United States is _____.

    a. 70 mph

    b. 75 mph

    c. 85 mph

    (d.) 90 mph

    Reference _R301.2(4)_

3. Based on Figure R301.2(5), what is the ground snow load for most of western Colorado?

    a. 30 psf

    b. 40psf

    c. 50 psf

    (d.) site-specific due to extreme local variations

    Reference _301.2(5)_

4. Unless local conditions warrant otherwise, what is the probability for termite infestation for dwellings constructed in most of Idaho?

    a. very heavy

    b. moderate to heavy

    (c.) light to moderate

    d. none to slight

    Reference _R301.2(6)_

5. For the purpose of determining the component and cladding loads on the roof surface of a building, the area at the ridge of a gable roof sloped at 20 degrees shall be considered as Pressure Zone _____ at other than the eaves.

   a. 0 b. 1

   (c.) 2 d. 3

   Reference 301.2 (2)

6. Where wood structural panels are used to protect windows in buildings located in windborne debris regions, #6 wood screws shall be located at a maximum of _____ on center to fasten those panels that span 8 feet.

   (a.) 9 inches b. 12 inches

   c. 16 inches d. 24 inches

   Reference TABLE R301.2.1.2

7. A "3-second gust" velocity of 90 miles per hour in the IRC is to be converted to a "fastest mile" wind speed of _____ for use in reference documents using the "fastest mile" criteria.

   (a.) 76 miles per hour b. 85 miles per hour

   c. 90 miles per hour d. 95 miles per hour

   Reference R301.2.1.3

8. What wind exposure category is appropriate for a dwelling located in a residential development in a suburban area?

   a. Exposure A (b.) Exposure B

   c. Exposure C d. Exposure D

   Reference R301.2.1.4

9. The seismic design category for a site having a calculated S_{DS} of 0.63g is _____.

   a. B (b.) D_0

   c. D_1 d. E

   Reference R301.2.2.1.1

10. For structures assigned to Seismic Design Category C, anchored masonry veneer is limited to _____ in thickness.

    a. 4 inches   b. 5 inches

    c. 6 inches   d. 8 inches

Reference _R703.7_

11. For a dwelling assigned to Seismic Design Category D_2 and constructed under the conventional provisions of the IRC, what is the maximum dead load permitted for 8-inch-thick masonry walls?

    a. 40 psf   b. 65 psf

    c. 80 psf   d. 85 psf

Reference _R301.2.2.2 / R301.2.2.2.1_

12. A portion of a building is considered irregular for seismic purposes when a floor opening, such as for a stairway, exceeds _____ of the least floor dimension.

    a. 15 percent   b. 25 percent

    c. $33^1/_3$ percent   d. 50 percent

Reference _____

13. Buildings constructed in regions where the ground snow load exceeds _____ must be designed in accordance with accepted engineering practice.

    a. 30 psf   b. 50 psf

    c. 70 psf   d. 90 psf

Reference _R301.2.3_

14. For rooms other than sleeping rooms, what is the minimum uniformly distributed live load that is to be used for the design of the floor system?

    a. 20 psf   b. 30 psf

    c. 40 psf   d. 50 psf

Reference _R301.5_

15. A minimum uniformly distributed live load of 10 psf is to be used for the design of attic areas having a maximum clear height of_____ inches.

    a. 30   b. 36

    c. 42   d. 60

Reference TABLE R301.5

16. A minimum uniformly distributed live load of _____ shall be used for the design of sleeping rooms.

    a. 10 psf   b. 20 psf

    c. 30 psf   d. 40 psf

Reference TABLE R301.5

17. Where no snow load is present, what is the minimum roof design live load for a 240 square foot tributary-loaded roof area having a slope of 8:12?

    a. 20 psf   b. 16 psf

    c. 14 psf   d. 12 psf

Reference TABLE R301.6

18. What is the maximum allowable deflection for floor joists?

    a. L/120   b. L/180

    c. L/240   **d.** L/360

TABLE
Reference R301.7

19. The maximum allowable deflection permitted for 7:12-sloped rafters with no ceiling attached is _____.

    a. L/120   **b.** L/180

    c. L/240   d. L/360

Reference TABLE R301.7

20. Except for walls perpendicular to the lot line, a minimum 1-hour fire-resistance rating is required for all exterior walls having a fire separation distance of less than _____ .

    a. 3 feet   b. 4 feet

    c. 5 feet   d. 10 feet

    Reference _TABLE R032.1_

21. What is the minimum distance a roof projection on a dwelling shall be located from an interior lot line?

    a. 0 feet; it may extend to the property line

    b. 12 inches

    c. 2 feet

    d. 4 feet

    Reference _R302_

22. Tool and storage sheds, playhouses and similar accessory structures having a maximum floor area of _____ are not required to have exterior wall protection based on location on the lot.

    a. 100 square feet   b. 120 square feet

    c. 150 square feet   d. 200 square feet

    Reference ___?___

23. Openings located less than 3 feet from an interior lot line shall be protected with what minimum fire protection rating?

    a. 0 hours; no fire protection rating is required

    b. 20 minutes

    c. 45 minutes

    d. openings are not permitted

    Reference _TABLE 302.1_

24. Where located in an exterior wall having a fire separation distance of less than 5 feet, penetrations protected with an approved penetration firestop system shall have a minimum fire-resistance rating of _____ .

    a. 20 minutes   b. 30 minutes

    c. 45 minutes   d. 1 hour

    Reference _____

25. Where a detached garage is located within 2 feet of a lot line, the maximum eave projection is limited to _____ inches.

a. 4   b. 6

c. 8   d. 12

Reference R302 1

26. For wind design purposes, a building located along the shoreline in a hurricane prone region is categorized as Exposure _____.

a. A   b. B

c. C   d. D

Reference R301.2.1.4

27. Which of the following buildings is exempt from the seismic requirements of the code?

a. a townhouse in Seismic Design Category C

b. a one-family dwelling in Seismic Design Categor C

c. a townhouse in Seismic Design Categor D_0

d. a two-family dwelling in Seismic Design Categor D_1

Reference R301.2.2

28. For a building located in Seismic Design Category D_1, the maximum average dead load for a floor assembly regulated by the conventional construction provisions is _____ psf.

a. 5   b. 10

c. 15   d. 20

Reference _____

29. Under the IRC conventional framing provisions, the maximum story height for steel wall framing is limited to _____, plus up to 16 inches for floor framing.

a. 10 feet   b. 12 feet

c. 16 feet   d. 22 feet

Reference R301.3

30. An in-fill panel for a guard shall be designed to withstand a minimum load of _____ applied horizontally on an area of 1 square foot.

    a. 50 pounds   b. 80 pounds

    c. 100 pounds   d. 200 pounds

Reference _Table 301.5 (F)_

31. Where structural wood panels are used to provide windborne debris protection for glazed openings, the fasteners shall be long enough to penetrate through the sheathing and into wood wall framing a minimum of _____ inch(es).

    a. $^3/_4$ b. 1

    c. $1^1/_4$ d. $1^1/_2$

Reference _Table R301.2.1.2_

32. For the purpose of addressing the minimum uniformly distributed live load, glazing used in handrail assemblies and guards shall be designed with a safety factor of _____.

    a. 0.5   b. 1.5

    c. 2.0   d. 4.0

Reference _Table 301.5 (i)_

33. Attic spaces served by a fixed stair shall be designed to support a minimum live load of _____.

    a. 10   b. 20

    c. 30   d. 40

Reference _Table R301.5 (H)_

34. Where the roof projection of a dwelling extends to a point two feet of an interior lot line, what minimum level of fire-resistance is required for the portion of the projection that is within 5 feet of the lot line?

    a. 30 minutes from both sides

    b. 1 hour on the underside

    c. no fire-resistance rating is required

    d. the projection is not permitted

Reference _Table R302.1_

35. The aggregate area of openings in an exterior wall located with a fire separation distance of four feet is limited to a maximum of _____ percent of the exterior wall area.

a. 10   b. 25

c. 50   d. 100

Reference _____

2006 IRC Sections R303 – R310
Building Planning II

OBJECTIVE: To develop an understanding of the health and safety criteria of the code, including light and ventilation; minimum room areas and ceiling height; sanitation; toilet, bath, and shower spaces; glazing, including safety glazing; carports and garages; and emergency escape and rescue openings.

REFERENCE: Sections R303 through R310, 2006 *International Residential Code*

- Where natural light is used to satisfy the minimum illumination requirements, how is the minimum required amount of glazing determined? Where artificial light is used, what illumination level is mandated?

- Where natural ventilation is used to satisfy the minimum ventilation requirements, what is the minimum required area of openable exterior openings?

- How must the ventilation and illumination of bathrooms occur?

- How must mechanical and gravity outside air intake openings be located in relationship to vents, chimneys, parking lots and other potential areas of a hazardous or noxious contaminant?

- Where must illumination be located in relationship to interior stairways? Exterior stairways?

- In what climatic areas must a heating system be provided? What performance level is mandated for the system?

- What is the minimum required size of the largest habitable room in a dwelling unit?

- What is the minimum dimension permitted for a habitable room other than a kitchen?

- What is the minimum ceiling height permitted for a living room or bedroom? A hallway? Bathroom? Basement? Where can a reduction in such heights be acceptable?

- How much clear floor space is required in front of a water closet? In front of a shower opening? What is the minimum distance needed between the centerline of a water closet and the nearest adjoining obstruction such as a wall or shower compartment?

- In what manner must safety glazing be identified? Multipane assemblies?

- Where is polished wired glass permitted to be used as safety glazing?

KEY POINTS:
(Cont'd)

- What specific locations in and adjacent to doors are subject to human impact and require safety glazing? In tub and shower areas? Adjacent to swimming pools, hot tubs, and spas? At stairways and stairway landings?

- When is sloped glazing considered a skylight? What glazing materials are permitted in skylights? When must a screen be installed below a skylight?

- How must a private garage be separated from a dwelling unit? What type of door is required? How should a duct penetration be addressed?

- How does a carport differ from a garage? What limitations are placed on carports?

- Where are escape and rescue openings required? What is the minimum size of such openings? Maximum sill height? What limitations are placed on the operation of the opening?

- When a window well serves an escape and rescue opening, what is its minimum size?

- How may a bulkhead enclosure be utilized as an escape and rescue opening?

- Under what conditions may an emergency escape window be located under a deck or porch?

Topic: Habitable Rooms

Category: Building Planning

Reference: IRC R303.1

Subject: Light, Ventilation and Heating

Code Text: *All habitable rooms shall be provided with aggregate glazing area of not less than 8 percent of the floor area of such rooms. Natural ventilation shall be through windows, doors, louvers or other approved openings to the outdoor air. The minimum openable area to the outdoors shall be 4 percent of the floor area being ventilated.* Exceptions allow the use of artificial light and mechanical ventilation.

Discussion and Commentary: A usable and sanitary interior environment depends on the inclusion of adequate light and ventilation for the habitable spaces within the dwelling unit. Traditionally, the use of natural light and, to some degree, natural ventilation has been mandated as the means for achieving such an environment. It has become increasingly more common to use artificial lighting and a mechanical ventilation system. These methods create additional design flexibility and functionality while maintaining a pleasant and sanitary living environment.

Interior room

"Borrowed" light

Glazed area to be minimum of 8% of floor area of both rooms

Plan view

50% of common wall to be open

Opening to be greater of 25 sq ft or 10% of interior room floor area

For SI: 1 square foot = 0.093 m²

Elevation

Borrowing natural light from adjoining room

Where the mechanical ventilation option is used for bathrooms and water-closet compartments, the minimum ventilation rates are set forth for either continuous ventilation or intermittent ventilation. In both cases, the air must be exhausted directly to the exterior of the dwelling.

Code Text: *Mechanical and gravity outside air intake openings shall be located a minimum of 10 feet (3048 mm) from any hazardous or noxious contaminant, such as vents, chimneys, plumbing vents, streets, alleys, parking lots and loading docks, except as otherwise specified in the IRC. Where a source of contaminant is located within 10 feet (3048 mm) of an intake opening, such opening shall be located a minimum of 2 feet (610 mm) below the contaminant source. Outside exhaust openings shall be located so as not to create a nuisance. Exhaust air shall not be directed onto walkways.*

Discussion and Commentary: In the context of this section, intake openings include windows, doors, combustion air intakes and similar openings that naturally or mechanically draw in air from the building exterior. The alternative to the 10-foot separation requirement, a 2-foot vertical separation distance, will allow noxious gases and contaminants to disperse into the atmosphere before they can be drawn into an air intake opening.

For SI: 1 foot = 304.8 mm.

Outdoor air exhaust and intake openings are regulated in the same manner as other openings in exterior walls. As such, they are not permitted in walls having a fire separation distance of less than 3 feet, except in exterior walls that are perpendicular to the lot line.

| **Topic:** Stairway Illumination | **Category:** Building Planning |
| **Reference:** IRC R303.6, R303.6.1 | **Subject:** Light, Ventilation and Heating |

Code Text: *Interior stairways shall be provided with an artificial light source located in the immediate vicinity of each landing of the stairway. See exception where light source is located over each stairway section. Exterior stairways shall be provided with an artificial light source located in the immediate vicinity of the top landing of the stairway. Where lighting outlets are installed in interior stairways, there shall be a wall switch at each floor level to control the lighting outlet where the stairway has six or more risers.* See exception for lights that are continuously illuminated or automatically controlled.

Discussion and Commentary: A stairway is one of the most hazardous areas of a dwelling unit. As such, the code highly regulates the design and construction of all stairways. In addition, adequate lighting must be provided to enable the stairway user to see the treads, their nosings and any obstructions that may be present. Stairway landings must also be adequately lighted.

Unless the light sources are on continuously or automatically activated, interior stairway lights must be controlled from both the top and bottom of each stairway consisting of six or more risers. For exterior stairway lighting, the control switch is to be located within the dwelling unit.

Code Text: *Every dwelling unit shall have at least one habitable room that shall have not less than 120 square feet (11.2 m²) of gross floor area. Other habitable rooms shall have a floor area of not less than 70 square feet (6.5 m²). Habitable rooms shall not be less than 7 feet (2134 mm) in any horizontal dimension.* See exceptions for kitchens regarding minimum size and horizontal dimensions.

Discussion and Commentary: Acceptable sizes for habitable rooms have been established. Because habitable rooms are expected to be those spaces within a dwelling unit where most activities take place, they are the only areas regulated. Most habitable rooms need be only 7 feet by 10 feet to comply with the provisions; however, at least one larger room must be provided. It is seldom that any habitable room in today's typical dwelling unit would be designed with such a small floor area.

A1 + A2 + A3 ≥ required room floor area per Section R304

A1 ≥ 50% of required room floor area per Section R304

For SI: 1 degree = 0.1745 rad

The minimum required floor area for any habitable room having a sloping ceiling, as would typically be encountered where an attic area is finished for use as a living or sleeping area, must be based on only those portions of the room with a ceiling height of at least 5 feet.

Topic: Minimum Height

Category: Building Planning

Reference: IRC R305.1

Subject: Ceiling Height

Code Text: *Habitable rooms, hallways, corridors, bathrooms, toilet rooms, laundry rooms and basements shall have a ceiling height of not less than 7 feet (2134 mm).* See four exceptions addressing ceilings with exposed beams, basements without habitable spaces, rooms with sloped ceilings, and bathrooms. *The required height shall be measured from the finish floor to the lowest projection from the ceiling.*

Discussion and Commentary: For both safety reasons and usability by the occupants, the minimum ceiling height throughout occupiable areas of a dwelling unit is regulated. Most rooms that are commonly used by the occupants are included, other than closets and storage areas. Where basements are used for habitable purposes, they too must comply with the minimum height requirement of 7 feet.

For SI: 1 inch = 25.4 mm, 1 foot = 304.8 mm.

Beams and girders spaced at least 4 feet on center may project not more than 6 inches below the required ceiling height.

Topic: Minimum Space Required

Reference: IRC R307

Category: Building Planning

Subject: Toilet, Bath and Shower Spaces

Code Text: *Fixtures shall be spaced as per Figure R307.1. Bathtub and shower floors and walls above bathtubs with installed shower heads and in shower compartments shall be finished with a nonabsorbent surface. Such wall surfaces shall extend to a height of not less than 6 feet (1829 mm) above the floor.*

Discussion and Commentary: It is necessary to provide adequate clearances at and around bathroom fixtures to allow for ease of use by the occupants of the dwelling unit. The code addresses clear floor space and clearances for lavatories, water closets, bathtubs, and showers. In addition, the minimum permitted size for a shower is 30 inches by 30 inches.

The need for nonabsorbent surfaces in shower areas subject to water splash is based on two concerns. For sanitation purposes, finish materials must be of a type that can be cleaned easily. Also, continued absorption of moisture will lead to deterioration of the structural components.

Topic: Safety Glazing Identification

Category: Building Planning

Reference: IRC R308.1

Subject: Glazing

Code Text: *Except as indicated in Section R308.1.1, each pane of glazing installed in hazardous locations as defined in Section R308.4 shall be provided with a manufacturer's designation specifying who applied the designation, designating the type of glass and the safety glazing standard with which it complies, which is visible in the final installation. The designation shall be acid etched, sandblasted, ceramic-fired, laser etched, embossed, or be of a type which once applied cannot be removed without being destroyed. A label shall be permitted in lieu of the manufacturer's designation.* See exceptions for tempered spandrel glass and where certifications of compliance are provided.

Discussion and Commentary: Improper glazing installed in areas subject to human impact can create a serious hazard. Accordingly, it is critical that glazing in such locations be appropriately identified to help ensure that the proper glazing is in place.

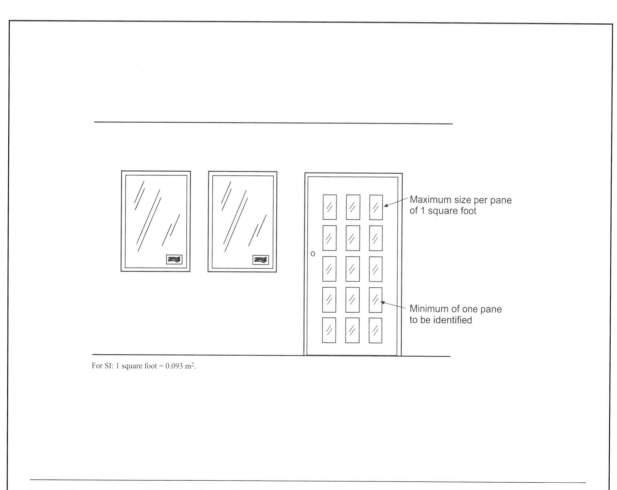

For SI: 1 square foot = 0.093 m².

In multipane assemblies, such as french doors, where the individual panes do not exceed one square foot in exposed area. Section R308.1.1 provides for a reduction in the required information on all but one pane. At least one pane must be fully identified.

Code Text: *Individual glazed areas, including glass mirrors in hazardous locations such as those indicated as defined in Section R308.4, shall pass the test requirements of CPSC 16 CFR, Part 1201. Glazing shall comply with CPSC 16 CFR, Part 1201 criteria for Category I or Category II as indicated in Table R308.3. See exception for complying jalousies and louvered windows.*

Discussion and Commentary: Glazing located in hazardous locations subject to human impact must pass the test requirements developed by the Consumer Product Safety Commission (CPSC). Two classifications of safety glazing have been established—Category I and Category II. Glazing identified as Category II is acceptable in all locations where safety glazing is mandated. Category I glazing is limited to use in small lites, 9 square feet or less, provided the glazing is not in a sliding glass patio door or in doors and enclosures for bathtubs, showers and similar areas.

TABLE R308.3
MINIMUM CATEGORY CLASSIFICATION OF GLAZING

EXPOSED SURFACE AREA OF ONE SIDE OF ONE LITE	GLAZING IN STORM OR COMBINATION DOORS (Category Class)	GLAZING IN DOORS (Category Class)	GLAZED PANELS REGULATED BY ITEM 7 OF SECTION R308.4 (Category Class)	GLAZED PANELS REGULATED BY ITEM 6 OF SECTION R308.4 (Category Class)	GLAZING IN DOORS AND ENCLOSURES REGULATED BY ITEM 5 OF SECTION R308.4 (Category Class)	SLIDING GLASS DOORS PATIO TYPE (Category Class)
9 sq ft or less	I	I	NR	I	II	II
More than 9 sq ft	II	II	II	II	II	II

For SI: 1 square foot = 0.0929 m².
NR means "No Requirement."

The use of polished wired glass complying with ANSI 97.1 in fire doors, fire windows and view panels in fire-resistance-rated walls is not permitted.

Code Text: *The following shall be considered specific hazardous locations for the purposes of glazing:* The code identifies eleven specific locations including glazing in and adjacent to doors, glazing in shower and bathtub enclosure walls, glazing in walls adjacent to pools and spas, and glazing adjacent to stairways and stairway landings.

Discussion and Commentary: A number of glazing locations have been identified over the years as being particular hazards due to their frequent impact by building occupants. Glazing in doors is of particular concern due to the increased likelihood of accidental impact by individuals operating or opening the doors. In addition, a person may push against a glazed portion of the door to gain leverage. Large pieces of glass may also create a hazard where located close to a travel path, because it is possible to impact glazing where no barrier is provided.

Safety glazing required

≤ 60"

For SI: 1 inch = 25.4 mm.

Glazing within a shower enclosure

There are a limited number of products and applications that are exempt from the requirements for hazardous locations, including small openings in doors through which a 3-inch-diameter sphere will not pass, and specific decorative assemblies, such as leaded, faceted or carved glass.

Code Text: *The following types of glazing may be used* (for skylights and sloped glazing)*: 1) laminated glass with a complying polyvinyl butyral interlayer; 2) fully tempered glass; 3) heat-strengthened glass; 4) wired glass; and 5) approved rigid plastics.*

Discussion and Commentary: A skylight or sloped glazing is considered *glass or other transparent or translucent glazing material installed at a slope of 15 degrees or more from vertical. Glazing materials in skylights, including unit skylights, solariums, sunrooms, roofs and sloped walls are included.* The requirements enhance the protection of a building's occupants from the possibility of falling glazing materials. Only certain glazing materials have the necessary characteristics to be permitted in an overhead installation. In addition, the provisions of the IRC address design loads normally attributed to roofs.

Glass area of any size

Slope of glass 30 degrees or less from vertical

Fully tempered glass w/o screening

10' max. to highest point of glass

Screen not required

For SI: 1 degree = 0.1745 rad 1 foot = 304.8 mm

Although fully-tempered and heat-strengthened glass are permitted as sloped glazing, it is often necessary that a protective retaining screen be located below the glass. Fully-tempered glass may spontaneously break into large chunks, whereas heat-strengthened glass can create falling shards.

Topic: Opening Protection

Reference: IRC R309.1

Category: Building Planning

Subject: Garages and Carports

Code Text: *Openings from a private garage directly into a room used for sleeping purposes shall not be permitted. Other openings between the garage and residence shall be equipped with solid wood doors not less than $1^3/_8$ inches (35 mm) in thickness, solid or honeycomb core steel doors not less than $1^3/_8$ inches (35 mm) thick, or 20-minute fire-rated doors.*

Discussion and Commentary: Although gypsum board or other approved material provides an adequate fire separation at the walls and/or ceiling between the garage and the dwelling unit, it is important that any other openings penetrating the separation be appropriately protected. Therefore, the type of door construction or fire rating of the door is mandated. Because the code covers only the door itself, it is not necessary to have a complete fire-rated door assembly, nor is it a requirement to provide a self-closing device. The level of separation provided by the door is consistent with that provided by the gypsum board installed on the garage side.

Cross section through garage

Ducts A - 0.019 in. (No. 26 Gage) galvanized steel with no openings into garage

Ducts B - any approved duct

For SI: 1 inch = 25.4 mm.

Where any duct passes through the gypsum board membrane located on the garage side of the separation, the duct material must be of minimum No. 26 gage sheet steel. There must be no openings in the duct within the garage area.

Code Text: *The garage shall be separated from the residence and its attic area by not less than $^1/_2$-inch (12.7 mm) gypsum board applied to the garage side. Garages beneath habitable rooms shall be separated from all habitable rooms above by not less than $^5/_8$-inch (15.9 mm) Type X gypsum board or equivalent. Where the separation is a floor-ceiling assembly, the structure supporting the separation shall also be protected by not less than $^1/_2$-inch (12.7 mm) gypsum board or equivalent.*

Discussion and Commentary: It is not uncommon for a fire to start in a private garage attached directly to a dwelling unit. In many cases, the fire may grow and go unnoticed for a period of time, becoming a distinct hazard to the residence and its occupants. The code mandates a minimum level of fire separation from the garage to the dwelling unit to allow the residents time for escape. In addition, the separation may be adequate to restrict the spread of fire to the dwelling unit until the fire can be controlled and extinguished.

The provisions regulating garages are applicable where the space is enclosed on at least three sides. Where enclosed on two sides or fewer, the parking area is considered a carport, and the separation requirements do not apply.

Code Text: *Garages located less than 3 feet (914 mm) from a dwelling unit on the same lot shall be protected with not less than $^1/_2$-inch (12.7 mm) gypsum board applied to the interior side of exterior walls that are within this area. Openings in these walls shall be regulated by Section R309.1. This provision does not apply to garage walls that are perpendicular to the adjacent dwelling unit wall.*

Discussion and Commentary: In order to protect a dwelling unit from a garage fire in a detached garage, a minimum degree of spatial separation is required. Where a separation of at least 3 feet is not provided, a minimum of one layer of $^1/_2$-inch gypsum board shall be applied on the interior side of the garage. Openings, such as doors or windows, that may occur in the garage wall that opposes the dwelling unit are regulated in a manner consistent with that for an attached garage.

Windows and other openings except complying doors not permitted

Minimum $^1/_2$" gypsum board

< 3'

Dwelling unit

Garage

Doors permitted if:
- Min $1^3/_8$" solid wood, or
- Min $1^3/_8$" steel, or
- 20-minute fire rated

Not applicable to walls or openings perpendicular to dwelling unit wall

For SI: 1 inch = 25.4 mm, 1 foot = 304.8 mm.

Although there is no appropriate method of measurement established for determining which portions of the garage wall must be protected, it is logical that the 3-foot measurement be taken at right angles from the garage wall insofar as the provision is based solely upon a fire spreading from the garage to the dwelling.

Code Text: *Basements and every sleeping room shall have at least one operable emergency escape and rescue opening.* See exception for small mechanical areas. *Such opening shall open directly into a public street, public alley, yard or court. Where basements contain one or more sleeping rooms, emergency egress and rescue openings shall be required in each sleeping room, but shall not be required in adjoining areas of the basement. Where emergency escape and rescue openings are provided they shall have a sill height of not more than 44 inches (1118 mm) above the floor.*

Discussion and Commentary: Because so many deaths and injuries from fire occur as the result of occupants of residential buildings being asleep at the time of a fire, the code requires that basements and all sleeping rooms have doors or windows that may be used for emergency escape or rescue. The concern is based on the fact that often a fire will have spread before the occupants are aware of the danger; thus, the normal means of escape will most likely be blocked.

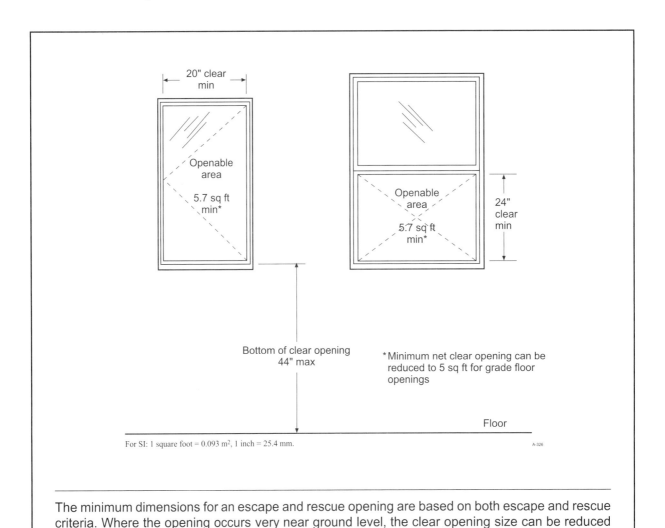

20" clear min

Openable area

5.7 sq ft min*

Openable area

5.7 sq ft min*

24" clear min

Bottom of clear opening 44" max

*Minimum net clear opening can be reduced to 5 sq ft for grade floor openings

Floor

For SI: 1 square foot = 0.093 m², 1 inch = 25.4 mm.

A-326

The minimum dimensions for an escape and rescue opening are based on both escape and rescue criteria. Where the opening occurs very near ground level, the clear opening size can be reduced because access from the exterior can be more easily accomplished without the use of a ladder.

Topic: Window Wells   **Category:** Building Planning
Reference: IRC R310.1, R310.2   **Subject:** Emergency Escape and Rescue Openings

Code Text: *Emergency escape and rescue openings with a finished sill height below the adjacent ground elevation shall be provided with a window well . . . . The minimum horizontal area of the window well shall be 9 square feet (0.9 m²), with a minimum horizontal projection and width of 36 inches (914 mm).* See exception for ladder encroachment into required dimensions. *The area of the window well shall allow the emergency and rescue opening to be fully opened.*

Discussion and Commentary: The window well provisions address those emergency escape and rescue openings that occur below grade. Simply applying the standard opening criteria to these window wells does not provide an adequate clear opening size to transition from the below-grade space to grade level. The increased cross-sectional dimension will be sufficient to provide for the escape of occupants or the entry of firefighters or other rescue personnel.

For SI: 1 foot = 304.8 mm, 1 inch = 25.4 mm.

Window Wells

It is acceptable for an emergency escape window to open into an area that is located under a porch or deck, provided the location of the porch or deck allows the window to be fully opened and an escape path at least 3 feet in height is available to a yard or court.

Quiz

Study Session 3
IRC Sections R303 – R310

1. When natural light is used to satisfy the minimum light requirements for habitable rooms, the aggregate glazing area shall be a minimum of _____ of the floor area.

   a. 4 percent   b. 5 percent

   (c.) 8 percent   d. 10 percent

   Reference R303.1

2. When natural ventilation is used to satisfy the minimum ventilation requirements for habitable rooms, the aggregate open area to the outdoors shall be a minimum of _____ of the floor area.

   (a.) 4 percent   b. 5 percent

   c. 8 percent   d. 10 percent

   Reference R303.1

3. Where a whole-house mechanical ventilation system is installed, the natural ventilation requirements are not applicable, provided the system is capable of supplying a minimum of _____ of outside ventilation air per computed occupant.

   (a.) 15 cfm   b. 20 cfm

   c. 0 cfm   d. 50 cfm

   Reference R303.1

4. Where an intermittent mechanical ventilation system is provided in a bathroom in lieu of exterior openings, the minimum required ventilation rate shall be _____.

   a. 15 cfm
   b. 20 cfm
   c. 30 cfm
   d. 50 cfm

   Reference R303.3

5. At least one habitable room with a minimum floor area of _____ square feet shall be provided in every dwelling unit.

   a. 100
   b. 120
   c. 150
   d. 200

   Reference R304.1

6. The floor area beneath a furred ceiling can be considered to be contributing to the minimum required habitable area for the room where it has a minimum height of _____ above the floor.

   a. 5 feet
   b. 6 feet, 4 inches
   c. 6 feet, 8 inches
   d. 7 feet

   Reference R304.4

7. Other than in a kitchen, what is the minimum permitted horizontal dimension of any habitable room?

   a. 6 feet
   b. 7 feet
   c. 8 feet
   d. 9 feet

   Reference R304.3

8. In general, the minimum ceiling height of all habitable rooms is _____.

   a. 6 feet, 8 inches
   b. 7 feet, 0 inches
   c. 7 feet, 6 inches
   d. 8 feet, 0 inches

   Reference R305.1

9. Ceilings in basements without habitable spaces shall have a minimum ceiling height of _____ from the finished floor to beams, ducts, or other obstructions.

    a. 6 feet, 4 inches
    b. 6 feet, 6 inches
    c. 6 feet, 8 inches
    d. 7 feet, 0 inches

    Reference R305.1 (2)

10. A minimum clearance of _____ shall be provided in front of a water closet.

    a. 21 inches
    b. 24 inches
    c. 30 inches
    d. 32 inches

    Reference Figure R307.1

11. The minimum distance between the centerline of a water closet and a side wall shall be _____.

    a. 15 inches
    b. 16 inches
    c. 18 inches
    d. 21 inches

    Reference Figure R307.1

12. Within a shower compartment, the walls shall be finished with a nonabsorbent surface to a minimum height of _____.

    a. 70 inches
    b. 72 inches
    c. 78 inches
    d. 84 inches

    Reference R307.2

13. Unless located a minimum of 60 inches above the floor or walking surface, glazing within a _____ arc of a door shall be considered to be installed in a hazardous location.

    a. 24-inch
    b. 30-inch
    c. 36-inch
    d. 60-inch

    Reference R308.4 / 6

14. Where located less than 5 feet above the walking surface, glazing in a wall enclosing an outdoor hot tub shall be considered to be installed in a hazardous location unless the glazing is a minimum of _____ horizontally from the water's edge.

    a. 3 feet   b. 4 feet

    c. 5 feet   d. 10 feet

Reference _R308.4/5_

15. A skylight is defined by the IRC as glass or other glazing material installed at a minimum slope of _____.

    a. 15 degrees from the vertical

    b. 30 degrees from the vertical

    c. 15 degrees from the horizontal

    d. 30 degrees from the horizontal

Reference _R308.6.1_

16. Curbs are not required for skylights installed on roofs having a minimum slope of _____.

    a. 2:12   b. 3:12

    c. 4:12   d. 5:12

Reference _R308.6.8_

17. Where a solid wood door is installed as a permitted opening between a garage and a residence, the minimum thickness of the door shall be _____.

    a. $1^1/_4$ inches   b. $1^3/_8$ inches

    c. $1^5/_8$ inches   d. $1^3/_4$ inches

Reference _R309.1_

18. Any duct penetrating the wall or ceiling separating a dwelling from a garage shall be constructed of a minimum _____ sheet steel or other approved material.

    a. No. 22 gage   b. No. 24 gage

    c. No. 26 gage   d. No. 28 gage

Reference _R309.1.1_

19. A garage shall be separated from the attic area of a residence by minimum _____ gypsum board applied to the garage side.

   a. $^1/_2$-inch   b. $^1/_2$-inch Type X

   c. $^5/_8$-inch   d. $^5/_8$-inch Type X

   Reference _R309.2_

20. Emergency escape and rescue openings shall have a maximum sill height of _____ above the floor.

   a. 40 inches   b. 42 inches

   c. 44 inches   d. 48 inches

   Reference _R310.1_

21. Emergency escape and rescue openings, when considered as grade floor openings, shall have a minimum net clear opening of _____.

   a. 4.0 square feet   b. 4.4 square feet

   c. 5.0 square feet   d. 5.7 square feet

   Reference _R310.1.1_

22. For an emergency escape and rescue opening, the minimum clear opening height shall be _____, and the minimum clear opening width shall be _____.

   a. 24 inches, 20 inches   b. 24 inches, 28 inches

   c. 28 inches, 22 inches   d. 28 inches, 24 inches

   Reference _R310.1.3_

23. Where a window well is required in conjunction with an escape and rescue opening, the window well shall have a minimum net clear area of _____ square feet with a minimum horizontal dimension of _____.

   a. 5.7, 20 inches   b. 5.7, 24 inches

   c. 9.0, 30 inches   d. 9.0, 36 inches

   Reference _R310.2_

24. A garage located beneath a habitable room shall be separated from the room by minimum _____ gypsum board applied to the garage side.

   a. $^1/_2$-inch

   b. $^1/_2$-inch Type X

   c. $^5/_8$-inch

   d. $^5/_8$-inch Type X ⟵ (circled)

   Reference _____

25. A 6-square-foot skylight of laminated glass shall have a minimum _____ polyvinyl butyral interlayer where located 14 feet above the walking surface.

   a.   0.015-inch ⟵ (circled)

   b.   0.024-inch

   c.   0.030-inch

   d.   0.044-inch

   Reference R308.6.2

26. Unless located at least 2 feet below the contaminant source, a mechanical outside air intake opening shall be located a minimum of _____ feet from a plumbing vent.

   a.   3

   b.   5

   c.   10 ⟵ (circled)

   d.   12

   Reference R303.4.1

27. The illumination source for interior stairs shall be capable of illuminating the treads and landings to a minimum level of _____ foot-candle(s).

   a.   1 ⟵ (circled)

   b.   5

   c.   8

   d.   10

   Reference R303.6

28. A minimum ceiling height of _____ is required at the front clearance area of bathroom fixtures.

   a.   6 feet, 4 inches

   b.   6 feet, 6 inches

   c.   6 feet, 8 inches ⟵ (circled)

   d.   7 feet, 0 inches

   Reference R305.1 /4

29. Glazing adjacent to a door to a storage closet is not required to be safety glazing provided the closet is a maximum of _____ in depth.

    a. 24 inches   b. 30 inches

    c. 36 inches   d. 42 inches

    Reference R308.4 /4
                   (EXCEPTIONS)

30. Glazing in a door shall be safety glazing where the opening allows the passage of a minimum _____ sphere.

    a. 3-inch   b. 3¹/₂-inch

    c. 4-inch   d. 6-inch

    Reference R308.4 / 1
                EXCEPTIONS

31. Where natural ventilation is utilized as the required ventilation for a bathroom, the minimum required openable area shall be _____ square feet.

    a. 1.5   b. 3.0

    c. 4.0   d. 5.7

    Reference R303.3

32. Where lighting outlets are installed in interior stairways, a wall switch shall be provided at each floor level to control the light outlet where the stairway has a minimum of _____ risers.

    a. two   b. three

    c. four   d. six

    Reference R303.6.1

33. A 4-square-foot glazed panel installed in an entry door shall have a minimum glazing category classification of Category _____.

    a. I   b. II

    c. III   d. IV

    Reference TABLE R308.3

34. Except for those basements with a maximum floor area of _____ square feet used only to house mechanical equipment, all basements shall be provided with at least one operable emergency escape and rescue opening.

    a. 120
                  b. 150

    (c.) 200
               d. 400

Reference __R310.1   EXCEPTION__

35. An emergency escape window may be installed under a deck or porch, provided a minimum height of _____ inches is maintained to a yard or court.

    a. 24
                (b.) 36

    c. 44
                d. 48

Reference __R310.5__

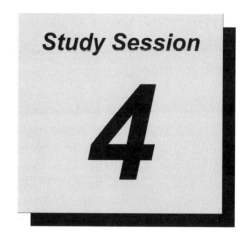

Study Session

2006 IRC Sections R311 – R324
Building Planning III

OBJECTIVE: To obtain an understanding of the fire and life safety criteria of the code, including stairways, ramps and landings; handrails and guards; smoke alarms; foam plastic and insulation; flame spread and smoke density of wall and ceiling finishes; separation of dwelling units, including townhouse separations; protection against decay and termite infestation; and flood-resistant construction.

REFERENCE: Sections R311 through R324, 2006 *International Residential Code*

KEY POINTS:
- How should enclosed space under stairs be addressed?
- What is the minimum required width of a hallway?
- How many exit doors from a dwelling are required? Under what conditions is an additional exit door required? What is the minimum required size of an exit door?
- What is the minimum size required for a landing? Under what conditions is a landing not required on each side of an exterior door?
- How wide must a stairway be? Where is this width measured? What type of encroachments into the minimum required width are permitted? How far may they encroach?
- What is the maximum allowable riser height in a stairway? Minimum allowable tread run? Within a flight of stairs, what is the maximum tolerance permitted between the smallest and greatest riser height? Between the smallest and greatest tread run?
- What are the various criteria for tread nosings? When are nosings not required?
- What is the minimum allowable headroom clearance at stairways and landings? How is this height measured?
- How is handrail height measured? What is the minimum height permitted? The maximum height? What sizes and shapes of handrails are acceptable?
- What types of special stairways are addressed? How do they differ in layout and size from traditional stairway design?
- What is the maximum slope permitted for a ramp? When is a handrail required for a ramp? What size landings are required?

KEY POINTS:
(Cont'd)

- When is a guard required? What is the minimum allowable height of the guard above the walking surface? How must intermediate rails be located? How do the provisions differ for guards on open sides of stairs?

- Where are smoke alarms required to be located? Do they need to be interconnected? How must they be powered? What must be done when alterations or repairs take place?

- How are foam plastics regulated? What are the restrictions on wall and ceiling finishes? Insulation materials?

- In a two-family dwelling, how must the units be separated? What minimum fire-resistive separation is mandated between townhouses? When is a common 2-hour fire-resistance-rated wall acceptable? When are parapets required? How are penetrations of dwelling unit separation assemblies to be protected?

- What areas are considered subject to decay damage? Termite damage? How should such conditions be abated?

- How must the premises be identified?

- What type of building constructed under the IRC requires some degree of accessibility?

- How are elevators and platform lifts regulated?

- What are the general requirements for buildings constructed in flood hazard areas?

Code Text: *Not less than one exit door conforming to the provisions of Section R311 shall be provided for each dwelling unit. The required exit door shall provide for direct access from the habitable portions of the dwelling to the exterior without requiring travel through a garage. The required exit door shall be a side-hinged door not less than 3 feet (914 mm) in width and 6 feet 8 inches (2032 mm) in height. Other doors shall not be required to comply with these minimum dimensions. All egress doors shall be readily openable from the side from which egress is to be made without the use of a key or special knowledge or effort.*

Discussion and Commentary: Regardless of the size of the dwelling unit, only one exterior exit door is required. Not only must the designated exit door be side-hinged, it must be at least 3 feet wide by 6 feet 8 inches in height. The other doors within the dwelling unit can be of any size and need not be of the swinging type.

Openable from inside without
- key
- special effort
- special knowledge

For SI: 1 inch = 25.4 mm, 1 foot = 304.8 mm.

Side-hinged 3 ft 0 in. x 6 ft 8 in. exit door

A-627

The designated exit door must be provided with latching and/or locking hardware that is under the control of and operable by any occupant seeking egress. The performance language of the code permits a variety of hardware options, provided they meet the intent of the provisions.

Topic: Landings at Doors

Category: Building Planning

Reference: IRC R311.4.3

Subject: Means of Egress

Code Text: *There shall be a floor or landing on each side of each exterior door. The floor or landing at the exterior door shall not be more than 1.5 inches (38 mm) lower than the top of the threshold.* See exceptions for 1) exterior landings at stairways of one or two risers located at other than the required exit door, 2) exterior landings at all exterior doorways where the door does not swing over the landing, and 3) the height of floors at exterior doors other than the required exit door.

Discussion and Commentary: As a general rule, a maximum elevation change of $1^1/_2$ inches is permitted at the exterior side of exterior doors, measured from the top of the threshold to the landing. A commonly utilized exception permits up to a $7^3/_4$-inch height difference, provided the exterior door, other than a screen or storm door, does not swing outward over the exterior landing. The user's typical familiarity with the change in elevation at the dwelling's exterior doors justifies such an allowance.

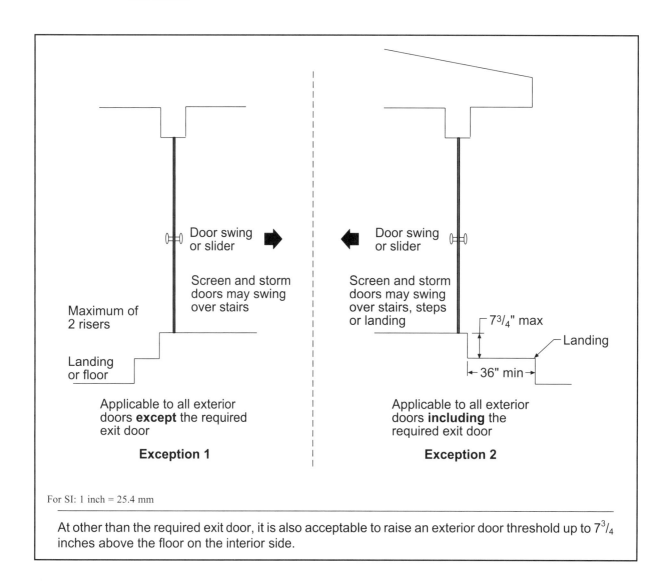

For SI: 1 inch = 25.4 mm

At other than the required exit door, it is also acceptable to raise an exterior door threshold up to $7^3/_4$ inches above the floor on the interior side.

Topic: Stairways	Category: Building Planning
Reference: IRC R311.5.1	Subject: Means of Egress

Code Text: *Stairways shall not be less than 36 inches (914 mm) in clear width at all points above the permitted handrail height and below the required headroom height. Handrails shall not project more than 4.5 inches (114 mm) on either side of the stairway and the minimum clear width of the stairway at and below the handrail height, including treads and landings, shall not be less than 31.5 inches (787 mm) where a handrail is installed on one side and 27 inches (698 mm) where handrails are provided on both sides.* See exception for the minimum width of spiral stairways.

Discussion and Commentary: Although a minimum width of 36 inches is required for stairways, the code is not as concerned about elements such as trim, stringers or other items that may be found below the level of the handrail, as long as they do not exceed the projection of the handrail(s). The width limitations are based on the body's movements as a person walks on a stairway.

Handrails

Riser

Tread

4¹/₂ in. max projection on each side into required 36 in. stairway width

Min 36 in. stairway width

Other projections of 4¹/₂ in. or less permitted below handrail height

For SI: 1 inch = 25.4 mm.

A-630

Typically, when a building code mandates a minimum width, it is expected that the width be the clear, net, usable, unobstructed width. In this case, however, limited projections into the minimum width are acceptable where located at or below the height of the handrail.

Code Text: *The maximum riser height shall be $7^3/_4$ inches (196 mm). The greatest riser height within any flight of stairs shall not exceed the smallest by more than $^3/_8$ inch (9.5 mm). The minimum tread depth shall be 10 inches (254 mm). The greatest tread depth within any flight of stairs shall not exceed the smallest by more than $^3/_8$ inch (9.5 mm).*

Discussion and Commentary: Although the rise and run configuration is important in regulating stairway safety, another significant safety factor relative to stairways is the uniformity of risers and treads in any flight (the section of a stairway leading from one landing to the next). It is important that any variation that would interfere with the rhythm of the stair user be avoided. Special allowances are provided for winder treads.

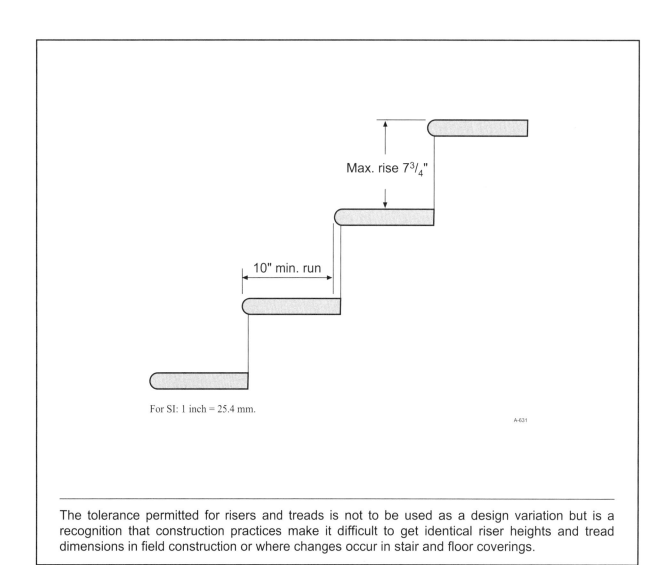

Max. rise $7^3/_4$"

10" min. run

For SI: 1 inch = 25.4 mm.

A-631

The tolerance permitted for risers and treads is not to be used as a design variation but is a recognition that construction practices make it difficult to get identical riser heights and tread dimensions in field construction or where changes occur in stair and floor coverings.

Topic: Special Stairways   **Category:** Building Planning
Reference: IRC R311.5.3.2, R311.5.8   **Subject:** Means of Egress

Code Text: *Winder treads shall have a minimum tread depth of 10 inches (254 mm) measured at a point 12 inches (305 mm) from the side where the treads are narrower. Winder treads shall have a minimum tread depth of 6 inches (152 mm) at any point. Spiral stairways are permitted, provided the minimum width shall be 26 inches (660 mm) with each tread having a 7 ¹/₂-inches (190 mm) minimum tread depth at 12 inches (305 mm) from the narrower edge. All treads shall be identical, and the rise shall be not more than 9 ¹/₂ inches (241 mm). Stairways serving bulkhead enclosures, not part of the required building egress, providing access from the outside grade level to the basement shall be exempt from the requirements of Sections R311.4.3 (landings at doors) and R311.5 (stairways) where the maximum height from the basement finished floor level to grade adjacent to the stairway does not exceed 8 feet (2438 mm).*

Discussion and Commentary: Three special types of stairways are allowed for use in dwelling units, provided they are constructed in compliance with the specific limitations mandated by the code.

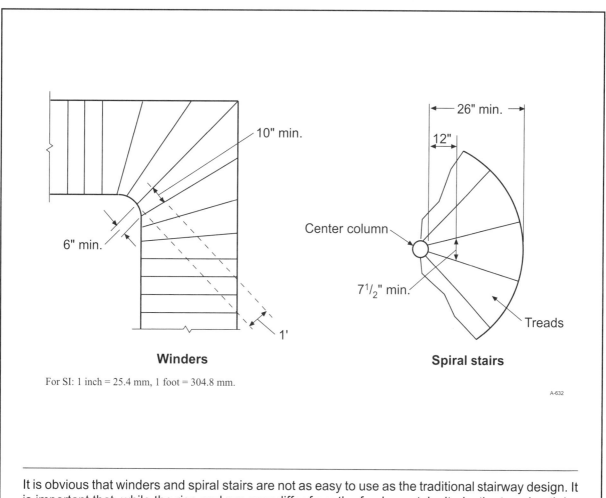

Winders   **Spiral stairs**

For SI: 1 inch = 25.4 mm, 1 foot = 304.8 mm.

A-632

It is obvious that winders and spiral stairs are not as easy to use as the traditional stairway design. It is important that, while the rise and run may differ from the fundamental criteria, the travel path be consistent in size and shape to reduce the hazard level.

Topic: Stairway Handrail	Category: Building Planning
Reference: IRC R311.5.6	Subject: Means of Egress

Code Text: *Handrails shall be provided on at least one side of each continuous run of treads or flight with four or more risers. Handrail height, measured vertically from the sloped plane adjoining the tread nosing, or finish surface of ramp slope, shall be not less than 34 inches (864 mm) and not more than 38 inches (965 mm). Handrails for stairways shall be continuous for the full length of the flight, from a point directly above the top riser of the flight to a point directly above the lowest riser of the flight. Handrail ends shall be returned or shall terminate in newel posts or safety terminals.* See exceptions.

Discussion and Commentary: Any stairway consisting of four or more risers must be provided with a handrail on at least one side. The rail is a proven safety feature for users of stairways, particularly those who fail to pay proper attention to stair use. Two types of handrail shapes are described, differing because of their perimeter measurement.

Type I circular handrail

1¹/₄" min.
2" max.

Section A-A

Handrail

38" max.
34" min.

Note: Vertical measurement made at the leading edge of the tread.

For SI: 1 inch = 25.4 mm.

A-633

An effective handrail must be of a size and shape that can be easily grasped by the majority of stair users. Whereas the minimum and maximum cross sections of 1¹/₄ inches and 2 inches are specific for circular rails, other shapes with equivalent grasping surfaces may be acceptable.

Topic: Ramps	**Category:** Building Planning
Reference: IRC R311.6	**Subject:** Means of Egress

Code Text: *Ramps shall have a maximum slope of one unit vertical in twelve units horizontal (8.3-percent slope). See exception where technically infeasible due to site constraints. A minimum 3-foot-by-3-foot (914 mm by 914 mm) landing shall be provided: at the top and bottom of ramps, where doors open onto ramps, and where ramps change direction. Handrails shall be provided on at least one side of all ramps exceeding a slope of one unit vertical in 12 units horizontal (8.33-percent slope).*

Discussion and Commentary: A ramp is defined as a walking surface that has a running slope steeper than 1 unit vertical in 20 units horizontal (5-percent slope). For general use purposes as well as exiting, the maximum permitted slope is 1:12. Ramps with slopes of 1:12 to 1:20 do not require a handrail, as the rise or descent is gradual enough to provide a safe travel path. Because ramps are generally limited to a 1:12 slope, the only ramps that will require a handrail are those permitted to be steeper than 1:12 by the exception for site constraints.

Handrail required when slope exceeds 1 unit vertical in 12 units horizontal.

12 max*

1

3 ft x 3 ft min landing required:
- At the top and bottom of ramp
- Where doors open onto ramp
- Where ramp changes direction

*1:8 maximum permitted where technically infeasible due to site constraints

For SI: 1 foot = 304.8 mm.

A-629

The ramp provisions in the IRC differ significantly from those in the IBC because of the building code's emphasis on accessibility. The IRC requirements are simply to allow safe movement through the dwelling unit and do not intend to address use by persons with physical disabilities.

Code Text: *Porches, balconies, ramps or raised floor surfaces located more than 30 inches (762 mm) above the floor or grade below shall have guards not less than 36 inches (914 mm) in height. Open sides of stairs with a total rise of more than 30 inches (762 mm) above the floor or grade below shall have guards not less than 34 inches (864 mm) in height measured vertically from the nosing of the treads.*

Discussion and Commentary: Some form of protection is necessary at elevated floor areas with a vertical drop of more than 30 inches, as a fall from that height can potentially result in injury. Where required, the guard must be of an adequate height to prevent someone from falling over the edge of the protected area.

For SI: 1 inch = 25.4 mm.

A-634

Required at balconies, porches, mezzanines and landings, guards are also needed on the open sides of stairs and ramps where the vertical distance to the surface below exceeds 30 inches. In such stairway situations, it is acceptable to use a complying handrail as the guard.

Topic: Opening Limitations
Reference: IRC R312.2

Category: Building Planning
Subject: Guards

Code Text: *Required guards on open sides of stairways, raised floor areas, balconies and porches shall have intermediate rails or ornamental closures which do not allow passage of a sphere 4 inches (102 mm) or more in diameter.* See exceptions for 1) triangular openings formed by riser, tread, and bottom rail, and 2) guards on open sides of stairs.

Discussion and Commentary: Guards must be constructed so that they not only prevent people from falling over them but also prevent children from crawling through them. The criteria spacing was chosen after many years of research and discussion. The chance of even a very small child being able to get through such a narrow opening is very low.

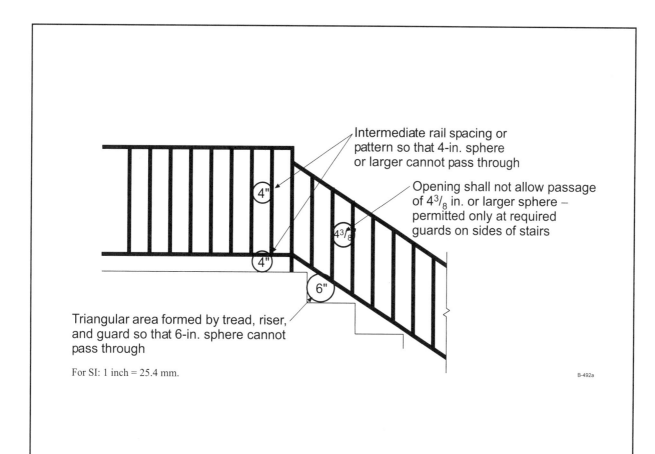

Intermediate rail spacing or pattern so that 4-in. sphere or larger cannot pass through

Opening shall not allow passage of 4³/₈ in. or larger sphere – permitted only at required guards on sides of stairs

Triangular area formed by tread, riser, and guard so that 6-in. sphere cannot pass through

For SI: 1 inch = 25.4 mm.

B-492a

In addition to limiting the opening size along the open sides of stairways, the code addresses the hazard created by open risers. Section R311.5.3.3 allows the use of open risers but only where the opening between the treads does not permit the passage of a 4-inch-diameter sphere.

Topic: Location and Interconnection	Category: Building Planning
Reference: IRC R313.2	Subject: Smoke Alarms

Code Text: *Smoke alarms shall be installed in the following locations: 1) in each sleeping room, 2) outside of each separate sleeping area in the immediate vicinity of the bedrooms, and 3) on each additional story of the dwelling, including basements but not including crawl spaces and unhabitable attics. When more than one smoke alarm is required to be installed within an individual dwelling unit the alarm devices shall be interconnected in such a manner that the actuation of one alarm will activate all of the alarms in the individual unit.*

Discussion and Commentary: A majority of deaths from fire in residential structures occur because of the delay in detecting and responding to a fire. Therefore, smoke alarms are mandated throughout the typically occupied areas of a dwelling unit to not only detect the products of combustion but also to alert the occupants of the fire.

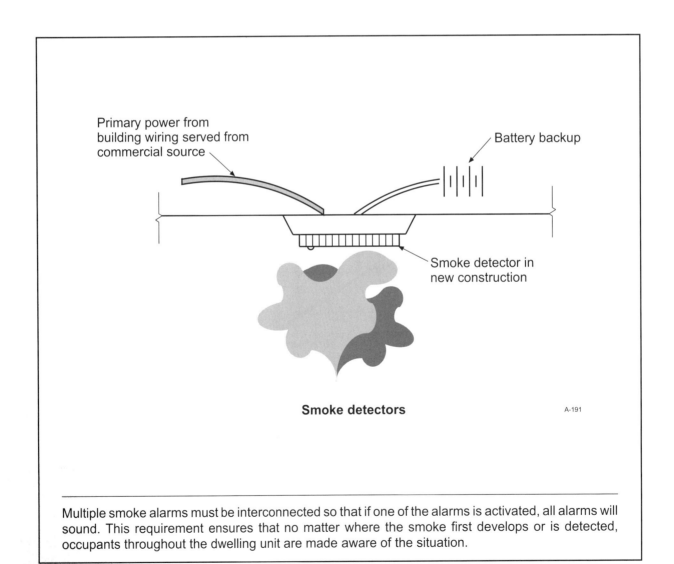

Primary power from building wiring served from commercial source

Battery backup

Smoke detector in new construction

Smoke detectors

A-191

Multiple smoke alarms must be interconnected so that if one of the alarms is activated, all alarms will sound. This requirement ensures that no matter where the smoke first develops or is detected, occupants throughout the dwelling unit are made aware of the situation.

Topic: Thermal Barrier

Category: Building Planning

Reference: IRC R314.1.4

Subject: Foam Plastic

Code Text: *Unless otherwise allowed in Section R314.5 or Section R314.6, foam plastic shall be separated from the interior of a building by an approved thermal barrier of minimum $^1/_2$-inch (12.7 mm) gypsum board or an approved finish material equivalent to a thermal barrier material that will limit the average temperature rise of the unexposed surface to no more than 250°F (139°C) after 15 minutes of fire exposure complying with the ASTM E 119 standard time temperature curve.*

Discussion and Commentary: Foam plastic must not be exposed to the building's interior during the early stages of a fire; rather, it must be isolated from the fire for a short period of time. The adequacy of the barrier is based on its ability to remain in place for 15 minutes.

Interior of building

Wall stud

Foam plastic insulation

Exterior sheathing

Min $^1/_2$ in. gypsum board or equivalent

For SI: 1 inch = 25.4 mm.

Optional methods of protection are provided where foam plastic insulation is placed in attic areas or crawl spaces. The level of separation is reduced, but not eliminated, where the attic or crawl space is accessed only for the service of utilities.

Code Text: *Wall and ceiling finishes shall have a flame-spread classification of not greater than 200. See exceptions for trim, doors and windows, and very thin materials cemented to a wall or ceiling surface. Wall and ceiling finishes shall have a smoke-developed index of not greater than 450. Tests shall be made in accordance with ASTM E 84.*

Discussion and Commentary: The control of interior finish materials is an important aspect of fire protection. The dangers of unregulated interior finish include both the rapid spread of fire and the contribution of additional fuel to the fire. Rapid fire spread presents a threat to the building occupants by limiting or denying their exit travel. This can be caused by either the rapid spread of the fire itself or by the production of large quantities of dense, black smoke that obscures the exit path or makes movement difficult.

Flame-spread ratings of typical construction materials

Material	Flame Spread
Ceilings:	
Glass-fiber sound-absorbing blankets	15 to 30
Mineral-fiber sound -absorbing panels	10 to 25
Sprayed cellulose fibers (treated)	20
Walls:	
Brick or concrete block	0
Cork	175
Gypsum board (with paper surface on both sides)	10 to 25
Southern pine (untreated)	130 to 190
Plywood paneling (untreated)	75 to 275
Red oak (untreated)	100

The installation of materials that will not significantly contribute to fire conditions is permitted. Trim, door and window frames, and baseboards are not regulated, because of their limited quantity. Thin materials such as wallpaper are allowed, as they tend to perform similar to their backing.

Topic: Two-family Dwellings	**Category:** Building Planning
Reference: IRC R317.1, R317.1.1	**Subject:** Dwelling Unit Separation

Code Text: *Dwelling units in two-family dwellings shall be separated from each other by wall and/or floor assemblies of not less than 1-hour fire-resistance rating when tested in accordance with ASTM E 119.* See exceptions for 1) a reduction in rating to $^1/_2$-hour for fully sprinklered buildings, and 2) a revised method of protection at the ceiling. *When floor assemblies are required to be fire-resistance-rated by Section R317.1, the supporting construction of such assemblies shall have an equal or greater fire-resistive rating.*

Discussion and Commentary: The code mandates a limited degree of fire separation to protect the occupants of one dwelling unit from the actions of their neighbor. Where the units are side by side, the separation wall must extend vertically to the underside of the roof deck. As an option to continuance of the 1-hour wall through the attic space, the ceiling can be protected with a minimum of one layer of $^5/_8$-inch Type X gypsum board, and the attic space need only be separated with a complying draftstop. Where one unit is located over the other, the fire-rated floor/ceiling assembly must extend to and be tight against the exterior walls. The required fire-resistance rating of one hour shall be verified by compliance with ASTM E 119, mandating an assembly constructed in accordance with its listing.

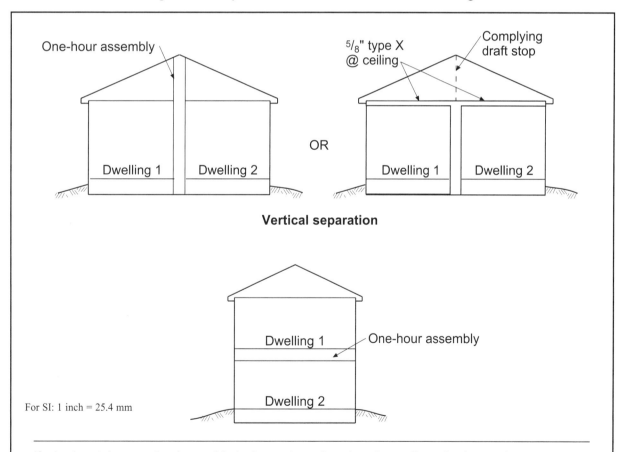

For SI: 1 inch = 25.4 mm

If a horizontal separation is provided, elements such as bearing walls and columns that support the separation must also have a 1-hour fire-resistance rating. It is of little value to provide a complying floor/ceiling assembly that quickly fails structurally because of unprotected supports.

Code Text: *Each townhouse shall be considered a separate building and shall be separated by fire-resistance-rated wall assemblies meeting the requirements of Section R302 for exterior walls. See exception for permitted use of common 2-hour fire-resistance-rated wall. The fire-resistance-rated wall or assembly separating townhouses shall be continuous from the foundation to the underside of the roof sheathing, deck or slab. The fire-resistance-rating shall extend the full length of the wall assembly, including wall extensions through and separating attached accessory enclosed structures.*

Discussion and Commentary: The application of townhouse provisions has its base in the exterior wall requirements of Section R302, which deal with the building's location on the lot. In general, because the "exterior wall" of the townhouse is essentially being constructed with no fire separation distance where one townhouse adjoins another, that wall must have a minimum 1-hour fire-resistance rating. The adjacent townhouse would have the same requirement; thus, two separate one-hour walls would be constructed side by side where one townhouse adjoins the other.

Construction type: gypsum wallboard, studs

Base layer $^5/_8$ in. Type X gypsum wallboard or veneer base applied at right angles to each side of 2 x 4 wood studs 24 in. on center with 6d coated nails, $1^7/_8$ in. long, 0.085 in. shank, $^1/_4$ in. heads, 24 in. on center. Face layer $^5/_8$ in. Type X gypsum wallboard or veneer base applied at right angles to studs over base layer with 8d coated nails, $2^3/_8$ in. long, 0.100 in. shank, $^1/_4$ in. heads, 8 in. on center. Stagger joints 24 in. on center each layer and side. Sound tested with studs 16 in. on center with nails for base layer spaced 6 in. on center (load-bearing).

2-hour wall assembly

(Fire Resistance/Sound Control Design Manual, 17th Edition, Gypsum Association)

For SI: 1 inch = 25.4 mm.

A-637

As an option, the construction of a single 2-hour fire-resistant wall is permitted in lieu of two separate 1-hour walls. This provides a similar level of separation between units. Where a 2-hour wall is constructed, no plumbing or mechanical elements are permitted within the wall.

Code Text: *Protection from decay shall be provided in the following locations by the use of naturally durable wood or wood that is preservative treated in accordance with AWPA U1 for the species, product, preservative and end use. Preservatives shall be listed in Section 4 of AWPA U1.* See seven specific locations where such lumber is required.

Discussion and Commentary: For those portions of a wood-framed structure that are subject to damage by decay, lumber must be pressure preservatively treated or of a species of wood having a natural resistance to decay. Such naturally durable wood includes the heartwood of decay-resistant redwood, cedars, black locust and black walnut. Crawl spaces and unexcavated areas under a building usually contain moisture-laden air. Foundation walls and floor slabs-on-grade absorb moisture from the ground and, by capillary action, move it to framing members to which they are in contact. Other locations also have similar conditions where damage to the wood is quite possible unless adequate clearance is maintained or the appropriate type of material is used.

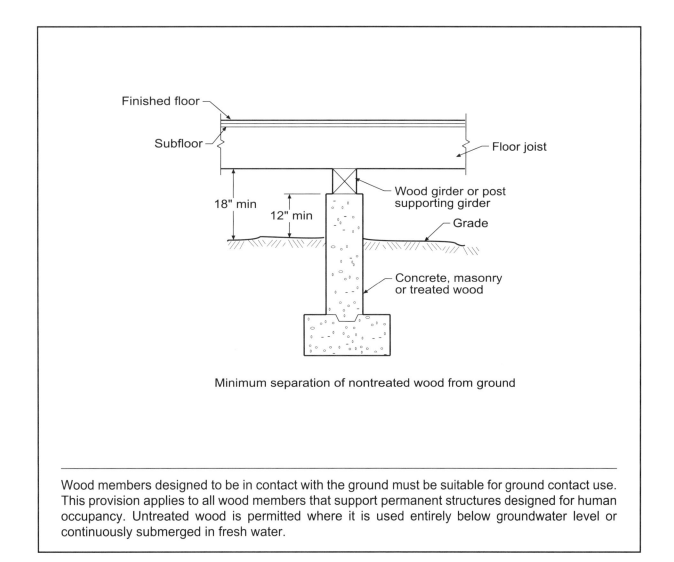

Minimum separation of nontreated wood from ground

Wood members designed to be in contact with the ground must be suitable for ground contact use. This provision applies to all wood members that support permanent structures designed for human occupancy. Untreated wood is permitted where it is used entirely below groundwater level or continuously submerged in fresh water.

Topic: Subterranean Control Methods
Reference: IRC R320.1

Category: Building Planning
Subject: Protection Against Termites

Code Text: *In areas subject to damage from termites as indicated by Table R301.2(1), methods of protection shall be one of the following methods or a combination of these methods: 1) chemical termiticide treatment, 2) termite baiting system, 3) pressure-preservative-treated wood, 4) naturally termite-resistant wood, and 5) physical barriers.*

Discussion and Commentary: Figure R301.2(6) depicts the geographical areas where termite damage to structures is probable. In those areas, the structure must be protected from termite damage in an appropriate manner. Although there are a number of approved methods, the most common is soil poisoning. Alternatives include the use of pressure preservatively treated wood and termite shields over perimeter walls. Often, a combination of these methods is necessary to establish the desired level of protection.

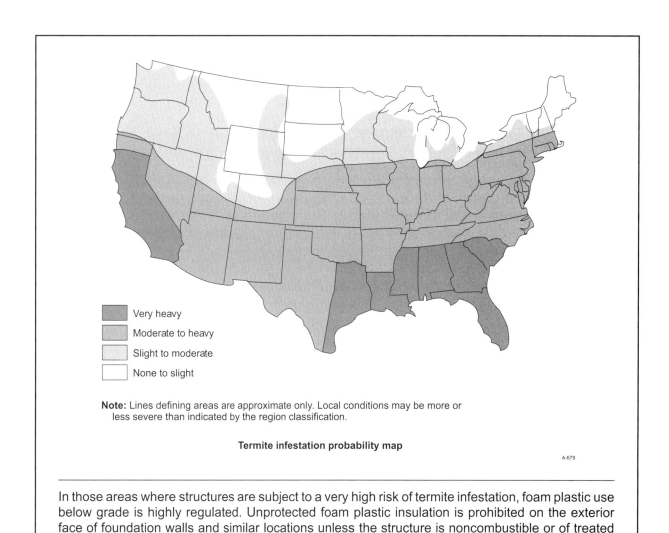

Very heavy

Moderate to heavy

Slight to moderate

None to slight

Note: Lines defining areas are approximate only. Local conditions may be more or less severe than indicated by the region classification.

Termite infestation probability map

A-679

In those areas where structures are subject to a very high risk of termite infestation, foam plastic use below grade is highly regulated. Unprotected foam plastic insulation is prohibited on the exterior face of foundation walls and similar locations unless the structure is noncombustible or of treated wood.

Code Text: *Approved numbers or addresses shall be provided for all new buildings in such a position as to be plainly visible and legible from the street or road fronting the property.*

Discussion and Commentary: It is important that buildings have plainly visible address numbers posted directly on the building or in such a place on the property that the building may be identified by emergency services such as fire, medical or police personnel. The primary concern is that emergency forces should be able to locate the building without going through a lengthy search procedure.

2903

If address not identifiable from street, it should posted on property to provide necessary visibility

Where an address identifier on the building will not be visible from the street, it is necessary that approved street numbers be placed in a location that would make them visible from the roadway fronting the property.

Code Text: *Where there are four or more dwelling units or sleeping units in a single structure, the provisions of Chapter 11 of the* International Building Code *for Group R-3 shall apply. In Group R-3 Occupancies where there are four or more dwelling units or sleeping units intended to be occupied as a residence in a single structure, every dwelling unit and sleeping unit intended to be occupied as a residence shall be a Type B unit.* See exception for reduction in number of Type B units.

Discussion and Commentary: Although most buildings constructed under the provisions of the *International Residential Code* are exempt from the accessibility provisions, structures containing four or more townhouses will be regulated. The required Type B units are described in ICC A117.1 and are consistent with the design and construction requirements of the federal Fair Housing Act.

A-327a

The technical requirements for Type B dwelling units are found in Section 1004 of ICC A117.1. The issues addressed by the accessibility standard include the primary entrance, accessible route, walking surfaces, doors, ramps, elevators and lifts, operable parts, bathrooms and kitchens.

Quiz

Study Session 4
IRC Sections R311 – R324

1. The minimum width of a hallway shall be _____.

   a. 32 inches   (b.) 36 inches

   c. 42 inches   d. 44 inches

   Reference R311.3

2. Every landing shall have a minimum length of _____ measured in the direction of travel.

   a. 30 inches   b. 32 inches

   (c.) 36 inches   d. 42 inches

   Reference R311.4.3

3. Unless technically infeasible due to site constraints, the maximum slope of a ramp shall be one unit vertical in _____ units horizontal.

   a. five   b. eight

   c. ten   (d.) twelve

   Reference R311.6.1

4. Handrails are not required on ramps having a maximum slope of _____.

   a. 1:5   b. 1:8

   (c.) 1:12   d. 1:15

   Reference R311.6.3

5. Where a door opens onto a ramp, a minimum _____ landing shall be provided.

   a. 32-inch-by-32-inch   (b.) 36-inch-by-36-inch

   c. 36-inch-by-60-inch   d. 60-inch-by-60-inch

   Reference _R311.6.2 / 2_

6. Stairways shall be a minimum _____ wide at all points above the permitted handrail height and below the required headroom height.

   a. 30 inches   b. 34 inches

   (c.) 36 inches   d. 38 inches

   Reference _R311.5.1_

7. Where a handrail is provided on one side of a stairway, what is the minimum required clear width at and below the handrail height?

   a. 27 inches   b. 29 inches

   (c.) $31^{1}/_{2}$ inches d. 36 inches

   Reference _R311.5.1_

8. Stairways shall have a maximum riser height of _____ and a minimum tread run of _____.

   a. $8^{1}/_{4}$ inches, 9 inches   b. 8 inches, 9 inches

   (c.) $7^{3}/_{4}$ inches, 10 inches   d. $7^{1}/_{2}$ inches, 10 inches

   Reference _R311.5.3.1 AND .2_

9. The maximum variation between the greatest riser height and the smallest riser height within any flight of stairs shall be _____.

   a. $^{1}/_{4}$ inch   (b.) $^{3}/_{8}$ inch

   c. $^{1}/_{2}$ inch   d. $^{5}/_{8}$ inch

   Reference _R311.5.3.1_

10. Unless a minimum tread depth of _____ is provided, a minimum $^{3}/_{4}$-inch nosing is required at the leading edge of all treads.

    a. $9^{1}/_{2}$ inches   b. 10 inches

    (c.) 11 inches   d. 12 inches

    Reference _R311.5.3.3_
    EXCEPTIONS / 1

11. Unless the stair has a total rise of no more than 30 inches, the opening between treads at open risers shall be such that a minimum _____ sphere shall not pass through.

   a. 3-inch

   b. 4-inch

   c. 6-inch

   d. no limitation is mandated

   Reference R311.5.3.3

12. Along all portions of a stairway, including at landings, the minimum headroom shall be _____.

   a. 6 feet, 6 inches

   b. 6 feet, 8 inches

   c. 6 feet, 9 inches

   d. 7 feet, 0 inches

   Reference R311.5.2

13. What is the maximum riser height permitted for spiral stairways?

   a. $7^3/_4$ inches

   b. 8 inches

   c. $8^1/_4$ inches

   d. $9^1/_2$ inches

   Reference R311.5.8.1

14. Stairway handrail height shall be measured vertically from the nosing of the treads, a minimum of _____ and a maximum of _____ .

   a. 30 inches, 34 inches

   b. 30 inches, 38 inches

   c. 34 inches, 38 inches

   d. 34 inches, 42 inches

   Reference R311.5.6.1

15. Where a stairway handrail is located adjacent to a wall, a minimum clearance of _____ shall be provided between the wall and the handrail.

   a. $1^1/_4$ inches

   b. $1^1/_2$ inches

   c. 2 inches

   d. $3^1/_2$ inches

   Reference R311.5.6.2

16. A Type I handrail having a circular cross section shall have a minimum outside diameter of _____ and a maximum outside diameter of _____ .

   a. $1^1/_4$ inches, 2 inches

   b. $1^1/_4$ inches, $2^5/_8$ inches

   c. $1^1/_2$ inches, 2 inches

   d. $1^1/_2$ inches, $2^5/_8$ inches

   Reference R311.5.6.3 / 1

17. A porch, balcony or similar raised floor surface more than 30 inches above the floor or grade below shall be provided with a guard having a minimum height of _____.

    a. 32 inches   b. 34 inches

    (c.) 36 inches   d. 42 inches

    Reference R312.1

18. Required guards on open sides of raised floor areas shall be provided with intermediate rails or ornamental closures such that a minimum _____ sphere cannot pass through.

    a. 3-inch   (b.) 4-inch

    c. 6-inch   d. 9-inch

    Reference R312.2

19. Which one of the following areas in a dwelling unit does not specifically require the installation of a smoke alarm?

    a. basement   (b.) kitchen

    c. habitable attic   d. sleeping room

    Reference R313.2

20. In general, foam plastic shall be separated from the interior of a dwelling by an approved thermal barrier of minimum _____.

    a. $^3/_8$-inch gypsum board   (b.) $^1/_2$-inch gypsum board

    c. $^5/_8$-inch gypsum board   d. $^5/_8$-inch Type X gypsum board

    Reference R314.4

21. Regulated wall and ceiling finishes shall have a maximum flame spread classification of _____.

    a. 25   b. 75

    (c.) 200   d. 450

    Reference R315.1

22. What is the minimum required separation for a wall between dwelling units in a nonsprinklered two-family dwelling?

    a. $^1/_2$-inch gypsum board on each side

    b. 30-minute fire-resistance rated

    (c.) 1-hour fire-resistance rated

    d. 2-hour fire-resistance rated

Reference R317.1

23. When a common wall is used to separate townhouses, it shall have a minimum _____ fire-resistance rating.

    a. 1-hour   (b.) 2-hour

    c. 3-hour   d. 4-hour

Reference R317.2 / EXCEPTION

24. Unless they are made of pressure preservatively treated or naturally durable wood, wood sill plates that rest on masonry or concrete exterior foundation walls shall be located a minimum of _____ from exposed ground.

    a. 6 inches   (b.) 8 inches

    c. 12 inches   d. 18 inches

Reference R319/2

25. In the establishment of a design flood elevation, the depth of peak elevation of flooding for a _____ flood is used.

    a. 10-year   b. 20-year

    c. 50-year   (d.) 100-year

Reference R324.1.3

26. Walls and the under stair surfaces that enclose accessible space under a stairway shall be protected on the enclosed side by minimum _____ gypsum board.

    (a.) $^1/_2$-inch   b. $^1/_2$-inch Type X

    c. $^5/_8$-inch   d. $^5/_8$-inch Type X

Reference R311.2.2

27. Where foam plastic is spray applied to a sill plate without a thermal barrier, it shall have a maximum density of _____ pcf.

    a. 1.0   b. 1.5

    c. 2.0   d. 3.0

Reference _R314.5.11/2_

28. The maximum width of a Type II handrail above the recess shall be _____ inches.

    a. $1^1/_2$ b. 2

    c. $2^5/_8$ d. $2^3/_4$

Reference _R311.5.6.3/2_

29. Handrails shall be provided on at least one side of a stairway having a minimum of _____ risers.

    a. 2   b. 3

    c. 4   d. 5

Reference _R311.5.6_

30. What is the minimum required size for an exit door from a dwelling unit?

    a. 32 inches by 78 inches   b. 32 inches by 80 inches

    c. 36 inches by 78 inches   d. 36 inches by 80 inches

Reference _R311.4.2_

31. Where the door does not swing over the landing, the exterior landing at an exterior doorway shall be located a maximum of _____ inches below the top of the threshold.

    a. 7   b. $7^1/_2$

    c. $7^3/_4$ d. 8

Reference _R311.4.3/2_

32. Where nosings are required on a stairway with solid risers, the nosings shall extend a minimum of _____ inch(es) and a maximum of _____ inch(es) beyond the risers.

   a. $^1/_2$, 1

   b. $^3/_4$, $1^1/_4$

   c. 1, $1^1/_2$

   d. $1^1/_4$, 2

   Reference _R311.5.3.3_

33. A flight of stairs shall have a maximum vertical rise of _____ feet between floor levels or landings.

   a. 8

   b. 9

   c. 10

   d. 12

   Reference _R311.5.4_ / _EXCEPTIONS_

34. Openings for required guards on the sides of stair treads shall be provided with intermediate rails or ornamental closures such that a minimum _____ sphere cannot pass through.

   a. $3^1/_2$

   b. 4

   c. $4^3/_8$

   d. 6

   Reference _R312.2_ / _2_ _EXCEPTIONS_

35. Where the wall assembly separating two dwelling units in a two-family dwelling does not extend through the attic space, the ceiling shall be protected by minimum _____ gypsum board.

   a. $^5/_8$-inch Type X

   b. $^5/_8$-inch

   c. $^1/_2$-inch Type

   d. $^1/_2$-inch

   Reference _R317.1_ / _2_ _EXCEPTIONS_

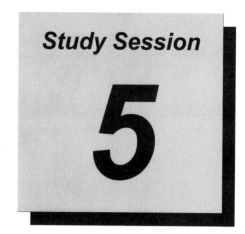

Study Session

5

2006 IRC Chapter 4
Foundations

OBJECTIVE: To gain an understanding of the requirements relating to foundation systems, including soil evaluation; footing size, depth and slope; foundation design and reinforcing; frost-protected shallow foundations; wood foundations; insulating form foundation walls; foundation drainage; waterproofing and dampproofing; and under-floor spaces.

REFERENCE: Chapter 4, 2006 *International Residential Code*

KEY POINTS:
- How must lots be graded? What is the minimum required slope of the grade away from the foundation?
- When are soils tests required? When are presumptive load-bearing values to be used?
- What criteria determines the minimum size of footings? What is the minimum required footing thickness? Minimum projection distance? Minimum depth?
- What additional conditions must be met for footings located in Seismic Design Categories D_0, D_1 and D_2? Where must any required reinforcing be located?
- What is the maximum slope permitted for the top surface of footings? The bottom surface?
- In what manner must the sill plate be connected to the foundation wall? What is the minimum size and spacing for anchor bolts? How must the connection be made in Seismic Design Categories D_0, D_1 and D_2?
- How shall buildings be located in respect to an adjacent ascending slope? Descending slope?
- How are foundations and floor slabs for buildings located on expansive soils to be regulated?
- Where a building is constructed on or adjacent to a slope, what is the minimum required height of a foundation above a street gutter or other point of drainage discharge?
- What are the provisions for the installation of wood foundation systems?
- How does a frost-protected shallow foundation work? What are the primary conditions for compliance?
- What criteria are necessary to address plain concrete and masonry foundation walls? Reinforced concrete and masonry foundation walls?

- What special conditions are required for foundation walls of buildings constructed in Seismic Design Categories D_0, D_1 and D_2? Where must reinforcement be located?
- When are pier foundations acceptable? What are the minimum requirements?
- What is an insulating concrete form (ICF) foundation wall system? What are the different types of ICF systems? How are they regulated?
- How must retaining walls be designed?
- When are drains required around foundation walls? What drainage methods are acceptable?
- How are soils classifed? What are the characteristics of the various soil groups?
- When must a foundation wall be dampproofed? Waterproofed? What methods are used for dampproofing and waterproofing?
- What is the minimum required size for under-floor ventilation openings? Where shall such openings be located? How may the required area of openings be reduced?
- Under what conditions is an unvented crawl space permitted?
- How shall access to an under-floor space be provided?
- When is a drainage system needed for an under-floor space?

Code Text: *Surface drainage shall be diverted to a storm sewer conveyance or other approved point of collection so as to not create a hazard. Lots shall be graded so as to drain surface water away from foundation walls. The grade shall fall a minimum of 6 inches (152 mm) within the first 10 feet (3048 mm).* See exception for use of drains or swales.

Discussion and Commentary: Proper site drainage is an important element in preventing wet basements, damp crawl spaces, eroded banks and the possible failure of a foundation system. Drainage patterns should result in adequate slopes to approved drainage devices that are capable of carrying concentrated runoff. It is often necessary to use gutters and downspouts to direct roof water to the appropriate drainage points.

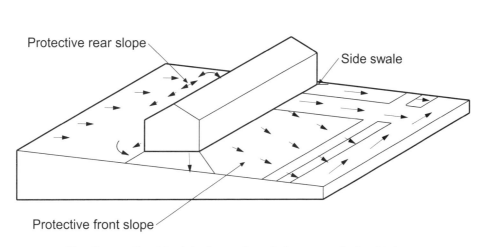

Grading method for lot where slope is from rear to front lot. Drainage swales are located at rear and sides of dwelling.

B-493

Drains or swales are effective measures where the site does not allow for the necessary fall away from the structure. Minimum slope gradients for swales are often based on ground frost and moisture conditions. The permeability of soil and type of drainage devices are also important.

Topic: Soil Tests

Category: Foundations

Reference: IRC R401.4, R401.4.1

Subject: General Requirements

Code Text: *In areas likely to have expansive, compressible, shifting or other unknown soil characteristics, the building official shall determine whether to require a soil test to determine the soil's characteristics at a particular location. This test shall be made by an approved agency using an approved method. In lieu of a complete geotechnical evaluation, the load-bearing values in Table R401.4.1 shall be assumed.*

Discussion and Commentary: Expansive soils, soil instability and increased lateral pressure due to a high water table or to surcharge loads from adjacent structures are special conditions that must be considered in the design of a foundation system. Where such conditions occur, they are beyond the scope of the prescriptive provisions of the code for foundation systems. In some cases, a soil test may be required to evaluate the soil's characteristics.

TABLE R401.4.1
PRESUMPTIVE LOAD–BEARING VALUES OF FOUNDATION MATERIALS[a]

CLASS OF MATERIAL	LOAD-BEARING PRESSURE (pounds per square foot)
Crystalline bedrock	12,000
Sedimentary and foliated rock	4,000
Sandy gravel and/or gravel (GW and GP)	3,000
Sand, silty sand, clayey sand, silty gravel and clayey gravel (SW, SP, SM, SC, GM and GC)	2,000
Clay, sandy clay, silty clay, clayey silt, silt and sandy silt (CL, ML, MH and CH)	1,500[b]

For SI: 1 pound per square foot = 0.0479 kPa.

a. When soil tests are required by Section R401.4, the allowable bearing capacities of the soil shall be part of the recommendations.

b. Where the building official determines that in-place soils with an allowable bearing capacity of less than 1,500 psf are likely to be present at the site, the allowable bearing capacity shall be determined by a soils investigation.

Where the bearing capacity of the soil has not been determined by geotechnical analysis such as borings, field load tests, laboratory tests and/or engineering analysis, it is a common practice to use presumptive bearing values for the design of the foundation system.

Code Text: *All lumber and plywood used in wood foundation systems shall be treated and dried after treatment in accordance with AWPA U1, and shall bear the label of an accredited agency. Concrete shall have a minimum specified compressive strength of f'$_c$, as shown in Table R402.2. Concrete subject to moderate or severe weathering as indicated in Table R301.2(1) shall be air entrained as specified in Table R402.2.*

Discussion and Commentary: Freezing and thawing cycles can be the most destructive weathering factors for concrete. For nonair entrained concrete, deterioration is caused by the water freezing in the cement matrix, in the aggregate, or both. Studies have documented that concrete provided with proper air entrainment is highly resistant to this deterioration.

TABLE R402.2
MINIMUM SPECIFIED COMPRESSIVE STRENGTH OF CONCRETE

TYPE OR LOCATION OF CONCRETE CONSTRUCTION	MINIMUM SPECIFIED COMPRESSIVE STRENGTH[a] (f'$_c$)		
	Weathering Potential[b]		
	Negligible	Moderate	Severe
Basement walls, foundations and other concrete not exposed to the weather	2,500	2,500	2,500[c]
Basement slabs and interior slabs on grade, except garage floor slabs	2,500	2,500	2,500[c]
Basement walls, foundation walls, exterior walls and other vertical concrete work exposed to the weather	2,500	3,000[d]	3,000[d]
Porches, carport slabs and steps exposed to the weather, and garage floor slabs	2,500	3,000[d,e,f]	3,500[d,e,f]

For SI: 1 pound per square inch = 6.895 kPa.
a. Strength at 28 days psi.
b. See Table R301.2(1) for weathering potential.
c. Concrete in these locations that may be subject to freezing and thawing during construction shall be air-entrained concrete in accordance with Footnote d.
d. Concrete shall be air-entrained. Total air content (percent by volume of concrete) shall be not less than 5 percent or more than 7 percent.
e. See Section R402.2 for maximum cementitious materials content.
f. For garage floors with a steel troweled finish, reduction of the total air content (percent by volume of concrete) to not less than 3 percent is permitted if the specified compressive strength of the concrete is increased to not less than 4,000 psi.

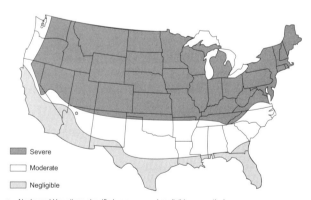

■ Severe
□ Moderate
▨ Negligible

a. Alaska and Hawaii are classified as severe and negligible, respectively.
b. Lines defining areas are approximate only. Local conditions may be more or less severe than indicated by region classification. A severe classification is where weather conditions result in significant snowfall combined with extended periods during which there is little or no natural thawing causing deicing salts to be used extensively.

Weathering probability map for concrete

A-680

Performance of wood foundation systems is dependent on the use of properly treated materials. Verification of the proper materials is provided by requiring identification showing the approval of an accredited inspection agency.

Code Text: *Minimum sizes for concrete and masonry footings shall be as set forth in Table R403.1 and Figure R403.3.1(1). The footing width, W, shall be based on the load-bearing value of the soil in accordance with Table R401.4.1. Spread footings shall be at least 6 inches (152 mm) in thickness. Footing projections, P, shall be at least 2 inches (51 mm) and shall not exceed the thickness of the footing.*

Discussion and Commentary: To avoid overstressing the footing, minimum size requirements are established based on construction type, number of stories and soil load-bearing value. The projection limitation is critical for 6-inch-thick footings supported by poor soil conditions. Excessive projections could result in the footing being cracked in the same plane as the foundation wall, which could occur if the allowable stress in the concrete is exceeded.

TABLE R403.1
MINIMUM WIDTH OF CONCRETE OR
MASONRY FOOTINGS
(inches)[a]

	LOAD-BEARING VALUE OF SOIL (psf)			
	1,500	2,000	3,000	≥4,000
Conventional light–frame construction				
1-story	12	12	12	12
2-story	15	12	12	12
3-story	23	17	12	12
4-inch brick veneer over light frame or 8-inch hollow concrete masonry				
1-story	12	12	12	12
2-story	21	16	12	12
3-story	32	24	16	12
8-inch solid or fully grouted masonry				
1-story	16	12	12	12
2-story	29	21	14	12
3-story	42	32	21	16

For SI: 1 inch = 25.4 mm, 1 pound per square foot = 0.0479 kPa.

a. Where minimum footing width is 12 inches, use of a single wythe of solid or fully grouted 12-inch nominal concrete masonry units is permitted.

Basement or crawl space
with concrete wall and
spread footing

Basement or crawl space
with foundation wall
bearing directly on soil

A-641a

In buildings constructed in Seismic Design Categories D_0, D_1 or D_2, footing reinforcement is mandated. Interconnection of the stem wall and its supporting footing is necessary to resist the tendency to slip during an earthquake. Various methods address different foundation systems.

Code Text: *All exterior footings shall be placed at least 12 inches (305 mm) below the undisturbed ground surface. Except where otherwise protected from frost, foundation walls, piers and other permanent supports of buildings and structures shall be protected from frost by one or more of the following methods: 1) extending below the frost line specified in Table R301.2(1); 2) constructing in accordance with Section R403.3 (frost protected shallow foundations); 3) constructing in accordance with ASCE 32 (Design and Construction of Frost Protected Shallow Foundations); and 4) erected on solid rock.* See exceptions for small freestanding accessory structures and decks not supported by a dwelling.

Discussion and Commentary: The volume changes (frost heave) that take place during freezing and thawing cycles produce excessive stresses in the foundations and create extensive damage to the supported walls. Thus, foundations must be extended below the depth of frost penetration.

Depth of footing

C-282

For SI: 1 inch = 25.4 mm.

Buildings should not be founded on or within frozen ground. This is necessary in order to prevent damage to the exterior walls and other walls bearing on frozen soil. Only where the soil is maintained in a permanently frozen condition can such construction be permitted.

Code Text: *The top surface of footings shall be level. The bottom surface of footings shall not have a slope exceeding one unit vertical in 10 units horizontal (10-percent slope). Footings shall be stepped where it is necessary to change the elevation of the top surface of the footings or where the slope of the bottom surface of the footings will exceed one unit vertical in ten units horizontal (10-percent slope).*

Discussion and Commentary: Although the code places no restriction on a stepped foundation, there is a recommended overlap of the top of the foundation wall beyond the step in the foundation. It should be larger than the vertical step in the foundation wall at that point. This recommendation is based on possible crack propagation at an angle of 45 degrees.

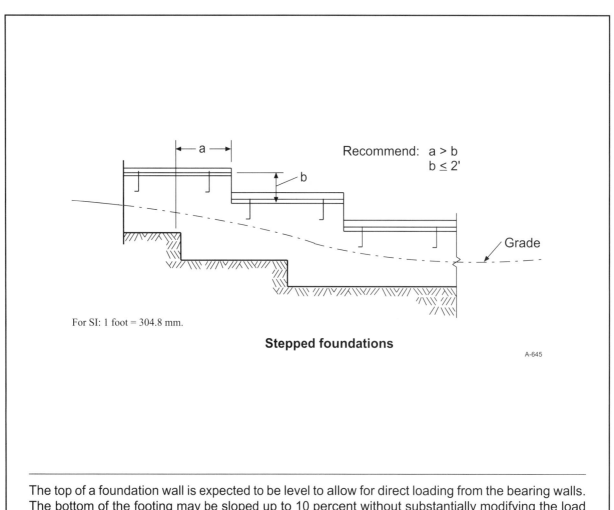

For SI: 1 foot = 304.8 mm.

Stepped foundations

A-645

The top of a foundation wall is expected to be level to allow for direct loading from the bearing walls. The bottom of the footing may be sloped up to 10 percent without substantially modifying the load path. If a steeper slope is encountered, the footing must be stepped.

Code Text: *When braced wall panels are supported directly on continuous foundations, the wall wood sill plate or cold-formed steel bottom track shall be anchored to the foundation in accordance with* Section R403.1.6. *The wood sole plate at exterior walls on monolithic slabs and wood sill plate shall be anchored to the foundation with anchor bolts spaced a maximum of 6 feet (1829 mm) on center. There shall be a minimum of two bolts per plate section with one bolt located not more than 12 inches (305 mm) or less than seven bolt diameters from each end of the plate section. Bolts shall be at least $^1/_2$ inch (13 mm) in diameter and shall extend a minimum of 7 inches (178 mm) into masonry or concrete.* See exceptions for foundation anchor straps and short walls connecting offset braced wall panels.

Discussion and Commentary: To prevent walls and floors from shifting under lateral loads, anchorage to the supporting foundation is needed. The minimum required connection is supplied by anchor bolts. Foundation straps are also acceptable if installed in accordance with the manufacturer's instructions and spaced to provide equivalent hold-down strength.

Within 12" of end of each piece but no closer to end than 7 bolt diameters

6' max.

Minimum of two bolts per piece

For SI: 1 inch = 25.4 mm, 1 foot = 304.8 mm.

Wood sill plate anchorage

A-120c

In Seismic Design Categories D_0, D_1 and D_2, large square plate washers are mandated to compensate for the practice of oversized bolt holes. These washers enable the nuts to be tightened enough to achieve an increased clamping action between the foundation wall and the sill plate. See Section R602.11.1 for anchorage details.

Code Text: *The placement of buildings and structures on or adjacent to slopes steeper than 1 unit vertical in 3 units horizontal (33.3-percent slope) shall conform to Sections R403.1.7.1 through R403.1.7.4. On graded sites, the top of any exterior foundation shall extend above the elevation of the street gutter at point of discharge or the inlet of an approved drainage device a minimum of 12 inches (305 mm) plus 2 percent. Alternate elevations are permitted subject to the approval of the building official, provided it can be demonstrated that required drainage to the point of discharge and away from the structure is provided at all locations on the site.*

Discussion and Commentary: Where natural drainage away from a building is not available, the site must be graded so that water will not drain toward, or accumulate at, the exterior foundation wall. A prescriptive elevation is set forth that will ensure positive drainage to a street gutter or other drainage point; however, any other method that moves water away from the building can be accepted by the building official.

Footing elevation on graded sites

Foundation clearances from slopes

For SI: 1 inch = 25.4 mm, 1 foot = 304.8 mm

When footings are located adjacent to a slope steeper than 1:3 (1 vertical to 3 horizontal), either at the top or the bottom, special clearances between the building and the sloping surfaces are required to protect against slope drainage, erosion and shallow failures.

Topic: Frost Protected Foundations

Category: Foundations

Reference: IRC R403.3

Subject: Footings

Code Text: *For buildings where the monthly mean temperature of the building is maintained at a minimum of 64°F (18°C), footings are not required to extend below the frost line when protected from frost by insulation in accordance with Figure R404.3(1) and Table R403.3.* Not permitted for unheated areas such as porches, garages and crawl spaces.

Discussion and Commentary: As a fundamental rule, footings must be placed below the frost line. However, the use of a frost protected foundation is an acceptable alternative, allowing placement of the footing/foundation above the frost line. This foundation system uses insulation to reduce the heat loss at the slab edge. By holding heat from the dwelling in the ground under the foundation, the insulation eliminates the potential for freezing, and thus the consequences of freeze/thaw conditions are avoided.

TABLE R403.3
MINIMUM INSULATION REQUIREMENTS FOR FROST-PROTECTED FOOTINGS IN HEATED BUILDINGS[a]

AIR FREEZING INDEX (°F-days)[b]	VERTICAL INSULATION R-VALUE[c,d]	HORIZONTAL INSULATION R-VALUE[c,e]		HORIZONTAL INSULATION DIMENSIONS PER FIGURE R403.3(1) (inches)		
		Along walls	At corners	A	B	C
1,500 or less	4.5	Not required	Not required	Not required	Not required	Not required
2,000	5.6	Not required	Not required	Not required	Not required	Not required
2,500	6.7	1.7	4.9	12	24	40
3,000	7.8	6.5	8.6	12	24	40
3,500	9.0	8.0	11.2	24	30	60
4,000	10.1	10.5	13.1	24	36	60

a. Insulation requirements are for protection against frost damage in heated buildings. Greater values may be required to meet energy conservation standards. Interpolation between values is permissible.
b. See Figure R403.3(2) for Air Freezing Index values.
c. Insulation materials shall provide the stated minimum R–values under long-term exposure to moist, below-ground conditions in freezing climates. The following R–values shall be used to determine insulation thicknesses required for this application: Type II expanded polystyrene—2.4R per inch; Type IV extruded polystyrene—4.5R per inch; Type VI extruded polystyrene—4.5R per inch; Type IX expanded polystyrene—3.2R per inch; Type X extruded polystyrene—4.5R per inch.
d. Vertical insulation shall be expanded polystyrene insulation or extruded polystyrene insulation.
e. Horizontal insulation shall be extruded polystyrene insulation.

Insulation detail

Flashing per Section R703.8

Insulation protection per Section R403.3.2

Slope final grade per Section R403.3.3

12" max

12" min

Slab-on-ground foundation floor per Sections R403.1 and R506

a. See Table R403.3 for required dimensions and R-values for vertical and horizontal insulation.

Vertical wall insulation[a]

Horizontal insulation[a]

Nominal 4 in. screened and washed gravel or crushed stone, drained per Section R403.3.3

For SI: 1 inch = 25.4 mm.

A-646a

It is often necessary to extend the insulation away from the foundation horizontally. By doing so, it is probable that the insulating material will be damaged by landscape work or other activities unless an approved method of protection is provided.

Code Text: *Concrete masonry and clay masonry foundation walls shall be constructed as set forth in Table R404.1.1(1)* (Plain Masonry), *R404.1.1(2)* (8-inch Reinforced Masonry), *R404.1.1(3)* (10-inch Reinforced Masonry) *or R404.1.1(4)* (12-inch Reinforced Masonry) *and shall also comply with the provisions of Section R404 and the applicable provisions of Sections R606, R607 and R608. In Seismic Design Categories D_0, D_1 or D_2, concrete masonry and clay masonry foundation walls shall also comply with Section R404.1.4.*

Discussion and Commentary: Foundation walls are typically designed and constructed to carry the vertical loads from the structure above, resist wind and any lateral forces transmitted to the foundation, and sustain earth pressures exerted against the wall, including any forces that may be imposed by frost action.

TABLE R404.1.1(2)
8-INCH MASONRY FOUNDATION WALLS WITH REINFORCING WHERE d > 5 INCHES[a]

WALL HEIGHT	HEIGHT OF UNBALANCED BACKFILL[e]	MINIMUM VERTICAL REINFORCEMENT[b,c] Soil classes and lateral soil load[d] (psf per foot below grade)		
		GW, GP, SW and SP soils 30	GM, GC, SM, SM-SC and ML soils 45	SC, ML-CL and inorganic CL soils 60
6 feet 8 inches	4 feet (or less)	#4 at 48″ o.c.	#4 at 48″ o.c.	#4 at 48″ o.c.
	5 feet	#4 at 48″ o.c.	#4 at 48″ o.c.	#4 at 48″ o.c.
	6 feet 8 inches	#4 at 48″ o.c.	#5 at 48″ o.c.	#6 at 48″ o.c.
7 feet 4 inches	4 feet (or less)	#4 at 48″ o.c.	#4 at 48″ o.c.	#4 at 48″ o.c.
	5 feet	#4 at 48″ o.c.	#4 at 48″ o.c.	#4 at 48″ o.c.
	6 feet	#4 at 48″ o.c.	#5 at 48″ o.c.	#5 at 48″ o.c.
	7 feet 4 inches	#5 at 48″ o.c.	#6 at 48″ o.c.	#6 at 40″ o.c.
8 feet	4 feet (or less)	#4 at 48″ o.c.	#4 at 48″ o.c.	#4 at 48″ o.c.
	5 feet	#4 at 48″ o.c.	#4 at 48″ o.c.	#4 at 48″ o.c.
	6 feet	#4 at 48″ o.c.	#5 at 48″ o.c.	#5 at 48″ o.c.
	7 feet	#5 at 48″ o.c.	#6 at 48″ o.c.	#6 at 40″ o.c.
	8 feet	#5 at 48″ o.c.	#6 at 48″ o.c.	#6 at 32″ o.c.
8 feet 8 inches	4 feet (or less)	#4 at 48″ o.c.	#4 at 48″ o.c.	#4 at 48″ o.c.
	5 feet	#4 at 48″ o.c.	#4 at 48″ o.c.	#5 at 48″ o.c.
	6 feet	#4 at 48″ o.c.	#5 at 48″ o.c.	#6 at 48″ o.c.
	7 feet	#5 at 48″ o.c.	#6 at 48″ o.c.	#6 at 40″ o.c.
	8 feet 8 inches	#6 at 48″ o.c.	#6 at 32″ o.c.	#6 at 24″ o.c.
9 feet 4 inches	4 feet (or less)	#4 at 48″ o.c.	#4 at 48″ o.c.	#4 at 48″ o.c.
	5 feet	#4 at 48″ o.c.	#4 at 48″ o.c.	#5 at 48″ o.c.
	6 feet	#4 at 48″ o.c.	#5 at 48″ o.c.	#6 at 48″ o.c.
	7 feet	#5 at 48″ o.c.	#6 at 48″ o.c.	#6 at 40″ o.c.
	8 feet	#6 at 48″ o.c.	#6 at 40″ o.c.	#6 at 24″ o.c.
	9 feet 4 inches	#6 at 40″ o.c.	#6 at 24″ o.c.	#6 at 16″ o.c.
10 feet	4 feet (or less)	#4 at 48″ o.c.	#4 at 48″ o.c.	#4 at 48″ o.c.
	5 feet	#4 at 48″ o.c.	#4 at 48″ o.c.	#5 at 48″ o.c.
	6 feet	#4 at 48″ o.c.	#5 at 48″ o.c.	#6 at 48″ o.c.
	7 feet	#5 at 48″ o.c.	#6 at 48″ o.c.	#6 at 32″ o.c.
	8 feet	#6 at 48″ o.c.	#6 at 32″ o.c.	#6 at 24″ o.c.
	9 feet	#6 at 40″ o.c.	#6 at 24″ o.c.	#6 at 16″ o.c.
	10 feet	#6 at 32″ o.c.	#6 at 16″ o.c.	#6 at 16″ o.c.

For SI: 1 inch = 25.4 mm, 1 foot = 304.8 mm, 1 pound per square foot per foot = 0.157 kPa/mm.

a. Mortar shall be Type M or S and masonry shall be laid in running bond.

b. Alternative reinforcing bar sizes and spacings having an equivalent cross-sectional area of reinforcement per lineal foot of wall shall be permitted provided the spacing of the reinforcement does not exceed 72 inches.

c. Vertical reinforcement shall be Grade 60 minimum. The distance from the face of the soil side of the wall to the center of vertical reinforcement shall be at least 5 inches.

d. Soil classes are in accordance with the Unified Soil Classification System and design lateral soil loads are for moist conditions without hydrostatic pressure. Refer to Table R405.1.

e. Unbalanced backfill height is the difference in height between the exterior finish ground level and the lower of the top of the concrete footing that supports the foundation wall or the interior finish ground level. Where an interior concrete slab-on-grade is provided and is in contact with the interior surface of the foundation wall, measurement of the unbalanced backfill height from the exterior finish ground level to the top of the interior concrete slab is permitted.

In lieu of an engineered foundation wall system, the IRC provides prescriptive tables for both plain (unreinforced) and reinforced masonry foundation walls. If reinforced walls are under consideration, three different wall widths (8-inch, 10-inch and 12-inch) are addressed.

Topic: Concrete Walls   **Category:** Foundations
Reference: IRC R404.1.2   **Subject:** Foundation Walls

Code Text: *Concrete foundation walls shall be constructed as set forth in Table R404.1.1(5) and shall also comply with the provisions of Section R404 and the applicable provisions of Section R402.2. In Seismic Design Categories D_0, D_1 and D_2, concrete foundation walls shall also comply with Section R404.1.4.*

Discussion and Commentary: Under the prescriptive provisions of the IRC, the minimum vertical reinforcement required in a concrete foundation wall is based upon multiple factors. The necessary information required for determining the proper reinforcement includes the 1) maximum wall height, 2) maximum unbalanced fill height, 3) type of soil, and 4) minimum wall thickness.

TABLE R404.1.1(5)
CONCRETE FOUNDATION WALLS[h, i, j, k]

MAXIMUM WALL HEIGHT (feet)	MAXIMUM UNBALANCED BACKFILL HEIGHT[b] (feet)	MINIMUM VERTICAL REINFORCEMENT SIZE AND SPACING[c, d, e, f, l]											
		Soil classes[a] and design lateral soil (psf per foot of depth)											
		GW, GP, SW and SP 30				GM, GC, SM, SM-SC and ML 45				SC, ML-CL and inorganic CL 60			
		Minimum wall thickness (inches)											
		5.5	7.5	9.5	11.5	5.5	7.5	9.5	11.5	5.5	7.5	9.5	11.5
5	4	PC	PC	PC	PC	PC	PC	PC	PC	PC	PC	PC	PC
	5	PC	PC	PC	PC	PC	PC	PC	PC	PC	PC	PC	PC
6	4	PC	PC	PC	PC	PC	PC	PC	PC	PC	PC	PC	PC
	5	PC	PC	PC	PC	PC	PCᵍ	PC	PC	#4@35″	PCᵍ	PC	PC
	6	PC	PC	PC	PC	#5@48″	PC	PC	PC	#5@36″	PC	PC	PC
7	4	PC	PC	PC	PC	PC	PC	PC	PC	PC	PC	PC	PC
	5	PC	PC	PC	PC	PC	PC	PC	PC	#5@47″	PC	PC	PC
	6	PC	PC	PC	PC	#5@42″	PC	PC	PC	#6@43″	#5@48″	PCᵍ	PC
	7	#5@46″	PC	PC	PC	#6@42″	#5@46″	PCᵍ	PC	#6@34″	#6@48″	PC	PC
8	4	PC	PC	PC	PC	PC	PC	PC	PC	PC	PC	PC	PC
	5	PC	PC	PC	PC	#4@38″	PCᵍ	PC	PC	#5@43″	PC	PC	PC
	6	#4@37″	PCᵍ	PC	PC	#5@37″	PC	PC	PC	#6@37″	#5@43″	PCᵍ	PC
	7	#5@40″	PC	PC	PC	#6@37″	#5@41″	PC	PC	#6@34″	#6@43″	PC	PC
	8	#6@43″	#5@47″	PCᵍ	PC	#6@34″	#6@43″	PC	PC	#6@27″	#6@32″	#6@44″	PC
9	4	PC	PC	PC	PC	PC	PC	PC	PC	PC	PC	PC	PC
	5	PC	PC	PC	PC	#4@35″	PCᵍ	PC	PC	#5@40″	PC	PCᶜ	PC
	6	#4@34″	PCᵍ	PC	PC	#6@48″	PC	PC	PC	#6@36″	#5@39″	PCᵍ	PC
	7	#5@36″	PC	PC	PC	#6@34″	#5@37″	PC	PC	#6@33″	#6@38″	#5@37″	PCᵍ
	8	#6@38″	#5@41″	PCᵍ	PC	#6@33″	#6@38″	#5@37″	PCᵍ	#6@24″	#7@39″	#6@39″	#4@48″ʰ
	9	#6@34″	#6@46″	PC	PC	#6@26″	#7@41″	#6@41″	PC	#6@19″	#7@31″	#7@41″	#6@39″

(continued)

The notes to Table R404.1.1(5) provide additional information necessary to the proper application of the provisions. Of considerable importance, the minimum concrete cover required for the reinforcement is established.

Topic: Insulating Concrete Form Foundation Walls	**Category:** Foundations
Reference: IRC R404.4	**Subject:** Foundation Walls

Code Text: *Insulating concrete form (ICF) foundation walls shall be designed and constructed in accordance with the provisions of Section R404.4 or in accordance with the provisions of ACI 318.*

Discussion and Commentary: An insulating concrete form (ICF) system uses stay-in-place forms of rigid foam plastic insulation, a hybrid of cement and foam insulation, a hybrid of cement and wood chips, or other insulating material for constructing cast-in-place concrete walls. Three types of ICF systems are prescriptively addressed in the IRC; waffle-grid, flat-wall and screen grid. The use of the prescriptive provisions of the IRC to regulate ICF construction is only applicable to buildings of limited size and height located in the lower seismic design categories and wind speed areas.

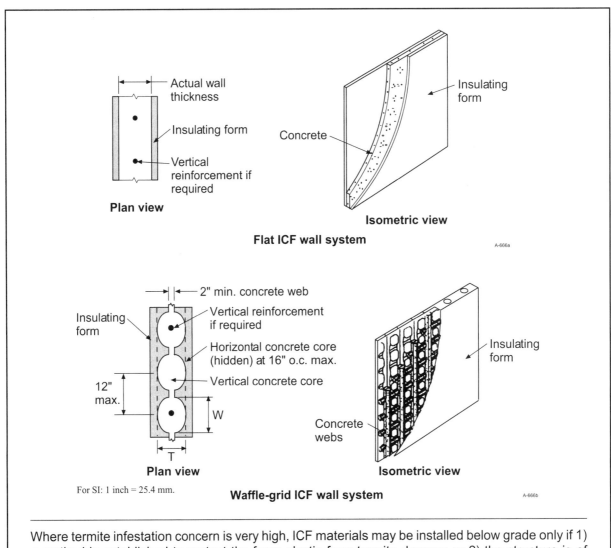

Where termite infestation concern is very high, ICF materials may be installed below grade only if 1) a method is established to protect the foam plastic from termite damage or 2) the structure is of noncombustible or pressure preservatively treated wood construction.

Topic: Retaining Walls
Reference: IRC R404.5

Category: Foundations
Subject: Foundation Walls

Code Text: *Retaining walls that are not laterally supported at the top and that retain in excess of 24 inches (610 mm) of unbalanced fill shall be designed to ensure stability against overturning, sliding, excessive foundation pressure and water uplift. Retaining walls shall be designed for a safety factor of 1.5 against lateral sliding and overturning.*

Discussion and Commentary: Retaining walls are a typical feature at many residential sites. As such, the IRC provides specific criteria for which a retaining wall must be designed.

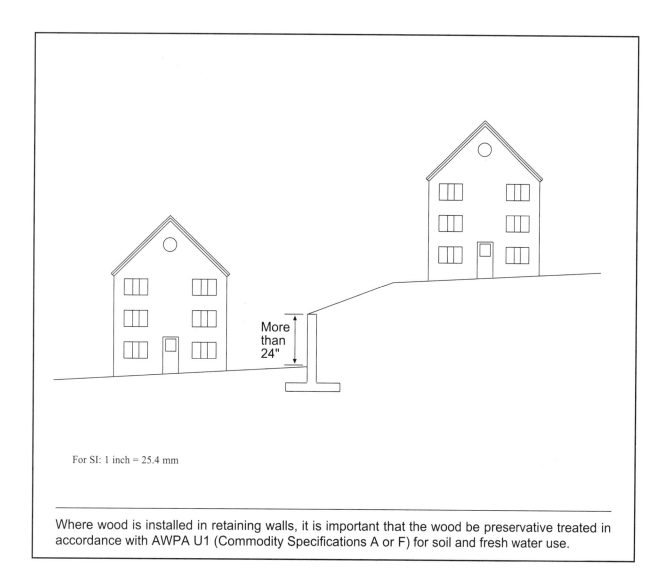

For SI: 1 inch = 25.4 mm

Where wood is installed in retaining walls, it is important that the wood be preservative treated in accordance with AWPA U1 (Commodity Specifications A or F) for soil and fresh water use.

Code Text: *Drains shall be provided around all concrete or masonry foundations that retain earth and enclose habitable or usable spaces located below grade. See exception for well-drained ground or sand-gravel mixture soils. Drainage tiles, gravel or crushed stone drains, perforated pipe or other approved systems or materials shall be installed at or below the area to be protected and shall discharge by gravity or mechanical means into an approved drainage system.*

Discussion and Commentary: To allow free groundwater that may be present adjacent to the foundation wall to be removed, drains are usually placed around houses. Such drains are intended to prevent leakage into interior spaces below grade. Drainage tiles are extremely important in areas having moderate to heavy rainfall and soils with a low percolation rate.

*Minimum limits of pea gravel

For SI: 1 inch = 25.4 mm.

B-494

A drainage system is not required if the foundation is in an area having well-drained ground or sand-gravel mixture soils (Group I). Such soils provide good drainage characteristics and include gravel sand mixtures, gravelly sands, silty gravels and silty sand.

Code Text: *Except where required by Section R406.2 to be waterproofed, foundation walls that retain earth and enclose interior spaces and floors below grade shall be dampproofed from the top of the footing to the finished grade. In areas where a high water table or other severe soil-water conditions are known to exist, exterior foundation walls that retain earth and enclose interior spaces and floors below grade shall be waterproofed from the top of the footing to the finished grade.*

Discussion and Commentary: Dampproofing installations generally consist of the application of one or more coatings of impervious compounds that are intended to prevent the passage of water vapor through walls under slight pressure. Waterproofing installations consist of the application of a combination of sealing materials and impervious coatings used to prevent the passage of moisture in either a vapor or liquid form under conditions of significant hydrostatic pressure.

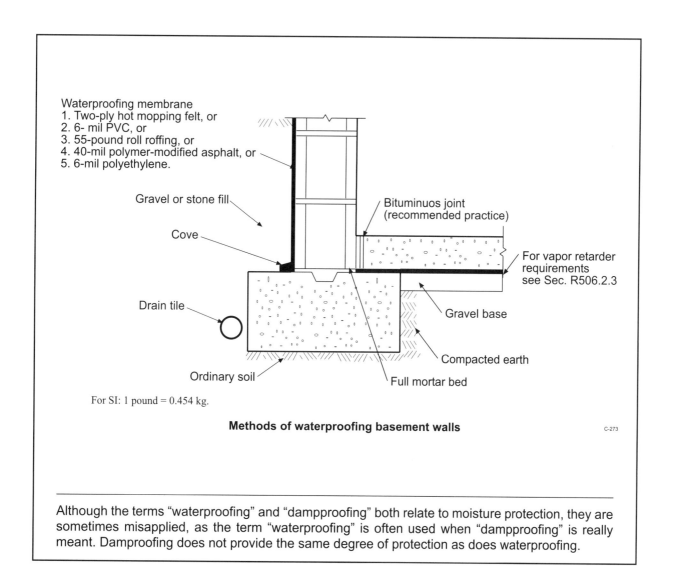

Waterproofing membrane
1. Two-ply hot mopping felt, or
2. 6- mil PVC, or
3. 55-pound roll roffing, or
4. 40-mil polymer-modified asphalt, or
5. 6-mil polyethylene.

Gravel or stone fill

Cove

Drain tile

Ordinary soil

Bituminuos joint
(recommended practice)

For vapor retarder
requirements
see Sec. R506.2.3

Gravel base

Compacted earth

Full mortar bed

For SI: 1 pound = 0.454 kg.

Methods of waterproofing basement walls

C-273

Although the terms "waterproofing" and "dampproofing" both relate to moisture protection, they are sometimes misapplied, as the term "waterproofing" is often used when "dampproofing" is really meant. Dampproofing does not provide the same degree of protection as does waterproofing.

Code Text: *The under-floor space between the bottom of the floor joists and the earth under any building (except space occupied by a basement) shall have ventilation openings through foundation walls or exterior walls. The minimum net area of ventilation openings shall not be less than 1 square foot (0.0929m^2) for each 150 square feet (14 m^2) of under-floor space area.* See two methods permitting the elimination of ventilation openings.

Discussion and Commentary: To control condensation in crawl space areas and thus reduce the chance of dry rot, natural ventilation of such spaces by reasonably distributed openings through exterior foundation walls is required. Susceptibility of a structure to condensation is a function of the geographical location and the climatic conditions.

16" x 24" min access hole through perimeter wall (18" x 24" if through floor)

3'

3'

3' max (typ. from corners)

Covered openings through foundation wall to crawl space

Minimum total net clear area of openings equal to the crawl space area divided by 150

For SI: 1 inch = 25.4 mm, 1 foot = 304.8 mm.

The installation of a covering material over the ventilation opening keeps rodents and vermin from entering the crawl space. The code lists a variety of materials including corrosion-resistant wire mesh having a least dimension of $^1/_8$ inch.

Quiz

Study Session 5
IRC Chapter 4

1. Unless grading is prohibited by physical barriers, lots shall be graded away from the foundation with a minimum fall of ___6___ within the first ___10___ .

   a. 6 inches, 5 feet   b. 6 inches, 10 feet

   c. 12 inches, 5 feet   d. 12 inches, 10 feet

   Reference _R401.3_

2. In the absence of a complete geotechnical evaluation to determine the soil's characteristics, clayey sand material shall be assumed to have a presumptive load-bearing value of ___2,000___ .

   a. 1,500 psf   b. 2,000 psf

   c. 3,000 psf   d. 4,000 psf

   Reference _R401.4.1_

3. Concrete used in a basement slab shall have a minimum compressive strength of ___2,500___ where a severe weathering potential exists.

   a. 2,000 psi   b. 2,500 psi

   c. 3,000 psi   d. 3,500 psi

   Reference _402.2_

4. Air entrainment for concrete subjected to weathering, when required for locations other than garage floors with a steel troweled finish, shall have a total air content of ___5___ minimum and __7____ maximum.

    a. 4 percent, 7 percent   b. 4 percent, 8 percent

    c. 5 percent, 7 percent   d. 5 percent, 8 percent

Reference R402.2 (d) dTable

5. A two-story dwelling of conventional light-frame construction is to be constructed in an area where the load-bearing value of the soil is 2,000 psf. What is the minimum required width of the concrete footings?

    a. 8 inches   b. 11 inches

    c. 12 inches   d. 15 inches

Reference Table R403.1

6. The minimum required thickness for concrete spread footings is _____.

    a. 4 inches   b. 6 inches

    c. 7 inches   d. 8 inches

Reference 403.1.1

7. What are the minimum and maximum footing projections for a concrete footing having a thickness of 8 inches?

    a. 2 inches, 4 inches   b. 2 inches, 8 inches

    c. 4 inches, 8 inches   d. 8 inches, 12 inches

Reference R403.1.1

8. Where a permanent wood foundation basement wall system is used, the wall shall be supported by a minimum __2×8___ footing plate resting on gravel or crushed stone fill a minimum of ___16___ in width.

    a. 2-inch-by-6-inch, 12 inches   b. 2-inch-by-8-inch, 16 inches

    c. 2-inch-by-12-inch, 16 inches   d. 2-inch-by-12-inch, 24 inches

Reference R403.1 (2)

9. For concrete foundation systems constructed in Seismic Design Category D$_2$, foundations with stem walls shall be provided with a minimum of __one #4__ bar(s) at the top of the wall and __one #4__ bar(s) at the bottom of the footing.

    ⓐ one #4, one #4   b. one #5, one #5

    c. one #5, two #4   d. two #4, two #5

Reference __403.1.3.1__

10. All exterior footings shall be placed a minimum of __12__ below the undisturbed ground surface.

    a. 6 inches   b. 8 inches

    c. 9 inches   ⓓ 12 inches

Reference __403.1.4__

11. The maximum permitted slope for the bottom surface of footings shall be __1:10__.

    a. 1:8   ⓑ 1:10

    c. 1:12   d. 1:20

Reference __403.1.5__

12. Anchor bolts used to attach wood sill plates to foundation walls shall be spaced a maximum of __6__ on center.

    a. 4 feet   b. 5 feet

    ⓒ 6 feet   d. 7 feet

Reference __403.1.6__

13. Anchor bolts used to attach a wood sole plate to a concrete foundation shall be a minimum of __1/2__ in diameter and extend a minimum of __7__ into the concrete.

    ⓐ $^1/_2$ inch, 7 inches   b. $^1/_2$ inch, 15 inches

    c. $^5/_8$ inch, 7 inches   d. $^5/_8$ inch, 15 inches

Reference __403.1.6__

14. On graded sites where structures are placed adjacent to slopes steeper than 1:3, the top of any exterior foundation wall shall extend a minimum of _____ above the elevation of the street gutter or other approved drainage discharge point.

   a. 6 inches plus 1 percent
   b. 6 inches plus 2 percent
   c. 12 inches plus 1 percent
   d. 12 inches plus 2 percent

   Reference _403 1.7 .3_

15. In order to use frost protected shallow foundations, the monthly mean temperature of the building must be maintained at a minimum of _____.

   a. 60°F
   b. 64°F
   c. 68°F
   d. 70°F

   Reference _403.3_

16. A dwelling located in an area with an air freezing index of 3,500 is constructed with a frost-protected shallow foundation. What is the minimum required horizontal insulation R-value at the corners of the foundation?

   a. 8.0
   b. 8.6
   c. 11.2
   d. 13.1

   Reference _Table R403.3_

17. A 9-foot-high plain masonry foundation wall is subjected to 7 feet of unbalanced backfill. If the soil class is GC, what is the minimum required nominal wall thickness?

   a. 8 inches
   b. 10 inches
   c. 12 inches
   d. 12 inches solid grouted

   Reference _Table R404.1.1 (1)/D_

18. An 8-foot-high, 9.5-inch-thick concrete foundation wall is subjected to 6 feet of unbalanced backfill. If the soil class is GP, what minimum vertical reinforcement is required?

   a. #4 at 38 inches on center
   b. #5 at 47 inches on center
   c. #6 at 43 inches on center
   d. no vertical reinforcement required

   Reference _Table 404.1.1(5)_

19. Where masonry veneer is used, concrete and masonry foundations shall extend a minimum of _____ above adjacent finished grade.

    a. 4 inches   b. 6 inches

    c. 8 inches   d. 12 inches

    Reference R404.1.6

20. Where the foundation wall supports a minimum of _____ of unbalanced backfill, backfill shall not be placed against the wall until the wall has sufficient strength and has been anchored to the floor above or has been sufficiently braced.

    a. 2 feet   b. 4 feet

    c. 6 feet   d. 8 feet

    Reference R404.1.7

21. Where wood studs having an F_b value of 1320 are used in a wood foundation wall, the minimum stud size shall be _____ and the maximum stud spacing shall be _____ on center.

    a. 2 inches by 4 inches, 24 inches

    b. 2 inches by 4 inches, 16 inches

    c. 2 inches by 6 inches, 24 inches

    d. 2 inches by 6 inches, 16 inches

    Reference R404.2.2

22. What is the maximum height of backfill permitted against a wood foundation wall not designed by AF&PA Report # 7?

    a. 3 feet   b. 4 feet

    c. 6 feet   d. 7 feet

    Reference R404.2.3

23. Required foundation drains of gravel or crushed stone shall extend a minimum of _____ beyond the outside edge of the footing and _____ above the top of the footing.

    a. 6 inches, 4 inches   b. 6 inches, 6 inches

    c. 12 inches, 4 inches   d. 12 inches, 6 inches

    Reference R405.1

24. As a general rule, the under-floor space between the bottom of floor joists and the earth shall be provided with ventilation openings sized for a minimum net area of 1 square foot for each _R408.1_ square feet of under-floor area.

   a. 100
   b. 120
   (c.) 150
   d. 200

   Reference _____

25. Access to an under-floor space through a perimeter wall shall be provided by a minimum _____ access opening.

   a. 16-inch-by-20-inch
   (b.) 16-inch-by-24-inch
   c. 18-inch-by-24-inch
   d. 18-inch-by-30-inch

   Reference _R408.4_

26. Frost protection is not required for a foundation that supports a freestanding accessory structure of light-framed construction that has a maximum floor area of _____ square feet and a maximum eave height of _____ feet.

   a. 200, 10
   b. 400, 10
   c. 400, 12
   (d.) 600, 10

   Reference _R403.1.4.1 #1_ _EXCEPTIONS_

27. Gravel used as a footing material for a wood foundation shall be washed, well graded and have a maximum stone size of _____ inch.

   a. $^1/_4$
   b. $^1/_2$
   (c.) $^3/_4$
   d. 1

   Reference _R403.2_

28. Horizontal insulation used in a frost protected shallow foundation system shall be protected against damage if it is located less than _____ inches below the ground surface.

   (a.) 12
   b. 15
   c. 18
   d. 24

   Reference _TABLE 403.3_

29. Where waterproofing of a masonry foundation wall is necessary due to the presence of a high water table, a minimum membrane thickness of _____ is required if the waterproofing material is polymer-modified asphalt.

    a. 6-mil   b. 16-mil

    c. 30-mil   (d.) 40-mil

Reference R406.2

30. Ventilating openings providing under-floor ventilation shall be located so that at least one such opening is installed a maximum of _____ feet of each corner of the building.

    a. 2   (b) 3

    c. 5   d. 6

Reference R408.2

31. A minimum of _____ anchor bolt(s) located _____ of the plate section is required to anchor maximum 24-inch-long walls connecting offset braced wall panels.

    (a.) 1, within the middle third

    b. 1, within the middle one-half

    c. 2, in the outer one-fourths

    d. 2, within 6 inches of each end

Reference R403.1.6 #2 EXCEPTIONS

32. What is the maximum slump permitted for ready-mixed concrete used in insulating concrete form (ICF) walls?

    a. 4 inches   b. $4\frac{1}{2}$ inches

    c. 5 inches   (d.) 6 inches

Reference R404.4.5

33. Where one coat of complying surface-bonding cement is used as the required dampproofing over the parging of a masonry foundation wall, the cement coating must be a minimum of _____ inch thick.

    a. $\frac{1}{8}$ b. $\frac{3}{16}$

    c. $\frac{1}{4}$ (d.) $\frac{3}{8}$

Reference R406.1

34. Steel columns used as a part of a foundation system shall be a minimum of _____ inches in diameter standard pipe size or approved equivalent.

a. 3　　　　　　　　　　　　b. 3 $^1/_2$

c. 4　　　　　　　　　　　　d. 5

Reference R407.3

35. The continuous vapor barrier required in an unvented crawl space shall extend a minimum of _____ inches up the stem wall.

a. 3　　　　　　　　　　　　b. 4

c. 6　　　　　　　　　　　　d. 8

Reference R408.3

2006 IRC Chapter 5
Floors

OBJECTIVE: To develop an understanding of the provisions regulating floor construction, including the sizing and installation of floor joists and girders; drilling and notching of floor framing members; engineered wood products such as trusses, glued-laminated members and I-joists; draftstopping; floor sheathing spans and installation; steel floor framing; and concrete slab-on-ground floors.

REFERENCE: Chapter 5, 2006 *International Residential Code*

KEY POINTS:
- How are the maximum allowable spans of floor joists determined? What design live loads are to be used for joists supporting sleeping rooms? For areas other than sleeping rooms? For attics accessed by a permanent stairway?
- What is the general limitation for cantilever spans of floor joists? What spans are permitted for cantilevered floor joists supporting a bearing wall and a roof? Supporting an exterior balcony?
- How shall bearing partitions be supported by floor joists? What method is acceptable where piping or vents are installed at the point of support?
- How must bearing partitions perpendicular to floor joists be located?
- What conditions regulate the sizing of girders for exterior bearing walls? For interior bearing walls?
- How much bearing is required for the ends of floor joists supported by wood or metal? Supported by concrete or masonry? What other methods are acceptable?
- What methods of connection are permitted where joists frame from opposite sides of a bearing wall or girder?
- How are floor joists to be restrained to resist twisting and lateral movement?
- What are the limitations for notches in floor joists? Bored holes?
- When are single headers and single trimmers permitted at floor openings? When are framing anchors or joist hangers mandated?
- How are wood floor trusses regulated? What type of information must be provided on the truss design drawings?

KEY POINTS:
(Cont'd)

- In what type of floor system is draftstopping required? At what intervals? What materials are permitted for draftstopping?
- How is wood structural panel sheathing identified? How does the panel span rating reflect the maximum allowable floor span? How is the allowable span for particleboard determined?
- What special concerns are addressed for pressure preservatively treated wood basement floors and floors on ground?
- What are the limitations for using the prescriptive provisions for steel floor framing?
- What materials are permitted as steel load-bearing members? How must they be identified? What are the fundamental fastening requirements for steel-to-steel floor connections? Floor to foundation or bearing wall connections? How are the maximum allowable spans determined?
- What is the minimum permitted thickness for concrete slab-on-ground floors? How must the site be prepared?
- How is any fill below a concrete slab-on-ground floor regulated? When is a base course needed? When is a vapor barrier required?
- When provided, where is reinforcement for slab-on-ground floors required to be located?

Topic: Framing at Braced Wall Lines	Category: Floors
Reference: IRC R502.2.1	Subject: Wood Floor Framing

Code Text: *A load path for lateral forces shall be provided between floor framing and braced wall panels located above or below a floor, as specified in Section R602.10.8 (Connections).*

Discussion and Commentary: The lateral forces induced by wind and earthquake are transferred to the building foundation through the roof and floor systems and the braced wall lines to which these framing members are attached. The braced wall lines in turn transfer such loads to the foundation by the sill plate and its connection to the foundation system. The proper and successful transfer of all loads from the building envelope system to the foundation is accomplished by proper creation of a complete load path.

Section view

The charging provisions for load path continuity are set forth in the floor requirements, whereas the details of such framing are established in the wall provisions of Chapter 6.

Code Text: *Table R502.3.1(1) shall be used to determine the maximum allowable span of floor joists that support sleeping areas and attics that are accessed by means of a fixed stairway in accordance with Section R311.5 (Stairways) provided that the design live load does not exceed 30 psf (1.44 kPa) and the design dead load does not exceed 20 psf (0.96 kPa). Table R502.3.1(2) shall be used to determine the maximum allowable span of floor joists that support all other areas of the building, other than sleeping rooms and attics, provided that the design live load does not exceed 40 psf (1.92 kPa) and the design dead load does not exceed 20 psf (0.96 kPa).*

Discussion and Commentary: The joist span tables are based on common lumber species and grades, joist spacing and design loads. Although a separate span table is available for sleeping rooms based on 30 psf, it is not uncommon to use the 40 psf table for the entire dwelling unit. Both tables assume a maximum design dead load of 10 psf.

TABLE R502.3.1(2)
FLOOR JOIST SPANS FOR COMMON LUMBER SPECIES
(Residential living areas, live load = 40 psf, L/D = 360)[b]

JOIST SPACING (inches)	SPECIES AND GRADE		DEAD LOAD = 10 psf				DEAD LOAD = 20 psf			
			2×6	2×8	2×10	2×12	2×6	2×8	2×10	2×12
			Maximum floor joist spans							
			(ft - in.)	(ft - in.)	(ft - in.)	(ft - in.)	(ft - in.)	(ft - in.)	(ft - in.)	(ft - in.)
12	Douglas fir-larch	SS	11- 4	15- 0	19- 1	23- 3	11- 4	15- 0	19- 1	23- 3
	Douglas fir-larch	#1	10-11	14- 5	18- 5	22- 0	10-11	14- 2	17- 4	20- 1
	Douglas fir-larch	#2	10- 9	14- 2	17- 9	20- 7	10- 6	13- 3	16- 3	18-10
	Douglas fir-larch	#3	8- 8	11- 0	13- 5	15- 7	7-11	10- 0	12- 3	14- 3
	Hem-fir	SS	10- 9	14- 2	18- 0	21-11	10- 9	14- 2	18- 0	21-11
	Hem-fir	#1	10- 6	13-10	17- 8	21- 6	10- 6	13-10	16-11	19- 7
	Hem-fir	#2	10- 0	13- 2	16-10	20- 4	10- 0	13- 1	16- 0	18- 6
	Hem-fir	#3	8- 8	11- 0	13- 5	15- 7	7-11	10- 0	12- 3	14- 3
	Southern pine	SS	11- 2	14- 8	18- 9	22-10	11- 2	14- 8	18- 9	22-10
	Southern pine	#1	10-11	14- 5	18- 5	22- 5	10-11	14- 5	18- 5	22- 5
	Southern pine	#2	10- 9	14- 2	18- 0	21- 9	10- 9	14- 2	16-11	19-10
	Southern pine	#3	9- 4	11-11	14- 0	16- 8	8- 6	10-10	12-10	15- 3
	Spruce-pine-fir	SS	10- 6	13-10	17- 8	21- 6	10- 6	13-10	17- 8	21- 6
	Spruce-pine-fir	#1	10- 3	13- 6	17- 3	20- 7	10- 3	13- 3	16- 3	18-10
	Spruce-pine-fir	#2	10- 3	13- 6	17- 3	20- 7	10- 3	13- 3	16- 3	18-10
	Spruce-pine-fir	#3	8- 8	11- 0	13- 5	15- 7	7-11	10- 0	12- 3	14- 3
16	Douglas fir-larch	SS	10- 4	13- 7	17- 4	21- 1	10- 4	13- 7	17- 4	21- 0
	Douglas fir-larch	#1	9-11	13- 1	16- 5	19- 1	9- 8	12- 4	15- 0	17- 5
	Douglas fir-larch	#2	9- 9	12- 7	15- 5	17-10	9- 1	11- 6	14- 1	16- 3
	Douglas fir-larch	#3	7- 6	9- 6	11- 8	13- 6	6-10	8- 8	10- 7	12- 4
	Hem-fir	SS	9- 9	12-10	16- 5	19-11	9- 9	12-10	16- 5	19-11
	Hem-fir	#1	9- 6	12- 7	16- 0	18- 7	9- 6	12- 0	14- 8	17- 0
	Hem-fir	#2	9- 1	12- 0	15- 2	17- 7	8-11	11- 4	13-10	16- 1
	Hem-fir	#3	7- 6	9- 6	11- 8	13- 6	6-10	8- 8	10- 7	12- 4
	Southern pine	SS	10- 2	13- 4	17- 0	20- 9	10- 2	13- 4	17- 0	20- 9
	Southern pine	#1	9-11	13- 1	16- 9	20- 4	9-11	13- 1	16- 4	19- 6
	Southern pine	#2	9- 9	12-10	16- 1	18-10	9- 6	12- 4	14- 8	17- 2
	Southern pine	#3	8- 1	10- 3	12- 2	14- 6	7- 4	9- 5	11- 1	13- 2
	Spruce-pine-fir	SS	9- 6	12- 7	16- 0	19- 6	9- 6	12- 7	16- 0	19- 6
	Spruce-pine-fir	#1	9- 4	12- 3	15- 5	17-10	9- 1	11- 6	14- 1	16- 3
	Spruce-pine-fir	#2	9- 4	12- 3	15- 5	17-10	9- 1	11- 6	14- 1	16- 3
	Spruce-pine-fir	#3	7- 6	9- 6	11- 8	13- 6	6-10	8- 8	10- 7	12- 4

(continued)

The span tables account for a uniform load condition. They also permit isolated concentrated loads such as nonbearing partitions. The use of additional joists and other adequate supports may be necessary where larger concentrated loads are encountered.

Topic: Floor Cantilevers	**Category:** Floors
Reference: IRC R502.3.3	**Subject:** Allowable Joist Spans

Code Text: *Floor cantilever spans shall not exceed the nominal depth of the wood floor joist. Floor cantilevers constructed in accordance with Table R502.3.3(1) shall be permitted when supporting a light-frame bearing wall and roof only. Floor cantilevers supporting an exterior balcony are permitted to be constructed in accordance with Table R502.3.3(2).*

Discussion and Commentary: Under the prescriptive limitations of the IRC, there are three conditions where the maximum allowable span of cantilevered floor joists can be determined. First, regardless of loading conditions, a maximum span equal to the depth of the floor joists is permitted. This requirement allows a very short cantilevered element. The second condition considers the support of both a single-story bearing wall and roof. The third situation is based on the cantilevered support of an exterior balcony with no roof load.

TABLE R502.3.3(2)
CANTILEVER SPANS FOR FLOOR JOISTS SUPPORTING EXTERIOR BALCONY[a, b, e, f]

Member Size	Spacing	Maximum Cantilever Span (Uplift Force at Backspan Support in lb)[c, d]		
		Ground Snow Load		
		≤ 30 psf	50 psf	70 psf
2 × 8	12″	42″ (139)	39″ (156)	34″ (165)
2 × 8	16″	36″ (151)	34″ (171)	29″ (180)
2 × 10	12″	61″ (164)	57″ (189)	49″ (201)
2 × 10	16″	53″ (180)	49″ (208)	42″ (220)
2 × 10	24″	43″ (212)	40″ (241)	34″ (255)
2 × 12	16″	72″ (228)	67″ (260)	57″ (268)
2 × 12	24″	58″ (279)	54″ (319)	47″ (330)

For SI: 1 inch = 25.4 mm, 1 pound per square foot = 0.0479 kPa.

a. Spans are based on No. 2 Grade lumber of Douglas fir-larch, hem-fir, southern pine, and spruce-pine-fir for repetitive (3 or more) members.

b. Ratio of backspan to cantilever span shall be at least 2:1.

c. Connections capable of resisting the indicated uplift force shall be provided at the backspan support.

d. Uplift force is for a backspan to cantilever span ratio of 2:1. Tabulated uplift values are permitted to be reduced by multiplying by a factor equal to 2 divided by the actual backspan ratio provided (2/backspan ratio).

e. A full-depth rim joist shall be provided at the cantilevered end of the joists. Solid blocking shall be provided at the cantilevered support.

f. Linear interpolation shall be permitted for ground snow loads other than shown.

The ratio of back span to cantilever span must be a minimum of 3:1 where the cantilevered floor joists support a bearing wall and a roof. Where only supporting an exterior balcony, the minimum ratio must be 2:1. It is important that connections be capable of resisting the indicated uplift force.

Code Text: *Joists under parallel bearing partitions shall be of adequate size to support the load. Double joists, sized to adequately support the load, that are separated to permit the installation of piping or vents shall be full depth solid blocked with lumber not less than 2 inches (51 mm) in nominal thickness spaced not more than 4 feet (1219 mm) on center. Bearing partitions perpendicular to joists shall not be offset from supporting girders, walls or partitions more than the joist depth unless such joists are of sufficient size to carry the additional load.*

Discussion and Commentary: It is often necessary to double or triple floor joists under bearing partitions in order to provide support for the additional load from above. The multiple joists create a beam-type member that is capable of properly transmitting the loads to the bearing points at the joist ends. Therefore, a beam of equivalent size may be substituted for the multiple joists.

For SI: 1 foot = 304.8 mm.

Spaced joists under bearing partition with blocking at penetration

A-651

Where piping or vents penetrate the floor system, the doubled joists may be separated to allow the pipe or vent to pass through. Solid blocking between the joists at 4-foot intervals is required so that the two joists will function as a combined member without twisting.

Topic: Allowable Girder Spans

Reference: IRC R502.5

Category: Floors

Subject: Wood Floor Framing

Code Text: *The allowable spans of girders fabricated of dimension lumber shall not exceed the values set forth in Tables R502.5(1) and R502.5(2).*

Discussion and Commentary: Table R502.5(1) provides the maximum span of girders in exterior walls based on several criteria. The loads supported by the header, building width, built-up member size and ground snow load are needed to use the prescriptive table. The built-up members listed vary as to depth and number of members. The more commonly encountered condition of girders for interior bearing walls is addressed in Table R502.5(2). In using this table, only the loading condition, building span, and built-up member size are needed to apply the table.

TABLE R502.5(2)
GIRDER SPANS[a] AND HEADER SPANS[a] FOR INTERIOR BEARING WALLS
(Maximum spans for Douglas fir-larch, hem-fir, southern pine and spruce-pine-fir[b] and required number of jack studs)

HEADERS AND GIRDERS SUPPORTING	SIZE	BUILDING WIDTH[c] (feet)					
		20		28		36	
		Span	NJ[d]	Span	NJ[d]	Span	NJ[d]
One floor only	2-2×4	3-1	1	2-8	1	2-5	1
	2-2×6	4-6	1	3-11	1	3-6	1
	2-2×8	5-9	1	5-0	2	4-5	2
	2-2×10	7-0	2	6-1	2	5-5	2
	2-2×12	8-1	2	7-0	2	6-3	2
	3-2×8	7-2	1	6-3	1	5-7	2
	3-2×10	8-9	1	7-7	2	6-9	2
	3-2×12	10-2	2	8-10	2	7-10	2
	4-2×8	9-0	1	7-8	1	6-9	1
	4-2×10	10-1	1	8-9	1	7-10	2
	4-2×12	11-9	1	10-2	2	9-1	2
Two floors	2-2×4	2-2	1	1-10	1	1-7	1
	2-2×6	3-2	2	2-9	2	2-5	2
	2-2×8	4-1	2	3-6	2	3-2	2
	2-2×10	4-11	2	4-3	2	3-10	3
	2-2×12	5-9	2	5-0	3	4-5	3
	3-2×8	5-1	2	4-5	2	3-11	2
	3-2×10	6-2	2	5-4	2	4-10	2
	3-2×12	7-2	2	6-3	2	5-7	3
	4-2×8	6-1	1	5-3	2	4-8	2
	4-2×10	7-2	2	6-2	2	5-6	2
	4-2×12	8-4	2	7-2	2	6-5	2

For SI: 1 inch = 25.4 mm, 1 foot = 304.8 mm.

a. Spans are given in feet and inches.

b. Tabulated values assume #2 grade lumber.

c. Building width is measured perpendicular to the ridge. For widths between those shown, spans are permitted to be interpolated.

d. NJ - Number of jack studs required to support each end. Where the number of required jack studs equals one, the header is permitted to be supported by an approved framing anchor attached to the full-height wall stud and to the header.

In addition to providing the maximum allowable girder spans, the table sets forth the number of jack studs. Typically, two jack studs are required to support the ends of the girder, providing at least 3 inches of bearing. Only under limited conditions is a single jack stud permitted.

Code Text: *The ends of each joist, beam or girder shall have not less than 1.5 inches (38 mm) of bearing on wood or metal and not less than 3 inches (76 mm) on masonry or concrete except where supported on a 1-inch-by-4-inch (25.4 mm by 102 mm) ribbon strip and nailed to the adjacent stud or by the use of approved joist hangers.*

Discussion and Commentary: Joists, beams and girders are typically provided with bearing surfaces. The required bearing length for masonry and concrete is twice that needed for wood or metal. As an alternative, bearing may be accomplished through the use of a ribbon strip, provided each member is directly connected to an adjoining stud. Approved joist hangers are also acceptable means of support in lieu of direct bearing.

Joist hanger at double top plate B-496

For SI: 1 inch = 25.4 mm.

Floor joist bearing on wood B-495

Additional criteria address conditions where joists frame over the top of a beam, girder or bearing wall. A minimum 3-inch (76 mm) lap is required unless a splice having equivalent strength is installed. Framing anchors or ledgers are mandated when framing into girder sides.

Topic: Lateral Restraint at Supports	**Category:** Floors
Reference: IRC R502.7	**Subject:** Wood Floor Framing

Code Text: *Joists shall be supported laterally at the ends by full-depth solid blocking not less than 2 inches (51 mm) nominal in thickness; or by attachment to a header, band, or rim joist, or to an adjoining stud; or shall be otherwise provided with lateral support to prevent rotation. See exception for Seismic Design Categories D_0, D_1 and D_2, where lateral restraint must also be provided at each intermediate support.*

Discussion and Commentary: The potential for rotation of floor joists is reduced by providing complying blocking at the end points. This is typically accomplished at exterior walls by nailing the rim joist directly to the joist ends. Full-depth blocking may be used where the joist ends bear at interior bearing walls. In addition, any other approved methods that can be shown to prevent the rotation of the joists are acceptable.

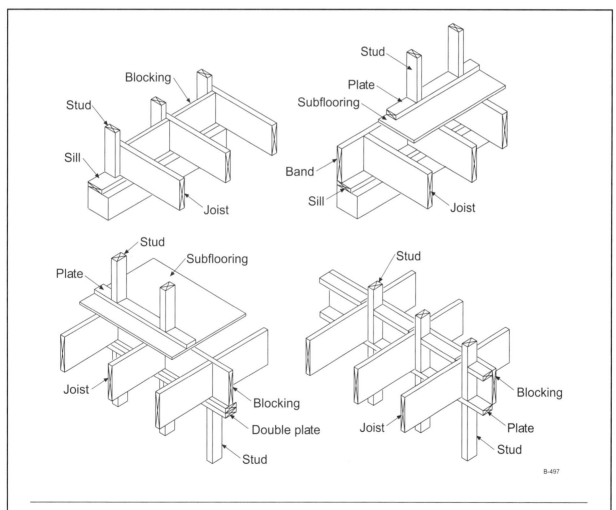

The potential for very deep members (2 inches by 14 inches or greater) to buckle laterally is mitigated by requiring solid blocking or diagonal bridging located at maximum intervals of 8 feet. As an alternative, a continuous 1-inch by 3-inch strip can be nailed to the bottoms of the joists at similar intervals.

Code Text: *Notches in solid lumber joists, rafters and beams shall not exceed one-sixth of the depth of the member, shall not be longer than one-third of the depth of the member and shall not be located in the middle one-third of the span. Notches at the ends of the member shall not exceed one-fourth the depth of the member. The tension side of members 4 inches (102 mm) or greater in nominal thickness shall not be notched except at the ends of the members.*

Discussion and Commentary: Notches potentially reduce the structural strength of framing members to unacceptable levels. Therefore, prescriptive limits on the size and location of any notches are set forth. The specified limitations are only applicable to solid sawn lumber members. Where engineered wood products such as trusses and I-joists are notched, the effects of the penetrations must be considered in the member design.

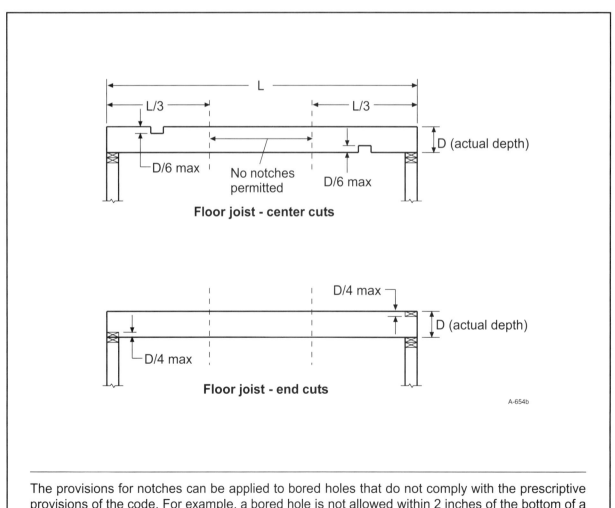

Floor joist - center cuts

Floor joist - end cuts

A-654b

The provisions for notches can be applied to bored holes that do not comply with the prescriptive provisions of the code. For example, a bored hole is not allowed within 2 inches of the bottom of a joist. If so located, it may be analyzed as a notch to determine compliance.

Code Text: *The diameter of holes bored or cut into members shall not exceed one-third the depth of the member. Holes shall not be closer than 2 inches (51 mm) to the top or bottom of the member, or to any other hole located in the member. Where the member is also notched, the hole shall not be closer than 2 inches (51 mm) to the notch.*

Discussion and Commentary: Where holes are drilled or bored in joists and other wood framing members, they tend to have the same effect as notches. The bored hole reduces the structural strength of the member. Holes cannot be oversized and must be located a minimum distance from the edges of the members. By placing any bored hole within the center portion of the structural member, the effect on the bending stresses is much less of a concern.

D/3 max

2 in. min from top and bottom of joist

For SI: 1 inch = 25.4 mm.

Pipe or conduit

D (actual depth)

A-654a

Although the code does not limit the number of bored holes within a member, it does restrict how far a bored hole can be located from a notch. Common sense should dictate that as few holes as necessary should be bored, with the holes well-distributed along the member span.

Topic: Framing of Openings
Reference: IRC R502.10

Category: Floors
Subject: Wood Floor Framing

Code Text: *Openings in floor framing shall be framed with a header and trimmer joists. When the header joist span does not exceed 4 feet (1219 mm), the header joist may be a single member the same size as the floor joist. When the header joist span exceeds 4 feet (1219 mm), the trimmer joists and the header joist shall be doubled and of sufficient cross section to support the floor joists framing into the header. Approved hangers shall be used for the header joist to trimmer joist connections when the header joist span exceeds 6 feet (1829 mm). Tail joists over 12 feet (3658 mm) long shall be supported at the header by framing anchors or on ledger strips not less than 2 inches by 2 inches (51 mm by 51 mm).*

Discussion and Commentary: Where larger floor openings are necessary to provide passage of such things as stairways or chases, the adequate transfer of floor loads to their supporting members must be provided.

For SI: 1 foot = 304.8 mm.

Floor framing for greater than 4 ft openings

A-655

In many situations, nails are insufficient to transfer vertical loads. Therefore, positive connections must be used when the span of the header or tail joists exceeds a specified length.

Code Text: *Wood trusses shall be designed in accordance with approved engineering practice. The design and manufacture of metal plate connected wood trusses shall comply with ANSI/TPI1. Trusses shall be braced to prevent rotation and provide lateral stability in accordance with the requirements specified in the construction documents for the building and on the individual truss design drawings. Truss members and components shall not be cut, notched, spliced or otherwise altered in any way without the approval of a registered design professional.*

Discussion and Commentary: There are no prescriptive provisions for wood trusses in the IRC. All trusses must be designed, and any modification or alteration must be approved by an architect or engineer.

Minimum Truss Design Drawing Information

1. Slope or depth, span, and spacing.
2. Location of all joints.
3. Required bearing widths.
4. Design loads as applicable.
   4.1 Top chord live load.
   4.2 Top chord dead load.
   4.3 Bottom chord live load.
   4.4 Bottom chord dead load.
   4.5 Concentrated loads and their points of application.
   4.6 Controlling wind and earthquake loads.
5. Adjustments to lumber and joint connector design values for conditions of use.
6. Each reaction force and direction.
7. Joint connector type and description (e.g., size, thickness or gage); and the dimensioned location of each joint connector except where symmetrically located relative to the joint interface.
8. Lumber size, species and grade for each member.
9. Connection requirements for:
   9.1 Truss-to-truss girder.
   9.2 Truss ply-to ply.
   9.3 Field splices.
10. Calculated deflection ratio and/or maximum description for live and total load.
11. Maximum axial compression forces in the truss members to enable the building designer to design the size, connections and anchorage of the permanent continuous lateral bracing. Forces shall be shown on the truss drawing or on supplemental documents.
12. Required permanent truss member bracing location.

Truss design drawings must be prepared by a registered design professional and submitted to the building official and approved prior to installation of the trusses. The drawings shall be provided with the shipment of trusses to the site and must contain at least the information specified in Section R502.11.4.

Code Text: *When there is usable space both above and below the concealed space of a floor/ceiling assembly, draftstops shall be installed so that the area of the concealed space does not exceed 1,000 square feet (92.9 m^2). Draftstopping shall divide the concealed space into approximately equal areas. Where the assembly is enclosed by a floor membrane above and a ceiling membrane below, draftstopping shall be provided in floor/ceiling assemblies under the following circumstances: 1) ceiling is suspended under the floor framing, or 2) floor framing is constructed of truss-type open-web or perforated members.*

Discussion and Commentary: Draftstopping, like fireblocking, is only mandated in combustible construction. Although the construction of draftstops is not as substantial as that for fireblocking, its integrity is just as critical and must be maintained.

Drawings courtesy
American Forest and Paper Association

For SI: 1 square foot = 0.093 m².

A-261

A variety of materials are specifically identified as acceptable for use as draftstopping. Such materials include minimum $^1/_2$-inch gypsum board, $^3/_8$-inch wood structural panels or Type 2-M-W particleboard, or any other approved materials adequately supported or secured in place.

Topic: Wood Structural Panels	**Category:** Floors
Reference: IRC R503.2.1.1, R503.2.2	**Subject:** Floor Sheathing

Code Text: *Where used as subflooring or combination subfloor underlayment, wood structural panels shall be of one of the grades specified in Table R503.2.1.1(1). The maximum allowable span for wood structural panels used as subfloor or combination subloor underlayment shall be as set forth in Table R503.2.1.1(1) or APA E30.*

Discussion and Commentary: The maximum spans for wood structural panel floor sheathing are limited by the stress and deflection imposed by the design live loads. The edges of the panels between floor supports are prevented from moving relative to each other by use of tongue and groove panel edges, the addition of wood blocking or by the use of an approved underlayment or structural finished floor system.

TABLE R503.2.1.1(1)
ALLOWABLE SPANS AND LOADS FOR WOOD STRUCTURAL PANELS FOR ROOF
AND SUBFLOOR SHEATHING AND COMBINATION SUBFLOOR UNDERLAYMENT[a, b, c]

SPAN RATING	MINIMUM NOMINAL PANEL THICKNESS (inch)	ALLOWABLE LIVE LOAD (psf)[h, l] SPAN @ 16² o.c.	SPAN @ 24² o.c.	MAXIMUM SPAN (inches) With edge support[d]	Without edge support	LOAD (pounds per square foot, at maximum span) Total load	Live load	MAXIMUM SPAN (inches)
Sheathing[e]					Roof[f]			Subfloor[j]
12/0	$5/16$	—	—	12	12	40	30	0
16/0	$5/16$	30	—	16	16	40	30	0
20/0	$5/16$	50	—	20	20	40	30	0
24/0	$3/8$	100	30	24	20[g]	40	30	0
24/16	$7/16$	100	40	24	24	50	40	16
32/16	$15/32, 1/2$	180	70	32	28	40	30	16[h]
40/20	$19/32, 5/8$	305	130	40	32	40	30	20[h,i]
48/24	$23/32, 3/48$	—	175	48	36	45	35	24
60/32	$7/8$	—	305	60	48	45	35	32
Underlayment, C-C plugged, single floor[e]					Roof[f]			**Combination subfloor underlayment[k]**
16 o.c.	$19/32, 5/8$	100	40	24	24	50	40	16[i]
20 o.c.	$19/32, 5/8$	150	60	32	32	40	30	20[i,j]
24 o.c.	$23/32, 3/4$	240	100	48	36	35	25	24
32 o.c.	$7/8$	—	185	48	40	50	40	32
48 o.c.	$1\,3/32, 1\,1/8$	—	290	60	48	50	40	48

For SI: 1 inch = 25.4 mm, 1 pound per square foot = 0.0479 kPa.

a. The allowable total loads were determined using a dead load of 10 psf. If the dead load exceeds 10 psf, then the live load shall be reduced accordingly.

b. Panels continuous over two or more spans with long dimension perpendicular to supports. Spans shall be limited to values shown because of possible effect of concentrated loads.

c. Applies to panels 24 inches or wider.

d. Lumber blocking, panel edge clips (one midway between each support, except two equally spaced between supports when span is 48 inches), tongue-and-groove panel edges, or other approved type of edge support.

e. Includes Structural 1 panels in these grades.

f. Uniform load deflection limitation: $1/180$ of span under live load plus dead load, $1/240$ of span under live load only.

g. Maximum span 24 inches for $15/32$-and $1/2$-inch panels.

h. Maximum span 24 inches where $3/4$-inch wood finish flooring is installed at right angles to joists.

i. Maximum span 24 inches where 1.5 inches of lightweight concrete or approved cellular concrete is placed over the subfloor.

j. Unsupported edges shall have tongue-and-groove joints or shall be supported with blocking unless minimum nominal $1/4$-inch thick underlayment with end and edge joints offset at least 2 inches or 1.5 inches of lightweight concrete or approved cellular concrete is placed over the subfloor, or $3/4$-inch wood finish flooring is installed at right angles to the supports. Allowable uniform live load at maximum span, based on deflection of $1/360$ of span, is 100 psf.

k. Unsupported edges shall have tongue-and-groove joints or shall be supported by blocking unless nominal $1/4$-inch-thick underlayment with end and edge joints offset at least 2 inches or $3/4$-inch wood finish flooring is installed at right angles to the supports. Allowable uniform live load at maximum span, based on deflection of $1/360$ of span, is 100 psf, except panels with a span rating of 48 on center are limited to 65 psf total uniform load at maximum span.

l. Allowable live load values at spans of 16" o.c. and 24" o.c are taken from reference standard APA E30, APA Engineered Wood Construction Guide. Refer to reference standard for allowable spans not listed in the table.

Wood structural panels are the class of panels manufactured with fully waterproof adhesive and include plywood, oriented strand board (OSB), and composite panels made up of a combination of wood veneers and reconstructed wood layers.

Code Text: *Load-bearing floor framing members shall comply with Figure R505.2(1) and with the dimensional and minimum thickness requirements specified in Tables R505.2(1) and R505.2(2). Screws for steel-to-steel connections shall be installed with a minimum edge distance and center-to-center spacing of 0.5 inch (12.7 mm), shall be self-drilling tapping, and shall conform to SAE J78.*

Discussion and Commentary: The provisions of Section R505 apply to the construction of floor systems using cold-formed steel framing. Cold-formed steel structural members have profiles with rounded corners and slender flat elements with large width-to-thickness ratios. Cold-formed steel structural members are produced from carbon or low alloy steel sheet, strip, plate or bars not more than one inch thick. Forming of the steel shapes is done at or near room temperature using bending brakes, press brakes or roll forming machines.

For SI: 1 inch = 25.4 mm, 1 foot = 304.8 mm.

Steel floor construction

A-656

The prescriptive provisions for cold-formed steel framing are only applicable to one- and two-story buildings with a maximum story height of 10 feet (plus the depth of the joist) and footprint of 40 feet by 60 feet. The wind speed, seismic design category and ground snow load are also limited.

Topic: Concrete Slab-on-ground Floors	**Category:** Floors
Reference: IRC R506.1, R506.2	**Subject:** Concrete Floors on Ground

Code Text: *Concrete slab-on-ground floors shall be a minimum 3.5 inches (89 mm) thick (for expansive soils, see Section R403.1.8). The specified compressive strength of concrete shall be as set forth in Section R402.2. The area within the foundation walls shall have all vegetation, top soil and foreign material removed. A 4-inch-thick (102 mm) base course consisting of clean graded sand, gravel, crushed stone or crushed blast-furnace slag passing a 2-inch (51 mm) sieve shall be placed on the prepared subgrade when the slab is below grade.* See exception for base course where slab is installed on well-drained or sand-gravel mixture soils.

Discussion and Commentary: Removal of construction debris and foreign materials, such as lumber formwork, stakes, tree stumps and other vegetation, limits the attraction of termites, insects and vermin. Top soil and vegetation should also be removed in that it is generally loosely compacted to such an extent as to allow soil settlement. Differential settlement can result in cracking of the floor slab and the interior wall/ceiling finishes.

Vapor retarder per Sect. R506.2.3

$3^1/_2$" concrete (min)

4" base per Sect. R506.2.2

For SI: 1 inch = 25.4 mm.

See Sect. R506.2.1 for limitations on fill material

B-498

Many of the moisture problems associated with enclosed slabs on ground can be minimized by installing a vapor retarder. The required vapor barrier, either single layer or multiple layer, must be properly installed with lapped joints, and the barrier cannot be punctured during construction.

Code Text: *Where provided in slabs on ground, reinforcement shall be supported to remain in place from the center to upper one third of the slab for the duration of the concrete placement.*

Discussion and Commentary: One of the major problems of slab-on-ground construction is the placement of reinforcement, whether it consists of welded wire mesh or reinforcing bars, prior to and during the placement of concrete. Although the IRC does not mandate such reinforcement, the proper installation is addressed in those cases where it is installed.

Reinforcing steel for slab on ground

Because the reinforcement must be supported in a manner that keeps it in place during the concrete placement, the practice of lifting reinforcement (that has been placed at the bottom of the slab) as the concrete is being placed is prohibited.

Quiz

Study Session 6
IRC Chapter 5

1. A minimum design live load of _____ shall be used for the determination of the maximum allowable floor joist spans in attics that are accessed by a fixed stairway.

   a. 10 psf   b. 20 psf

   c. 30 psf   d. 40 psf

   Reference R502.3.1

2. Where 2-inch by 10-inch floor joists of #2 hem-fir are spaced at 16 inches on center, what is the maximum span where such joists support a sleeping room and a dead load of 10 psf?

   a. 16 feet, 0 inches   b. 16 feet, 10 inches

   c. 17 feet, 8 inches   d. 19 feet, 8 inches

   Reference TABLE R.502.3.1. (1)

3. Where 2-inch by 10-inch floor joists of #1 spruce-pine-fir are spaced at 16 inches on center, what is the maximum span where such joists support a living room and a dead load of 20 psf?

   a. 14 feet, 1 inch   b. 15 feet, 5 inches

   c. 16 feet, 0 inches   d. 16 feet, 9 inches

   Reference TABLE R502.3.1. (1)

4. Where the ground snow load does not exceed 30 psf, what is the maximum span of a built-up spruce-pine-fir girder consisting of three 2 by 12s when it is located at an exterior bearing wall and supports a roof, a ceiling and one center-bearing floor in a building having a width of 28 feet?

    a. 5 feet, 2 inches   b. 7 feet, 8 inches

    c. 8 feet, 0 inches   (d.) 8 feet, 11 inches

Reference _TABLE R502.5 (1)_

5. Where a built-up girder of three 2 by 10s is used for an exterior bearing wall, what is the minimum number of jack studs required to support a girder that carries a roof, ceiling and one clear span floor?

    a. 1   (b.) 2

    c. 3   d. 4

Reference _Table R505.5 (1)_

6. What is the maximum girder span for an interior bearing wall using a built-up girder consisting of three 2 by 10s and supporting one floor in a building having a width of 36 feet?

    a. 5 feet, 5 inches   (b.) 6 feet, 9 inches

    c. 7 feet, 7 inches   d. 8 feet, 10 inches

Reference _TABLE R502.5 (2)_

7. Doubled joists under parallel bearing partitions that are separated to permit the installation of piping or vents shall be blocked at maximum intervals of _____ on center.

    a. 16 inches   b. 32 inches

    (c.) 48 inches   d. 60 inches

Reference _R502.4_

8. The ends of floor joists shall bear a minimum of _____ on wood or metal.

    a. 1 inch   (b.) $1^{1}/_{2}$ inches

    c. $2^{1}/_{2}$ inches   d. 3 inches

Reference _R 502.6_

9. Where floor joists frame from opposite sides across the top of a wood girder and are lapped, the minimum lap shall be _____.

    (a.) 3 inches   b. 4 inches

    c. 6 inches   d. 8 inches

Reference R502.6.1

10. Bridging shall be provided to support floor joists laterally where the minimum 2x joist depth exceeds _____ nominal size.

    a. 6 inches   b. 8 inches

    c. 10 inches   (d.) 12 inches

Reference R502.7.1

11. What is the maximum permitted length of a notch in a floor joist?

    a. 2 inches

    b. twice the notch depth

    (c.) $^1/_3$ the depth of the joist

    d. $^1/_6$ the depth of the joist

Reference R502.8.1

12. A hole bored through a floor joist shall have a maximum diameter of _____.

    a. 2 inches

    b. $^1/_6$ the depth of the joist

    c. $^1/_4$ the depth of the joist

    (d.) $^1/_3$ the depth of the joist

Reference R502.8.1

13. Tail joists exceeding _____ in length shall be supported by framing anchors or on complying ledger strips.

    a. 4 feet   b. 6 feet

    c. 8 feet   (d.) 12 feet

Reference R502.10

14. Where a ceiling is suspended below the floor framing, draftstops shall be installed so that the maximum area of any concealed space is _____ square feet.

   a. 100
   b. 400
   c. 1,000
   d. 1,500

   Reference R502.12

15. Where wood structural panels are used for required draftstops in concealed floor/ceiling assemblies, what is the minimum thickness mandated?

   a. $^5/_{16}$ inch
   b. $^3/_8$ inch
   c. $^{15}/_{32}$ inch
   d. $^{23}/_{32}$ inch

   Reference R502.12.1

16. Wood structural panels $^{15}/_{32}$-inch thick are to be used as subfloor sheathing and are to be covered with $^3/_4$-inch wood finish flooring installed at right angles to the joists. What is the maximum allowable span of the wood structural panels if the span rating of the panels is $^{32}/_{16}$?

   a. 0 inches; the panels may not be used as subfloor sheathing

   b. 16 inches

   c. 20 inches

   d. 24 inches

   Reference R503.2.1.1 (1) ǂǂ

17. Where Species Group 2 sanded plywood is used as combination subfloor underlayment, what is the minimum required plywood thickness where the joists are spaced at 16 inches on center?

   a. $^1/_2$ inch
   b. $^5/_8$ inch
   c. $^3/_4$ inch
   d. $^7/_8$ inch

   Reference TABLE R503.2.1.1. (2)

18. Steel floor framing constructed in accordance with the IRC is limited to buildings a maximum of _____ stories with each story having a maximum stud height of _____ feet.

   a. 2, 10
   b. 2, 12
   c. 3, 10
   d. 3, 12

   Reference R505.1.1.1 /
   R301.3 #2

19. Where complying screws are used for the connection of steel floor framing members, screws shall extend through the steel a minimum of _____.

   a. $\frac{1}{8}$ inch   b. $\frac{3}{8}$ inch

   c. three exposed threads   d. five exposed threads

   Reference R505.2.4

20. In a steel floor framing system, No. 8 screws spaced at a maximum of _____ on center on the edges and _____ on center at the intermediate supports shall be used to fasten the subfloor to the floor joists.

   a. 6 inches, 10 inches   b. 6 inches, 12 inches

   c. 8 inches, 12 inches   d. 8 inches, 14 inches

   Reference Table R505.3.1 (2)

21. Cold-formed 33 ksi steel joists are installed as single spans at 16 inches on center in the floor framing system for a dwelling. If the nominal joist size is 800S162-43, the maximum span for a 40 psf live load is _____.

   a. 10 feet, 5 inches   b. 12 feet, 3 inches

   c. 14 feet, 1 inch   d. 15 feet, 6 inches

   Reference Table R505.3.2 (1)

22. Concrete slab-on-ground floors shall be a minimum of _____ thick.

   a. 3 inches   b. $3\frac{1}{2}$ inches

   c. 4 inches   d. 5 inches

   Reference R506.1

23. The maximum fill depth when preparing a site for construction of a concrete slab-on-ground floor is _____ for earth and _____ for clean sand or gravel.

   a. 6 inches, 12 inches   b. 8 inches, 16 inches

   c. 8 inches, 24 inches   d. 12 inches, 24 inches

   Reference R506.2.1

24. A base course is not required for the prepared subgrade for a concrete floor slab below grade where the soil is classified as _____.

    a. GM   b. SC

    c. CH   d. OH

Reference R506.2.2 EXCEPTION TABLE R405.1

25. Where a vapor retarder is required between a concrete floor slab and the prepared subgrade, the joints of the vapor retarder shall be lapped a minimum of _____.

    a. 2 inches   b. 4 inches

    c. 6 inches   d. 8 inches

Reference R506.2.3

26. Where supporting only a light-frame exterior bearing wall and roof, what is the maximum cantilever span for 2 by 10 floor joists spaced at 16 inches on center, provided the roof has a width of 32 feet and the ground snow load is 30 psf?

    a. 18 inches   b. 21 inches

    c. 22 inches   d. 26 inches

Reference TABLE R 502.3.3.(1)

27. Where supporting an exterior balcony in an area having a ground snow load of 50 psf, what is the maximum cantilever span for 2 by 12 floor joists spaced at 16 inches on center?

    a. 49 inches   b. 57 inches

    c. 67 inches   d. 72 inches

Reference TABLE R502.3.3(2)

28. Unless the joists are of sufficient size to carry the load, bearing partitions perpendicular to floor joists shall be offset a maximum of _____ from supporting girders, walls or partitions.

    a. 6 inches

    b. 12 inches

    c. the depth of the floor sheathing

    d. the depth of the floor joists

Reference R502.4

29. Where an opening in floor framing is framed with a single header joist the same size as the floor joist, the maximum header joist span shall be _____ feet.

  a. 3   b. 4

  c. 6   d. 12

Reference _R502.10_

30. If a base course is required for a concrete floor slab installed below grade, the minimum thickness of the base course shall be _____ inches.

  a. 3   b. 4

  c. 6   d. 8

Reference _R506.2.2_

31. Unless supported by other approved means, the ends of each joist, beam or girder shall have a minimum of _____ inches of bearing on masonry or concrete.

  a. $1^1/_2$ b. 2

  c. 3   d. 4

Reference _R502.6_

32. Notches at the ends of solid lumber joists, rafters and beams shall have a maximum depth of _____.

  a. 2 inches

  b. 3 inches

  c. one-half the depth of the member

  d. one-fourth the depth of the member

Reference _R502.8.1_

33. Holes bored into solid lumber joists, rafter and beams shall be located a minimum of _____ from the top or bottom of the member.

  a. 1 inch

  b. 2 inches

  c. one-sixth the member depth

  d. one-third the member depth

Reference _R502.8.1_

34. Where the roof framing members are spaced at 24 inches on center, what is the maximum allowable live load for roof sheathing rated at $^{32}/_{16}$?

    a. 30 psf   b. 40 psf

    c. 70 psf   d. 180 psf

    Reference Table R503.2.1.1(1)

35. Where reinforcement is provided in concrete slab-on-ground floors, the reinforcement shall be supported to remain within what section of the slab during the concrete placement?

    a. the middle one-third

    b. the middle one-half

    c. the center to the upper one-third

    d. the center to the upper one-fourth

    Reference R506.2.4

Study Session

7

2006 IRC Chapters 6 and 7
Wall Construction and Wall Covering

OBJECTIVE: To obtain an understanding of the requirements dealing with wall construction and wall covering, including wood wall framing details; headers in both exterior and interior walls; fireblocking; braced wall panels and braced wall lines; cripple walls; steel wall framing; masonry wall construction; exterior windows and exterior doors; gypsum board plaster, ceramic tile and other interior wall finishes; and exterior wall finishes such as siding, plaster and veneer.

REFERENCE: Chapters 6 and 7, 2006 *International Residential Code*

KEY POINTS:
- 602.3.1
- How is the maximum allowable spacing of wood studs determined? How is a double top plate to be installed? A single top plate? 602.3.2 Exception   602.3.2 —
- What is the appropriate number and type of fasteners for connecting wood structural members? What is the maximum allowable spacing of such fasteners? What alternative fastening methods are acceptable?
- Where must a bored hole in a stud be located? What is the maximum permitted size of a bored hole located in a bearing wall? In a nonbearing wall? 602.6 (1 & 2)   602.6 —
- What is the maximum notch size allowed in a bearing stud wall? In a nonbearing wall? How are notches and bored holes in a top plate to be addressed?
- What is the method for determining the maximum header span in an interior bearing wall? In an exterior bearing wall? In a nonbearing wall? 602.7 —
- What is a wood structural panel box header? How is a box header to be constructed?— P.135   602.7.1 —
- Where is fireblocking required? What materials can be used as fireblocking? — 602.8.1   602.8 —
- How are foundation cripple studs to be sized? When must cripple walls be sheathed? 602.9   602.9 —
- What methods are acceptable for the construction of braced wall panels? What percentage of a braced wall line must consist of braced wall panels? How close to the end of a braced wall line is a braced wall panel required? What is minimum required panel length? 602.10 +   602.10 —
- What alternate methods of construction are provided for braced wall panels? What are the code benefits of continuous structural panel sheathing? 602.10.3 •
- How do braced wall panels need to be connected to the floor framing and top plates? What special conditions apply in Seismic Design Categories D_0, D_1 and D_2? 602.10.11   602.10.8 •

159

KEY POINTS: • What are the general provisions for load-bearing steel wall framing? How are such walls to
(Cont'd) be fastened? How are connections to be made?

• What is the minimum required thickness of masonry walls? Parapet walls?

• How must masonry walls be laterally supported? What is the minimum spacing between lateral supports? How are walls to be anchored at floors and roofs?

• How shall glass masonry be constructed?

• What is the minimum window sill height required for operable exterior windows? Under what conditions may the opening height be lower?

• What is the performance level required for exterior windows and doors? How must such windows and doors be tested and identified? How should they be attached to the structure?

• What limitations are placed on insulating concrete form wall construction?

• How must gypsum board be applied on an interior wall? How must it be fastened?

• Where is the use of water-resistant gypsum backing board prohibited?

• When is weather-resistant sheathing paper required on an exterior wall?

• What is the criteria for the installation of weep screeds in exterior plaster applications?

• How shall masonry veneer be supported? What method of anchorage is mandated?

• How must fiber cement siding be installed? Vinyl siding?

Code Text: *Components of exterior walls shall be fastened in accordance with Tables R602.3(1) through R602.3(4).*

Discussion and Commentary: The fastener schedules provide the minimum nailing requirements for the connection of structural wood framing members, wood structural panels, and other types of wall sheathing materials. For dimensional lumber, the number of nails and maximum nail spacing is specified. The use of staples is permitted in specific locations. Wood structural panels and other sheathing materials are regulated by thickness for nail size, number and spacing. Staples and other alternate attachments are also permitted when in compliance with the code. Although many of the listed building elements are to some degree related to wall construction, a number of the connection locations are specific to floor and roof framing.

TABLE R602.3(1)
FASTENER SCHEDULE FOR STRUCTURAL MEMBERS

DESCRIPTION OF BUILDING ELEMENTS	NUMBER AND TYPE OF FASTENER[a,b,c]	SPACING OF FASTENERS
Joist to sill or girder, toe nail	3-8d ($2\text{-}^1/_2'' \times 0.113''$)	—
1″ × 6″ subfloor or less to each joist, face nail	2-8d ($2^1/_2'' \times 0.113''$) 2 staples, $1^3/_4''$	— —
2″ subfloor to joist or girder, blind and face nail	2-16d ($3^1/_2'' \times 0.135''$)	—
Sole plate to joist or blocking, face nail	16d ($3^1/_2'' \times 0.135''$)	16″ o.c.
Top or sole plate to stud, end nail	2-16d ($3^1/_2'' \times 0.135''$)	—
Stud to sole plate, toe nail	3-8d ($2^1/_2'' \times 0.113''$) or 2-16d ($3^1/_2'' \times 0.135''$)	—
Double studs, face nail	10d ($3'' \times 0.128''$)	24″ o.c.
Double top plates, face nail	10d ($3'' \times 0.128''$)	24″ o.c.
Sole plate to joist or blocking at braced wall panels	3-16d ($3^1/_2'' \times 0.135''$)	16″ o.c.
Double top plates, minimum 24-inch offset of end joints, face nail in lapped area	8-16d ($3^1/_2'' \times 0.135''$)	—
Blocking between joists or rafters to top plate, toe nail	3-8d ($2^1/_2'' \times 0.113''$)	—
Rim joist to top plate, toe nail	8d ($2^1/_2'' \times 0.113''$)	6″ o.c.
Top plates, laps at corners and intersections, face nail	2-10d ($3'' \times 0.128''$)	—
Built-up header, two pieces with $^1/_2''$ spacer	16d ($3^1/_2'' \times 0.135''$)	16″ o.c. along each edge
Continued header, two pieces	16d ($3^1/_2'' \times 0.135''$)	16″ o.c. along each edge
Ceiling joists to plate, toe nail	3-8d ($2^1/_2'' \times 0.113''$)	—
Continuous header to stud, toe nail	4-8d ($2^1/_2'' \times 0.113''$)	—
Ceiling joist, laps over partitions, face nail	3-10d ($3'' \times 0.128''$)	—
Ceiling joist to parallel rafters, face nail	3-10d ($3'' \times 0.128''$)	—
Rafter to plate, toe nail	2-16d ($3^1/_2'' \times 0.135''$)	—
1″ brace to each stud and plate, face nail	2-8d ($2^1/_2'' \times 0.113''$) 2 staples, $1^3/_4''$	— —
1″ × 6″ sheathing to each bearing, face nail	2-8d ($2^1/_2'' \times 0.113''$) 2 staples, $1^3/_4''$	— —
1″ × 8″ sheathing to each bearing, face nail	2-8d ($2^1/_2'' \times 0.113''$) 3 staples, $1^3/_4''$	— —

(continued)

The code allows the use of built-up members as headers and girders in various applications. The fastening schedules require staggered nailing with minimum 10d nails on built-up members so they may act as a single member. Thus, the loads are properly transferred to all elements of the built-up member.

Topic: Design and Construction
Reference: IRC R602.3.1, R602.3.2

Category: Wall Construction
Subject: Wood Wall Framing

Code Text: *The size, height and spacing of studs shall be in accordance with Table R602.3(5). See exceptions for limits on use of utility grade studs and studs exceeding 10 feet in height. Wood stud walls shall be capped with a double top plate installed to provide overlapping at corners and intersections with bearing partitions. End joints in top plates shall be offset at least 24 inches (610 mm). Joints in plates need not occur over studs. Plates shall be not less than 2 inches (51 mm) nominal thickness and have a width at least equal to the width of the studs.* See exception for use of a single top plate.

Discussion and Commentary: Unless the studs are supporting two floors and a roof, it is possible to use 2-inch by 4-inch studs for the framing of bearing walls up to 10 feet in height. The center-to-center spacing varies based on the loading conditions and the lumber grade. If the stud height exceeds 10 feet, it is likely that larger studs and closer spacing will be required.

TABLE R602.3(5)
SIZE, HEIGHT AND SPACING OF WOOD STUDS[a]

STUD SIZE (inches)	BEARING WALLS					NONBEARING WALLS	
	Laterally unsupported stud height[a] (feet)	Maximum spacing when supporting roof and ceiling only (inches)	Maximum spacing when supporting one floor, roof and ceiling (inches)	Maximum spacing when supporting two floors, roof and ceiling (inches)	Maximum spacing when supporting one floor only (inches)	Laterally unsupported stud height[a] (feet)	Maximum spacing (inches)
2 × 3[b]	—	—	—	—	—	10	16
2 × 4	10	24	16	—	24	14	24
3 × 4	10	24	24	16	24	14	24
2 × 5	10	24	24	—	24	16	24
2 × 6	10	24	24	16	24	20	24

For SI: 1 inch = 25.4 mm.
a. Listed heights are distances between points of lateral support placed perpendicular to the plane of the wall. Increases in unsupported height are permitted where justified by analysis.
b. Shall not be used in exterior walls.

The installation of double top plates is mandated for a variety of reasons. They tie the building together at corners and at wall intersections, serve as beams by supporting joists, rafters and trusses that do not bear above studs, and work as chords for floor and roof diaphragms.

Code Text: *Any stud in an exterior wall or bearing partition may be cut or notched to a depth not exceeding 25 percent of its width. Studs in nonbearing partitions may be notched to a depth not to exceed 40 percent of a single stud width. Any stud may be bored or drilled, provided that the diameter of the resulting hole is no greater than 60 percent of the stud width, the edge of the hole is no closer than $^5/_8$ inch (15.9 mm) to the edge of the stud, and the hole is not located in the same section as a cut or notch. Studs located in exterior walls or bearing partitions drilled over 40 percent and up to 60 percent shall also be doubled with no more than two successive doubled studs bored.* See exceptions for the use of approved stud shoes.

Discussion and Commentary: In order to maintain the structural integrity of stud wall systems, limits are imposed on the notching or drilling of studs. Where a bored hole is oversized, it must be considered a notch for compliance purposes.

Top plates

Bored hole max
diameter 40 percent
of stud depth

Stud

$^5/_8$" min to edge

$^5/_8$" min to edge

Notch must not exceed
25 percent of stud depth

Bored holes shall not be located
in the same cross section of cut
or notch in stud

For SI: 1 inch = 25.4 mm.

If hole is between 40 percent and
60 percent of stud depth, then stud
must be double. No more than two
successive studs can be doubled
and so bored.

Note: Condition for exterior and bearing walls

B-499

Where nonmetallic sheathed (NM) cable is installed through bored holes, the nearest edge of the hole must be located at least $1^1/_4$ inches from the edge of the stud. Otherwise, a minimum $^1/_{16}$-inch steel plate, shoe or sleeve must be installed to provide the necessary physical protection.

Code Text: *When piping or ductwork is placed in or partly in an exterior wall or interior load-bearing wall, necessitating cutting, drilling or notching of the top plate by more than 50 percent of its width, a galvanized metal tie of not less than 0.054 inch thick (1.37 mm)(16 ga) and $1^1/_2$ inches (38 mm) wide shall be fastened across and to the plate at each side of the opening with not less than eight 16d nails at each side or equivalent.* See exception for conditions where entire side of wall with notch is covered by wood structural panel sheathing.

Discussion and Commentary: It is not uncommon for the top plate to be cut or notched to allow for the placement of plumbing, heating or other pipes in exterior walls and load-bearing interior walls. Where removal of the top plate is extensive (more than one-half of its width), a substantial tie is necessary to provide for plate continuity and retain the structural integrity of the wall system as a whole.

Exterior or bearing wall

Notch greater than 50 percent of the plate width

16 gage (0.054 in.) and 1.5 in. wide metal tie across and to each side of the notch with 8-16d nails each side

Top plates

Pipe

For SI: 1 inch = 25.4 mm

Where a double top plate is provided, as is typically the case, only the upper of the two plates needs to have a strap installed as described. However, protection of the piping may still be required to avoid potential damage caused by nails or screws.

Topic: Headers   **Category:** Wall Construction
Reference: IRC R602.7, R602.7.2   **Subject:** Wood Wall Framing

Code Text: *For header spans see Tables R502.5(1) and R502.5(2). Load-bearing headers are not required in interior or exterior nonbearing walls. A single flat 2-inch-by-4-inch (51 mm by 102 mm) member may be used as a header in interior or exterior nonbearing walls for openings up to 8 feet (2438 mm) in width if the vertical distance to the parallel nailing surface above is not more than 24 inches (610 mm). For such nonbearing headers, no cripples or blocking are required above the header.*

Discussion and Commentary: Where it is necessary to transfer floor, ceiling, wall and roof loads to the foundation, headers must be installed over openings created for doors, windows and similar items. The maximum allowable spans are provided for various conditions, with the minimum permitted size of built-up header and supporting jack studs identified.

TABLE R502.5(1)—continued
GIRDER SPANS[a] AND HEADER SPANS[a] FOR EXTERIOR BEARING WALLS
(Maximum spans for Douglas fir-larch, hem-fir, southern pine and spruce-pine-fir[b] and required number of jack studs)

GIRDERS AND HEADERS SUPPORTING	SIZE	GROUND SNOW LOAD (psf)[e]																
		30						50						70					
		Building width[c] (feet)																	
		20		28		36		20		28		36		20		28		36	
		Span	NJ[d]	Span	NJ[d]	Span	NJ[d]	Span	NJ[d]	Span	NJ[d]	Span	NJ[d]	Span	NJ[d]	Span	NJ[d]	Span	NJ[d]
Roof, ceiling, and two clear span floors	2-2×4	2-1	1	1-8	1	1-6	2	2-0	1	1-8	1	1-5	2	2-0	1	1-8	1	1-5	2
	2-2×6	3-1	2	2-8	2	2-4	2	3-0	2	2-7	2	2-3	2	2-11	2	2-7	2	2-3	2
	2-2×8	3-10	2	3-4	2	3-0	3	3-10	2	3-4	2	2-11	3	3-9	2	3-3	2	2-11	3
	2-2×10	4-9	2	4-1	3	3-8	3	4-8	2	4-0	3	3-7	3	4-7	3	4-0	3	3-6	3
	2-2×12	5-6	3	4-9	3	4-3	3	5-5	3	4-8	3	4-2	3	5-4	3	4-7	3	4-1	4
	3-2×8	4-10	2	4-2	2	3-9	2	4-9	2	4-1	2	3-8	2	4-8	2	4-1	2	3-8	2
	3-2×10	5-11	2	5-1	2	4-7	3	5-10	2	5-0	2	4-6	3	5-9	2	4-11	2	4-5	3
	3-2×12	6-10	2	5-11	3	5-4	3	6-9	2	5-10	3	5-3	3	6-8	2	5-9	3	5-2	3
	4-2×8	5-7	2	4-10	2	4-4	2	5-6	2	4-9	2	4-3	2	5-5	2	4-8	2	4-2	2
	4-2×10	6-10	2	5-11	2	5-3	2	6-9	2	5-10	2	5-2	2	6-7	2	5-9	2	5-1	2
	4-2×12	7-11	2	6-10	2	6-2	3	7-9	2	6-9	2	6-0	3	7-8	2	6-8	2	5-11	3

For SI: 1 inch = 25.4 mm, 1 pound per square foot = 0.0479 kPa.
a. Spans are given in feet and inches.
b. Tabulated values assume #2 grade lumber.
c. Building width is measured perpendicular to the ridge. For widths between those shown, spans are permitted to be interpolated.
d. NJ - Number of jack studs required to support each end. Where the number of required jack studs equals one, the header is permitted to be supported by an approved framing anchor attached to the full-height wall stud and to the header.
e. Use 30 psf ground snow load for cases in which ground snow load is less than 30 psf and the roof live load is equal to or less than 20 psf.

Box headers are permitted when minimum $^{15}/_{32}$-inch wood structural panels are used. Header depths of both 9 inches and 15 inches are addressed, with panels attached to both sides or one side only. Adequate attachment to the framing is critical, with nails spaced at 3 inches on center.

Code Text: *Fireblocking shall be provided to cut off all concealed draft openings (both vertical and horizontal) and to form an effective barrier between stories, and between a top story and the roof space. Fireblocking shall be provided in wood-frame construction in the following locations: 1) in concealed spaces of stud walls and partitions, including furred spaces, at the ceiling and floor level and at 10 foot (3048 mm) intervals both vertical and horizontal; 2) at all interconnections between concealed vertical and horizontal spaces; 3) in concealed spaces between stair stringers at the top and bottom of the run; 4) at openings around vents, pipes and ducts at ceiling and floor level; 5) at chimneys and fireplaces per Section R1001.16; and 6) at cornices of a two-family dwelling at the line of dwelling unit separation.*

Discussion and Commentary: Fireblocking is defined as the installation of building materials to resist the free passage of flame to other areas of the building through concealed spaces.

Fireblocking—Furred soffit

A-661a

The technique of wood-frame platform construction provides natural fireblocking in many locations to restrict fire movement between horizontal and vertical concealed spaces. Where platform framing and vertical openings occur, a more prescriptive approach is necessary.

Topic: Braced Wall Lines
Reference: IRC R602.10.1

Category: Wall Construction
Subject: Wood Wall Framing

Code Text: *Braced wall lines shall consist of braced wall panel construction methods in accordance with Section R602.10.3. The amount and location of bracing shall be in accordance with Table R602.10.1 and the amount of bracing shall be the greater of that required by the seismic design category or the design wind speed.*

Discussion and Commentary: Braced wall lines are the lateral-resisting elements in conventional construction that are similar to shear walls in engineered structures. In order to properly transfer the lateral loads through the floor or roof diaphragms to the braced wall panels in the braced wall lines, the structure must be adequately connected. There are eight different types of braced wall panels that may be used.

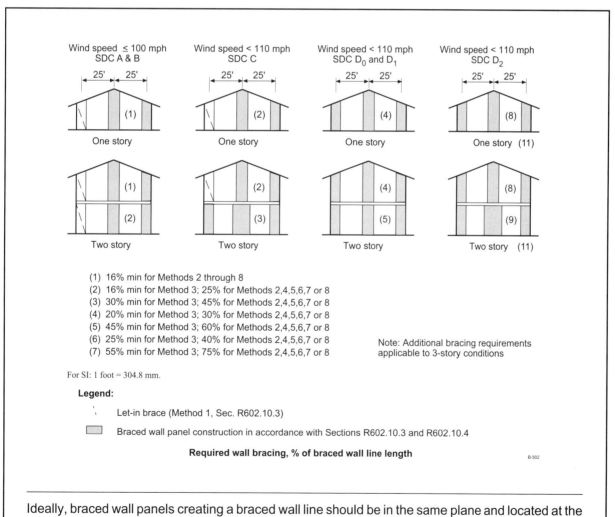

(1) 16% min for Methods 2 through 8
(2) 16% min for Method 3; 25% for Methods 2,4,5,6,7 or 8
(3) 30% min for Method 3; 45% for Methods 2,4,5,6,7 or 8
(4) 20% min for Method 3; 30% for Methods 2,4,5,6,7 or 8
(5) 45% min for Method 3; 60% for Methods 2,4,5,6,7 or 8
(6) 25% min for Method 3; 40% for Methods 2,4,5,6,7 or 8
(7) 55% min for Method 3; 75% for Methods 2,4,5,6,7 or 8

Note: Additional bracing requirements applicable to 3-story conditions

For SI: 1 foot = 304.8 mm.

Legend:

     Let-in brace (Method 1, Sec. R602.10.3)

     Braced wall panel construction in accordance with Sections R602.10.3 and R602.10.4

Required wall bracing, % of braced wall line length

B-502

Ideally, braced wall panels creating a braced wall line should be in the same plane and located at the ends of each braced wall line. However, it is possible to offset the panels up to 4 feet on either side of the braced wall line, with the first panel up to $12^{1}/_{2}$ feet from the end of the wall.

Code Text: *Load-bearing steel wall framing members shall comply with Figure R603.2(1) and the dimensional and minimum thickness requirements specified in Tables R603.2(1) and R603.2(2). All exterior steel framed walls and interior load-bearing steel framed walls shall be constructed in accordance with the provisions of Section R603.3 and Figure R603.3.*

Discussion and Commentary: The use of load-bearing steel members is regulated prescriptively not only for wall framing but also for floor and roof system framing. In order to verify that the steel members are in compliance with appropriate materials standards, the framing members must have a legible label, stencil, stamp or embossment with the manufacturer's identification as well as the minimum steel thickness, coating designation and yield strength.

For SI: 1 inch = 25.4 mm.

Steel wall construction

A-664

A variety of tables, figures and details are provided in the code to assist the user in understanding the construction techniques and structural limitations of steel-framed structures. When in compliance with the prescriptive provisions, the building is considered structurally sound.

Code Text: *The minimum thickness of masonry bearing walls more than one story high shall be 8 inches (203 mm). Solid masonry walls of one-story dwellings and garages shall not be less than 6 inches (152 mm) in thickness when not greater than 9 feet (2743 mm) in height, provided that when gable construction is used, an additional 6 feet (1829 mm) is permitted to the peak of the gable. Masonry walls shall be laterally supported in either the horizontal or vertical direction at intervals as required by Section R606.9.*

Discussion and Commentary: The minimum required width of masonry units is based directly on the height of the wall, both in number of stories and vertical dimension. Lateral support may be provided horizontally at complying intervals along the length of masonry walls, determined by the walls' length-to-thickness ratio, or vertically at the floors and/or the roof.

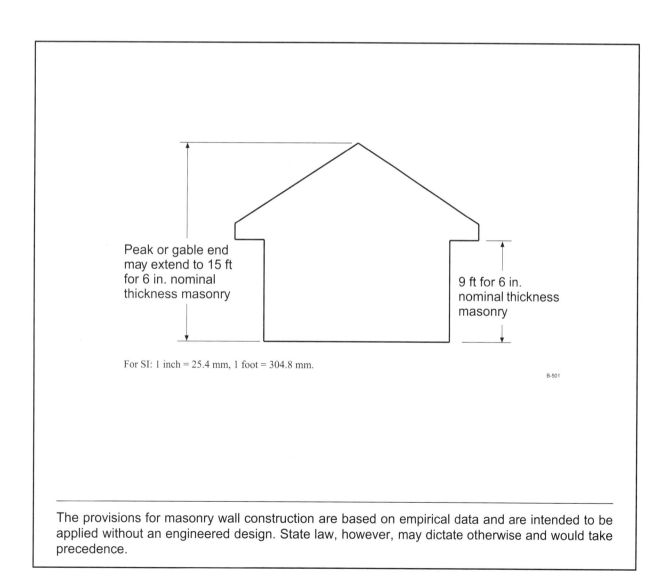

Peak or gable end may extend to 15 ft for 6 in. nominal thickness masonry

9 ft for 6 in. nominal thickness masonry

For SI: 1 inch = 25.4 mm, 1 foot = 304.8 mm.

B-501

The provisions for masonry wall construction are based on empirical data and are intended to be applied without an engineered design. State law, however, may dictate otherwise and would take precedence.

Code Text: *Masonry walls shall be laterally supported in either the horizontal or the vertical direction. The maximum spacing between lateral supports shall not exceed the distances in Table R606.9. Lateral support shall be provided by cross walls, pilasters, buttresses or structural frame members when the limiting distance is taken horizontally, or by floors or roofs when the limiting distance is taken vertically.*

Discussion and Commentary: The limitations on the maximum unsupported height or length of masonry walls specified in Table R606.9 provide reasonable performance. At the base of the wall, footings are a lateral support point. Thus, the unsupported height from the footing to the anchorage point at the floor or roof is the unsupported height, which must be limited to the values in Table R606.9.

TABLE R606.9
SPACING OF LATERAL SUPPORT FOR MASONRY WALLS

CONSTRUCTION	MAXIMUM WALL LENGTH TO THICKNESS OR WALL HEIGHT TO THICKNESS[a,b]
Bearing walls:	
Solid or solid grouted	20
All other	18
Nonbearing walls:	
Exterior	18
Interior	36

For SI: 1 foot = 304.8 mm.

a. Except for cavity walls and cantilevered walls, the thickness of a wall shall be its nominal thickness measured perpendicular to the face of the wall. For cavity walls, the thickness shall be determined as the sum of the nominal thicknesses of the individual wythes. For cantilever walls, except for parapets, the ratio of height to nominal thickness shall not exceed 6 for solid masonry, or 4 for hollow masonry. For parapets, see Section R606.2.4.

b. An additional unsupported height of 6 feet is permitted for gable end walls.

In lieu of unsupported dimensional limitations being measured vertically from footing to supporting floor or roof, the span limitations may be met with the use of pilasters, columns, piers, cross walls or similar elements whose relative stiffness is greater than the wall.

Code Text: *Masonry walls in Seismic Design Category A, B or C shall be anchored to roof structures with metal strap anchors spaced in accordance with the manufacturer's instructions, $^1/_2$-inch (13 mm) bolts spaced not more than 6 feet (1829 mm) on center, or other approved anchors. Anchors shall be embedded at least 16 inches (406 mm) into the masonry, or be hooked or welded to bond beam reinforcement placed not less than 6 inches (152 mm) from the top of the wall.*

Discussion and Commentary: All anticipated lateral forces must have a path of tranfer from the roof and floor diaphragms to the masonry walls for transfer to the building's foundation. It is necessary to provide adequate connections to maintain the required load path. Where the building is located in one of the higher seismic design categories, additional conditions are mandated.

See table for bolt size and spacing

For SI: 1 inch = 25.4 mm, 1 foot = 304.8 mm.

Ledger bolt size and spacing

Joist span	Bolt size and spacing	
	Roof	Floor
10'	$^1/_2$ at 2' 6" $^7/_8$ at 3' 6"	$^1/_2$ at 2' 0" $^7/_8$ at 2' 9"
10-15'	$^1/_2$ at 1' 9" $^7/_8$ at 2' 6"	$^1/_2$ at 1' 4" $^7/_8$ at 2' 0"
15-20'	$^1/_2$ at 1' 3" $^7/_8$ at 2' 0"	$^1/_2$ at 1' 0" $^7/_8$ at 1' 6"

A-665a

Where the floor diaphragm is connected to a masonry wall, a ledger is used. The ledger fasteners transfer shear forces when loading is in the plane of the wall and prevent separation of the wall and flooring system when the forces occur out of plane.

Code Text: *In dwelling units, where the opening of an operable window is located more than 72 inches (1829 mm) above the finished grade or surface below, the lowest part of the clear opening of the window shall be a minimum of 24 inches (610 mm) above the finished floor of the room in which the window is located. Glazing between the floor and 24 inches (610 mm) shall be fixed or have openings through which a 4-inch-diameter (102 mm) sphere cannot pass.* See exceptions for windows that 1) do not open far enough to allow passage of a 4-inch sphere, or 2) are protected by complying window guards.

Discussion and Commentary: Historical data have shown that each year a considerable number of children fall from windows. It has been estimated that a sizable percentage of those falls occurred through windows with a low sill height. The restrictions on window sill height are intended to raise the height of the opening at the sill above the center of gravity of a small child, thus reducing the number of falls.

For SI: 1 inch = 25.4 mm

The measurements on both the interior and exterior sides of the residence are taken from the lowest part of the clear window opening, providing for consistent application of the provision. Where the lower window panel is inoperable, the measurement is to be taken from the lowest point of the lowest operable panel.

Code Text: *Exterior windows and doors shall be designed to resist the design wind loads specified in Table R301.2(2) adjusted for height and exposure per Table R301.2(3). Window and glass door assemblies shall be anchored in accordance with the published manufacturer's recommendations to achieve the design pressure specified. Substitute anchoring systems used for substrates not specified by the fenestration manufacturer shall provide equal or greater anchoring performance as demonstrated by accepted engineering practice.*

Discussion and Commentary: Glass doors and windows are components of the exterior wall. As such, they are regulated through the specification of performance criteria for such components, as well as their supporting elements, in order to protect against glass breakage due to high wind pressure.

Through the frame anchoring method. Anchors shall be provided to transfer load from the window or door frame into the rough opening substrate.

Apply frame clip to window or door frame in accordance with published manufacturer's recommendations. Anchors shall be provided to transfer load from the frame clip Into the rough opening substrate

Figure R613.8(3)
Through the frame

Figure R613.8(4)
Frame clip

Figure R613.8(6)
Through the frame

Figure R613.8(7)
Frame clip

C-274

In hurricane-prone regions, the exterior windows and glass doors shall be protected from wind-borne debris in accordance with Section R301.2.1.2. The installation of impact-resistant glazing or precut wood structural panels are options.

Code Text: *Maximum spacing of supports and the size and spacing of fasteners used to attach gypsum board shall comply with Table R702.3.5. Gypsum board shall be applied at right angles or parallel to framing members. All edges and ends of gypsum board shall occur on the framing members, except those edges and ends that are perpendicular to the framing members. Interior gypsum board shall not be installed where it is directly exposed to the weather or to water.*

Discussion and Commentary: Table R702.3.5 is a comprehensive table identifying the minimum thickness and fastening requirements for gypsum wallboard. The wallboard thickness ($^3/_8$-inch, $^1/_2$-inch, or $^5/_8$-inch), location of the wallboard (wall or ceiling), orientation of the wallboard to the framing members (parallel or perpendicular), spacing of framing members (16 inches or 24 inches on center), type of fasteners (nails or screws) and use of adhesive (with or without) are all set forth in the table.

TABLE R702.3.5
MINIMUM THICKNESS AND APPLICATION OF GYPSUM BOARD

THICKNESS OF GYPSUM BOARD (inches)	APPLICATION	ORIENTATION OF GYPSUM BOARD TO FRAMING	MAXIMUM SPACING OF FRAMING MEMBERS (inches o.c.)	MAXIMUM SPACING OF FASTENERS (inches) Nails[a]	MAXIMUM SPACING OF FASTENERS (inches) Screws[b]	SIZE OF NAILS FOR APPLICATION TO WOOD FRAMING[c]
Application without adhesive					
$^3/_8$	Ceiling[d]	Perpendicular	16	7	12	13 gage, $1^1/_4$″ long, $^{19}/_{64}$″ head; 0.098″ diameter, $1^1/_4$″ long, annular-ringed; or 4d cooler nail, 0.080″ diameter, $1^3/_8$″ long, $^7/_{32}$″ head.
$^3/_8$	Wall	Either direction	16	8	16	
$^1/_2$	Ceiling	Either direction	16	7	12	13 gage, $1^3/_8$″ long, $^{19}/_{64}$″ head; 0.098″ diameter, $1^1/_4$″ long, annular-ringed; 5d cooler nail, 0.086″ diameter, $1^5/_8$″ long, $^{15}/_{64}$″ head; or gypsum board nail, 0.086″ diameter, $1^5/_8$″ long, $^9/_{32}$″ head.
$^1/_2$	Ceiling[d]	Perpendicular	24	7	12	
$^1/_2$	Wall	Either direction	24	8	12	
$^1/_2$	Wall	Either direction	16	8	16	
$^5/_8$	Ceiling	Either direction	16	7	12	13 gage, $1^5/_8$″ long, $^{19}/_{64}$″ head; 0.098″ diameter, $1^3/_8$″ long, annular-ringed; 6d cooler nail, 0.092″ diameter, $1^7/_8$″ long, $^1/_4$″ head; or gypsum board nail, 0.0915″ diameter, $1^7/_8$″ long, $^{19}/_{64}$″ head.
$^5/_8$	Ceiling[e]	Perpendicular	24	7	12	
$^5/_8$	Wall	Either direction	24	8	12	
$^5/_8$	Wall	Either direction	16	8	16	
Application with adhesive					
$^3/_8$	Ceiling[d]	Perpendicular	16	16	16	Same as above for $^3/_8$″ gypsum board
$^3/_8$	Wall	Either direction	16	16	24	
$^1/_2$ or $^5/_8$	Ceiling	Either direction	16	16	16	Same as above for $^1/_2$″ and $^5/_8$″ gypsum board, respectively
$^1/_2$ or $^5/_8$	Ceiling[d]	Perpendicular	24	12	16	
$^1/_2$ or $^5/_8$	Wall	Either direction	24	16	24	
Two $^3/_8$ layers	Ceiling	Perpendicular	16	16	16	Base ply nailed as above for $^1/_2$″ gypsum board; face ply installed with adhesive
Two $^3/_8$ layers	Wall	Either direction	24	24	24	

For SI: 1 inch = 25.4 mm.

a. For application without adhesive, a pair of nails spaced not less than 2 inches apart or more than $2^1/_2$ inches apart may be used with the pair of nails spaced 12 inches on center.

b. Screws shall be Type S or W per ASTM C 1002 and shall be sufficiently long to penetrate wood framing not less than $^5/_8$ inch and metal framing not less than $^3/_8$ inch.

c. Where metal framing is used with a clinching design to receive nails by two edges of metal, the nails shall be not less than $^5/_8$ inch longer than the gypsum board thickness and shall have ringed shanks. Where the metal framing has a nailing groove formed to receive the nails, the nails shall have barbed shanks or be 5d, $13^1/_2$ gage, $1^5/_8$ inches long, $^{15}/_{64}$-inch head for $^1/_2$-inch gypsum board; or 6d, 13 gage, $1^7/_8$ inches long, $^{15}/_{64}$-inch head for $^5/_8$-inch gypsum board.

d. Three-eighths-inch-thick single-ply gypsum board shall not be used on a ceiling where a water-based textured finish is to be applied, or where it will be required to support insulation above a ceiling. On ceiling applications to receive a water-based texture material, either hand or spray applied, the gypsum board shall be applied perpendicular to framing. When applying a water-based texture material, the minimum gypsum board thickness shall be increased from $^3/_8$ inch to $^1/_2$ inch for 16-inch on center framing, and from $^1/_2$ inch to $^5/_8$ inch for 24-inch on center framing or $^1/_2$-inch sag-resistant gypsum ceiling board shall be used.

e. Type X gypsum board for garage ceilings beneath habitable rooms shall be installed perpendicular to the ceiling framing and shall be fastened at maximum 6 inches o.c. by minimum $1^7/_8$ inches 6d coated nails or equivalent drywall screws.

Where screws are used for attaching gypsum board to wood framing, they shall be either Type W or Type S and must penetrate the wood at least $^5/_8$ inch. Type S screws are to be used where attachment is made to light-gage steel framing, with a minimum penetration of $^3/_8$ inch.

Code Text: *One layer of No. 15 asphalt felt, free from holes and breaks, complying with ASTM D 226 for Type 1 felt or other approved water-resistive barrier shall be applied over studs or sheathing of all exterior walls. Such felt or material shall be applied horizontally, with the upper layer lapped over the lower layer not less than 2 inches (51 mm). Where joints occur, felt shall be lapped not less than 6 inches (152 mm).*

Discussion and Commentary: Structural members are adversely affected by moisture, particularly under cyclic conditions of wetting and drying. Due to the potential for water accumulation within the wall cavity, a water-resistant membrane is mandated behind any exterior siding or veneer.

Under certain conditions, the value of asphalt felt or a similar weather-resistant barrier is questionable or its use unnecessary. It is permissible to eliminate the membrane in detached acessory buildings or where the siding, finish materials or lath provides the needed protection.

Topic: Masonry Veneer Anchorage   **Category:** Wall Covering
Reference: IRC R703.7.4, R703.7.4.1   **Subject:** Exterior Covering

Code Text: *Masonry veneer shall be anchored to the supporting wall with corrosion-resistant metal ties. Where veneer is anchored to wood backings by corrugated sheet metal ties, the distance separating the veneer from the sheathing material shall be a maximum of a nominal 1 inch (25 mm). Each tie shall be spaced not more than 24 inches (610 mm) on center horizontally and vertically and shall support not more than 2.67 square feet (0.25 m^2) of wall area. See exception for closer spacing in high-wind or high-seismic areas.*

Discussion and Commentary: Two types of ties are approved for the attachment of masonry veneer to wood construction: corrugated sheet metal ties and metal strand wire ties. The clearance between the veneer and the wood backing is limited and varies based upon which type of tie is used.

Clearance of $^3/_4$-in. min air space

Building paper or approved water-repellent sheathing[b]

Veneer

Metal tie[b]

Roofing

Wallboard

1 in.-air space or
1 in.-mortared space[c]

Building paper or approved water-repellent sheathing[b]

Metal tie[b]

Flashing[a]

Steel lintel[d]

Weephole[a]

Sealant

Wallboard

a. See Sections R703.7.5 and R703.7.6.
b. See Sections R703.2 and R703.7.4.
c. See Sections R703.7.4.2 and R703.7.4.3.
d. See Section R703.7.2.1

For SI: 1 inch = 25.4 mm.

B-505

Where veneer is attached to a wall as facing material, it is intended to provide ornamentation, protection, insulation or a combination of these features. It does not provide any structural strength to the wall, nor is it intended to carry any load other than its own weight.

Quiz

Study Session 7
IRC Chapters 6 and 7

1. What is the maximum center-to-center stud spacing permitted for a 2-inch by 6-inch, 8-foot-high wood stud bearing wall supporting two floors, roof and ceiling?

   a. 8 inches   b. 12 inches

   (c) 16 inches   d. 24 inches

   Reference _Table R602.3(5)_

2. An exterior bearing wall consists of 2-inch by 4-inch wood studs at 12 inches on center. If located in Seismic Design Category C and exposed to a wind speed of 90 mph, what is the maximum stud height when the wall supports a roof only?

   a. 10 feet   (b.) 12 feet

   c. 14 feet   d. 16 feet

   Reference _Table R602.3.1_

3. The minimum offset for end joints in a double top plate shall be ___24"___.

   (a) 24 inches   b. 36 inches

   c. 48 inches   d. 60 inches

   Reference _602.3.2_

4. A single top plate is permitted in a wood stud bearing wall where the rafters or joists are located within _____ of the center of the studs.

    a. 0 inches; no tolerance is permitted

    b. 1 inch

    c. $1^1/_2$ inches

    d. 5 inches

Reference 602.3.2 Exception

5. Where the wall sheathing is nailed directly to the studs, what is the maximum stud spacing permitted for $^3/_8$-inch wood structural panel wall sheathing with a span rating of 12/0.

    a. 12 inches   b. 16 inches

    c. 20 inches   d. 24 inches

Reference Table 602.3 (3)

6. In a nonbearing partition, a wood stud may be notched a maximum of _____to_____.

    a. $^5/_8$ inch

    b. $1^3/_8$ inches

    c. 40 percent of the stud width

    d. 60 percent of the stud width

Reference 602.6 #1

7. A bored hole shall not be located with _____$5/8''$_____ of the edge of a wood stud.

    a. $^3/_8$ inch   b. $^1/_2$ inch

    c. $^5/_8$ inch   d. 1 inch

Reference 602.6 #2

8. In a nonbearing interior partition, what is the maximum diameter permitted for a bored hole in a wood stud?

    a. $1^1/_2$ inches

    b. 25 percent of the stud depth

    c. 40 percent of the stud depth

    d. 60 percent of the stud depth

Reference 602.6 #2

9. A 15-inch-deep box header in an exterior wall is constructed with wood structural panels on both sides. What is the maximum allowable header span for a condition where the header supports a clear-span roof truss with a span of 26 feet?

    a. 4 feet   b. 5 feet

    c. 7 feet   (d.) 8 feet

Reference *Table R602.7.2*

10. Which of the following materials is not specifically identified by the IRC as a fireblocking material?

    a. $^{1}/_{4}$-inch cement-based millboard

    b. $^{1}/_{2}$-inch gypsum board

    (c.) $^{15}/_{32}$-inch wood structural panel

    d. $^{3}/_{4}$-inch particleboard

Reference *602.8.1*

11. Where unfaced fiberglass batt insulation is used as a fireblocking material in the wall cavity of a wood stud wall system, the insulation shall be installed with a minimum vertical height of *16* .

    (a.) 16 inches   b. 3 feet

    c. 4 feet   d. the entire stud space

Reference *602.8.1.1*

12. A foundation cripple wall shall be considered an additional story for stud sizing requirements where the wall height exceeds *48* .

    a. 14 inches   b. 30 inches

    c. 36 inches   (d.) 48 inches

Reference *602.9*

13. For one-story buildings located in Seismic Design Category B with a wind speed of 90 mph, particleboard wall sheathing panels used as wall bracing shall be provided for a minimum of *16%* of the braced wall line.

    (a.) 16 percent   b. 25 percent

    c. 30 percent   d. 45 percent

Reference *Table R602.10.1*

14. Braced wall panels shall begin a maximum of _12.5_ from the end of a braced wall line.

    a. 2 feet, 8 inches   b. 4 feet

    c. 8 feet   (d.) 12 feet, 6 inches

    Reference _602.10.1_

15. In steel-framed wall construction, rafters may be offset a maximum of _3/4"_ from the centerline of the load-bearing steel studs.

    a. 0 inches; no tolerance is permitted

    (b.) $^3/_4$ inch

    c. 1 inch

    d. 5 inches

    Reference _603.1.2_

16. Holes in load-bearing steel wall framing members shall have a minimum distance of _10_ between the edge of the bearing surface and the edge of the hole.

    a. $1^1/_2$ inches   b. 3 inches

    c. 4 inches   (d.) 10 inches

    Reference _603.2 #6_

17. In unreinforced masonry construction, what is the maximum vertical spacing of longitudinal reinforcement where units are laid in stack bond?

    a. 12 inches o.c.   (b.) 16 inches o.c.

    c. 20 inches o.c.   d. 24 inches o.c.

    Reference _606.8_

18. Where ledgers are used at the connection between a masonry wall and a wood floor system having floor joists spanning 16 feet, $^1/_2$-inch ledger bolts shall be located at a maximum of _1'0"_ on center, provided the building is in Seismic Design Category A, B or C and the wind loads are less than 30 psf.

    (a.) 1 foot, 0 inches   b. 1 foot, 3 inches

    c. 1 foot, 9 inches   d. 2 feet, 0 inches

    Reference _Fig. R606.11(i)_

19. When $1/2$-inch gypsum board is used as an interior wall covering and installed perpendicular to framing members at 16 inches on center, the maximum spacing of nails is _____16_____ on center where adhesive is used.

    a. 7 inches   b. 8 inches

    c. 12 inches   d. 16 inches

    Reference _Table 702.3.5_

20. Screws for attaching gypsum board to light-gage steel framing shall penetrate the steel a minimum of _____3/8"_____ .

    a. $1/4$ inch   b. $3/8$ inch

    c. $1/2$ inch   d. $5/8$ inch

    Reference _702.3.6_

21. One-half-inch thick water-resistant gypsum backing board is permitted to be used on a ceiling where the framing members are spaced a maximum of _____12_____ on center.

    a. 12 inches   b. 16 inches

    c. 20 inches   d. 24 inches

    Reference _702.3.8_

22. Where $3/8$-inch particleboard is used as an exterior wall covering, what type of fasteners are required if the particleboard is attached directly to the studs?

    a. 0.120 nail, 2 inches long

    b. 6d box nail

    c. 8d box nail

    d. direct attachment to the studs is prohibited

    Reference _Table R703.4_

23. In Seismic Design Category D_1, metal ties for anchoring masonry veneer to a supporting wall shall support a maximum of _____2 sq ft_____ of wall area.

    a. 2 square feet   b. $2^2/3$ square feet

    c. $3^1/2$ square feet   d. $4^1/2$ square feet

    Reference _703.7.4.1 Exception_

24. An Exterior Insulation Finish System (EIFS) shall terminate a minimum of _____ above the finished ground level.

    a. 1 inch   b. 2 inches

    c. 6 inches   d. 8 inches

Reference _703.9_

25. What is the minimum required size of a steel angle spanning 8 feet used as a lintel supporting one story of masonry veneer above?

    a. $3 \times 3 \times {}^{1}/_{4}$ b. $4 \times 3 \times {}^{1}/_{4}$

    c. $5 \times 3^{1}/_{2} \times {}^{5}/_{16}$ d. $6 \times 3^{1}/_{2} \times {}^{5}/_{16}$

Reference _Table R703.7.3_

26. Utility grade studs, where used in loadbearing walls supporting only a roof and a ceiling, shall have a maximum height of _____ feet.

    a. 8   b. 10

    c. 12   d. 14

Reference _602.3.1 Exc 1_

27. Double top plates shall be face-nailed together in the required lapped area with _____ fasteners.

    a. four 16d   b. six 10d

    c six 16d   d. eight 16d

Reference _Table R602.3(1)_

28. ${}^{15}/_{32}$-inch wood structural panels attached to wall framing with 15 gage staples shall be fastened at a maximum of _____ inches at the panel edges and _____ inches at the intermediate supports.

    a. 3, 6   b. 4, 8

    c. 5, 10   d. 6, 12

Reference _Table R602.3(2)_

29. Where the top plate of an interior load-bearing wall is notched by more than __50__ of its width to accommodate piping, a complying metal tie shall be installed.

    a. 25 percent   b. 40 percent

    c. 50 percent   d. 60 percent

Reference _602.6.1_

30. A weep screed installed on exterior stud walls in an exterior plaster application shall be placed a minimum of _2_ inch(es) above paved areas.

    a. 1   b. 2

    c. 4   d. 6

Reference _703.6.2.1_

31. Where an exterior wall top plate is notched to the extent that a metal tie is required across the opening, the tie shall be fastened to the plate at each side of the opening with a minimum of _8_ 16d nails at each side, or equivalent.

    a. two   b. four

    c. six   d. eight

Reference _602.6.1_

32. For a building located is Seismic Design Category D_0, plate washers used in the connection of braced wall line sills to a concrete foundation shall be a minimum of _3_ in size.

    a. 0.188 inch by 2 inches by 2 inches

    b. 0.229 inch by 2 inches by 2 inches

    c. 0.229 inch by 3 inches by 3 inches

    d. 0.375 inch by 3 inches by 3 inches

Reference _602.11.1_

33. What is the minimum required thickness of a vertical core in an 8-inch waffle-grid ICF wall?

    a. 5 inches   b. 5.5 inches

    c. 6.25 inches   d. 7 inches

Reference _Table R611.2_

34. Where the opening of an operable window is located more than 72 inches above the surface below, the lowest part of the clear opening of the window shall be a minimum of _____24_____ inches above the finished floor.

    a. 24   b. 30

    c. 36   d. 44

Reference _____613.2_____

35. The water-resistive vapor permeable barrier required to be applied over wood-based wall sheathing shall have a performance level equivalent to that of _____2_____ layer(s) of Grade _____D_____ paper.

    a. one, A   b. one, B

    c. two, C   d. two, D

Reference _____703.6.3_____

Study Session

8

2006 IRC Chapters 8 and 9
Roof/Ceiling Construction and Roof Assemblies

OBJECTIVE: To gain an understanding of the requirements for ceiling construction, roof construction and roof coverings, including ceiling joist and rafter sizing; cutting and notching of framing members; framing at openings; purlins; steel roof framing; roof ventilation and attic access; and roof covering materials such as asphalt shingles, concrete or clay tiles, wood shingles and wood shakes.

REFERENCE: Chapters 8 and 9, 2006 *International Residential Code*

KEY POINTS:
- How must roof drainage water be discharged? ~ 901.3
  802.1 -- How must wood framing members be identified? What special concerns must be addressed when using fire-retardant-treated wood?
  802 • What are the acceptable framing methods at the roof ridge? At valleys and hips? Under what condition are ridges, valleys and hips required to be designed as beams?
  802.3.1 • How are rafters to be tied together? At what maximum intervals must collar ties be located?
  802.4 -- How is the maximum allowable ceiling joist span determined? Roof rafter span? *802.5* How does the location of rafter ties affect the maximum allowable rafter span? Where are collar ties to be located? *802.3.1*
  802.9 • What methods are prescribed for framing around ceiling and roof openings? When is lateral support mandated? ~ 802.7
  802.5.1 • What is the purpose of a purlin system? How are purlins sized? What is the maximum spacing of purlin braces? The maximum unbraced length?
  803.22 -- How is the maximum allowable span for roof sheathing determined? How must roof sheathing be installed? *803.2.3 d Tables 12602-3(1)*
  802.10.5 • How are trusses to be connected to the wall plates to address uplift pressures?
  804.1.1 • What are the general limitations for the prescriptive use of steel roof framing systems? What are the general requirements? *804.1*
  806.1 • When is roof ventilation required? What is minimum required ventilating area? *(806.2)* How can this area be reduced? *806.2*

185

KEY POINTS:

(handwritten: 806.4)

(handwritten: 807.1) **(Cont'd)**

- Under what circumstances are unvented conditioned attic assemblies permitted?
- When is attic access required? What is the minimum size needed for an attic access opening? How much headroom clearance must be provided over the access opening?
- *(handwritten: 808.1)* At what distance must combustible insulation be separated from recessed lighting fixtures, fan motors, and other heat-producing devices?
- What is Class A, Class B or Class C roofing? When is such roofing required? What specific materials are considered Class A roof coverings?
- Where is flashing required? How is roof drainage to be addressed? *(handwritten: 903.4)*
- How must roofing materials be identified? *(handwritten: 904.4)*
- *(handwritten: 905.2)* What are the installation requirements for asphalt shingles? Clay and concrete tile? *(handwritten: 905.3)* Wood shingles? Wood shakes? *(handwritten: 905.8)*
- What other types of roof coverings are regulated by the code?
- What level of work is considered reroofing? What is a roof repair? Roof replacement? When must the existing roof coverings be removed prior to any reroofing work?

186

Topic: Framing Details
Reference: IRC R802.3

Category: Roof-Ceiling Construction
Subject: Wood Roof Framing

Code Text: *Rafters shall be framed to ridge board or to each other with a gusset plate as a tie. Ridge board shall be at least 1-inch (25 mm) nominal thickness and not less in depth than the cut end of the rafter. At all valleys and hips there shall be a valley or hip rafter not less than 2-inch (51 mm) nominal thickness and not less in depth than the cut end of the rafter.*

Discussion and Commentary: For the roof loads to be carried to the exterior walls, a continuous tie must be created at the roof ridge. This is typically accomplished by installing the rafters in direct opposition to each other and nailing them directly to a ridge board having a depth at least that of the rafter ends. Another option is the use of a gusset plate or similar element that ties the roof system together at the ridge line. Continuity from rafter to rafter is the key requirement.

Ridge board

A-670a

Section A-A

For SI: 1 inch = 25.4 mm.

Ridge alternative - gusset plate tie

A-670b

Where the roof has a relatively shallow slope (less than 3 units vertical in 12 units horizontal), the ridge must be designed as a beam in order to carry the necessary tributary loading. Positive bearing for the rafters must be provided unless an approved alternative method is used.

Code Text: *Ceiling joists and rafters shall be nailed to each other in accordance with Table R802.5.1(9), and the rafter shall be nailed to the top wall plate in accordance with Table R602.3(1). Ceiling joists shall be continuous or securely joined in accordance with Table R802.5.1(9) where they meet over interior partitions and are nailed to adjacent rafters to provide a continuous tie across the building when such joists are parallel to the rafters.*

Discussion and Commentary: Where ceiling joists are located parallel to roof rafters, they must be tied together. Adequate attachment is mandated to transfer the thrust from the rafters to the joists. Where the ceiling joists do not run parallel to the rafters, it is necessary to use other means, such as minimum 2-inch by 4-inch rafter ties located near the top plate. Collar ties or ridge straps are also required in order to resist wind uplift. The ties shall be at least 1-inch by 4-inch members spaced at a maximum of 4 feet on center.

1 x 4 collar ties at 4 ft o.c. to resist wind uplift and located within upper $\frac{1}{3}$ of attic space

Ridge board

2 x 4 rafter ties

For SI: 1 inch = 25.4 mm.

Roof framing with ceiling joists not parallel to rafters

B-506

There are many locations where no ceiling joists exist to tie the roof together at the top plate line. Under such conditions, the ridge must be designed to act as a girder and support the necessary tributary roof load. Both the girder and its supporting elements must be designed.

Code Text: *Spans for ceiling joists shall be in accordance with Tables R802.4(1) and R802.4(2).*

Discussion and Commentary: Two tables are provided for the more common ceiling joist conditions. The maximum allowable spans are provided for uninhabitable attics with no storage (10 psf live load) and limited storage (20 psf). Table R301.4 indicates that the 10 psf criterion is to be used only where the attic height is limited to 42 inches. It is assumed that higher attics create conditions that allow for some degree of storage, thus mandating a higher design load. Criteria for truss roof systems is also provided. Similar to the listings for floor joists and rafter spans, four common lumber species are listed, along with a variety of lumber grades. Four center-to-center spacing conditions are included, as are four sizes of dimension lumber. The combination of these criteria provides all the information necessary to determine the maximum allowable ceiling joist spans.

TABLE R802.4(1)
CEILING JOIST SPANS FOR COMMON LUMBER SPECIES
(Uninhabitable attics without storage, live load = 10 psf, L/Δ = 240)

CEILING JOIST SPACING (inches)	SPECIES AND GRADE		DEAD LOAD = 5 psf			
			2 × 4	2 × 6	2 × 8	2 × 10
			Maximum ceiling joist spans			
			(feet - inches)	(feet - inches)	(feet - inches)	(feet - inches)
12	Douglas fir-larch	SS	13-2	20-8	Note a	Note a
	Douglas fir-larch	#1	12-8	19-11	Note a	Note a
	Douglas fir-larch	#2	12-5	19-6	25-8	Note a
	Douglas fir-larch	#3	10-10	15-10	20-1	24-6
	Hem-fir	SS	12-5	19-6	25-8	Note a
	Hem-fir	#1	12-2	19-1	25-2	Note a
	Hem-fir	#2	11-7	18-2	24-0	Note a
	Hem-fir	#3	10-10	15-10	20-1	24-6
	Southern pine	SS	12-11	20-3	Note a	Note a
	Southern pine	#1	12-8	19-11	Note a	Note a
	Southern pine	#2	12-5	19-6	25-8	Note a
	Southern pine	#3	11-6	17-0	21-8	25-7
	Spruce-pine-fir	SS	12-2	19-1	25-2	Note a
	Spruce-pine-fir	#1	11-10	18-8	24-7	Note a
	Spruce-pine-fir	#2	11-10	18-8	24-7	Note a
	Spruce-pine-fir	#3	10-10	15-10	20-1	24-6
16	Douglas fir-larch	SS	11-11	18-9	24-8	Note a
	Douglas fir-larch	#1	11-6	18-1	23-10	Note a
	Douglas fir-larch	#2	11-3	17-8	23-0	Note a
	Douglas fir-larch	#3	9-5	13-9	17-5	21-3
	Hem-fir	SS	11-3	17-8	23-4	Note a
	Hem-fir	#1	11-0	17-4	22-10	Note a
	Hem-fir	#2	10-6	16-6	21-9	Note a
	Hem-fir	#3	9-5	13-9	17-5	21-3
	Southern pine	SS	11-9	18-5	24-3	Note a
	Southern pine	#1	11-6	18-1	23-1	Note a
	Southern pine	#2	11-3	17-8	23-4	Note a
	Southern pine	#3	10-0	14-9	18-9	22-2
	Spruce-pine-fir	SS	11-0	17-4	22-10	Note a
	Spruce-pine-fir	#1	10-9	16-11	22-4	Note a
	Spruce-pine-fir	#2	10-9	16-11	22-4	Note a
	Spruce-pine-fir	#3	9-5	13-9	17-5	21-3

(continued)

In a limited number of applications, the span tables may not address the loading conditions under consideration. Therefore, it may be necessary to refer to another source to determine the maximum allowable joist spans. AF&PA provides span tables for such situations.

Code Text: *Spans for rafters shall be in accordance with Tables R802.5.1(1) through R802.5.1(8). The span of each rafter shall be measured along the horizontal projection of the rafter.*

Discussion and Commentary: Eight span tables are available in the code to determine the maximum allowable span of rafters in light-frame wood construction. The tables are specific for snow loads of 30 psf, 50 psf and 70 psf, as well as a live load of 20 psf. Each of these loading conditions includes two different deflection criteria. Where a ceiling is attached to the rafters, such as a cathedral ceiling, the deflection is limited to L/240. Where there is no ceiling material supported by the rafters, a maximum deflection of L/180 is permitted. The tables' format is similar to that for floor joists and ceiling joists.

TABLE R802.5.1(1)—continued
RAFTER SPANS FOR COMMON LUMBER SPECIES
(Roof live load=20 psf, ceiling not attached to rafters, $L/\Delta = 180$)

RAFTER SPACING (inches)	SPECIES AND GRADE		DEAD LOAD = 10 psf					DEAD LOAD = 20 psf				
			2 × 4	2 × 6	2 × 8	2 × 10	2 × 12	2 × 4	2 × 6	2 × 8	2 × 10	2 × 12
			Maximum rafter spans[a]									
			(feet - inches)	(feet - inches)	(feet - inches)	(feet - inches)	(feet - inches)	(feet - inches)	(feet - inches)	(feet - inches)	(feet - inches)	(feet - inches)
24	Douglas fir-larch	SS	9-1	14-4	18-10	23-4	Note b	8-11	13-1	16-7	20-3	23-5
	Douglas fir-larch	#1	8-7	12-6	15-10	19-5	22-6	7-5	10-10	13-9	16-9	19-6
	Douglas fir-larch	#2	8-0	11-9	14-10	18-2	21-0	6-11	10-2	12-10	15-8	18-3
	Douglas fir-larch	#3	6-1	8-10	11-3	13-8	15-11	5-3	7-8	9-9	11-10	13-9
	Hem-fir	SS	8-7	13-6	17-10	22-9	Note b	8-7	12-10	16-3	19-10	23-0
	Hem-fir	#1	8-4	12-3	15-6	18-11	21-11	7-3	10-7	13-5	16-4	19-0
	Hem-fir	#2	7-11	11-7	14-8	17-10	20-9	6-10	10-0	12-8	15-6	17-11
	Hem-fir	#3	6-1	8-10	11-3	13-8	15-11	5-3	7-8	9-9	11-10	13-9
	Southern pine	SS	8-11	14-1	18-6	23-8	Note b	8-11	14-1	18-6	22-11	Note b
	Southern pine	#1	8-9	13-9	17-9	21-1	25-2	8-3	12-3	15-9	18-3	21-9
	Southern pine	#2	8-7	12-3	15-10	18-11	22-2	7-5	10-8	13-9	16-5	19-3
	Southern pine	#3	6-5	9-6	12-1	14-4	17-1	5-7	8-3	10-6	12-5	14-9
	Spruce-pine-fir	SS	8-5	13-3	17-5	21-8	25-2	8-4	12-2	15-4	18-9	21-9
	Spruce-pine-fir	#1	8-0	11-9	14-10	18-2	21-0	6-11	10-2	12-10	15-8	18-3
	Spruce-pine-fir	#2	8-0	11-9	14-10	18-2	21-0	6-11	10-2	12-10	15-8	18-3
	Spruce-pine-fir	#3	6-1	8-10	11-3	13-8	15-11	5-3	7-8	9-9	11-10	13-9

Check sources for availability of lumber in lengths greater than 20 feet.

For SI: 1 inch = 25.4 mm, 1 foot = 304.8 mm, 1 pound per square foot = 0.0479 kPa.

a. The tabulated rafter spans assume that ceiling joists are located at the bottom of the attic space or that some other method of resisting the outward push of the rafters on the bearing walls, such as rafter ties, is provided at that location. When ceiling joists or rafter ties are located higher in the attic space, the rafter spans shall be multiplied by the factors given below:

H_C/H_R	Rafter Span Adjustment Factor
1/3	0.67
1/4	0.76
1/5	0.83
1/6	0.90
1/7.5 or less	1.00

where:

H_C = Height of ceiling joists or rafter ties measured vertically above the top of the rafter support walls.

H_R = Height of roof ridge measured vertically above the top of the rafter support walls.

b. Span exceeds 26 feet in length.

It is important that the horizontal thrust created by roof loads be resisted by the connection of the rafters to ceiling joists or rafter ties. This connection must occur very near the top plate line. Where the horizontal ties are higher in the space, a reduction in the rafter span is required.

Topic: Purlins
Reference: IRC R802.5.1

Category: Roof-Ceiling Construction
Subject: Wood Roof Framing

Code Text: *Installation of purlins to reduce the span of rafters is permitted as shown in Figure R802.5.1. Purlins shall be sized no less than the required size of the rafters that they support. Purlins shall be continuous and shall be supported by 2-inch by 4-inch (51 mm by 102 mm) braces installed to bearing walls at a slope not less than 45 degrees from the horizontal. The braces shall be spaced not more than 4 feet (1219 mm) on center and the unbraced length of braces shall not exceed 8 feet (2438 mm).*

Discussion and Commentary: Purlins provide an effective method for reducing rafter size by reducing the rafter span. The span is measured horizontally from the exterior wall to the purlin and from the purlin to the ridge, with the controlling span determined by the greater length.

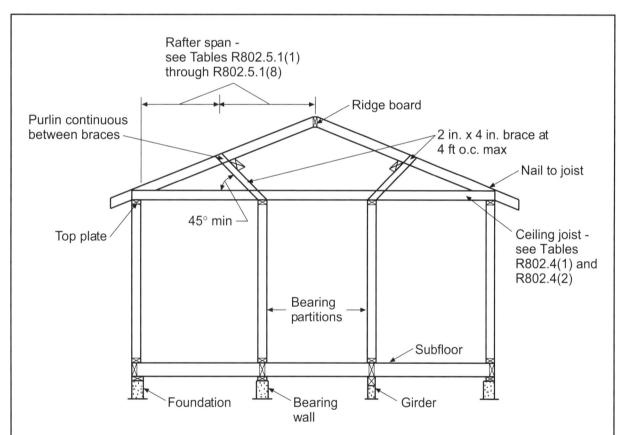

Note: Where ceiling joists run perpendicular to the rafters, rafter ties shall be nailed to the rafter near the plate line and spaced not more than 4 ft on center.

For SI: 1 inch = 25.4 mm, 1 foot = 304.8 mm, 1 degree = 0.01745 rad.

Braced rafter construction

A-669

It is critical that purlin braces be supported by bearing walls. The use of 2-inch by 4-inch braces is based on a maximum span of 4 feet between such purlin supports. Although there is no specific distance limit between bearing points at the bearing wall, the 45-degree criteria must be met.

Code Text: *Roof assemblies which are subject to wind uplift pressures of 20 pounds per square foot (960 kN/m2) or greater shall have roof rafters or trusses attached to their supporting wall assemblies by connections capable of providing the resistance required in Table R802.11. Wind uplift pressures shall be determined using an effective wind area of 100 square feet (9.3 m2) and Zone 1 in Table R301.2(2), as adjusted for height and exposure per Table R301.2(3).*

Discussion and Commentary: Because roof uplift caused by high winds can be a significant factor in roof-system damage, the code requires roof-to-wall connections capable of resisting this force. In addition, it is necessary to maintain the integrity of the continuous load path by providing a complying means to transmit the uplift forces from the rafter or truss ties to the foundation.

TABLE R802.11
REQUIRED STRENGTH OF TRUSS OR RAFTER CONNECTIONS TO RESIST WIND UPLIFT FORCES[a, b, c, e, f]
(Pounds per connection)

BASIC WIND SPEED (mph) (3–second gust)	ROOF SPAN (feet)							OVERHANGS[d] (pounds/foot)
	12	20	24	28	32	36	40	
85	-72	-120	-145	-169	-193	-217	-241	-38.55
90	-91	-151	-181	-212	-242	-272	-302	-43.22
100	-131	-218	-262	-305	-349	-393	-436	-53.36
110	-175	-292	-351	-409	-467	-526	-584	-64.56

For SI: 1 inch = 25.4 mm, 1 foot = 305 mm, 1 mph = 0.447 m/s, 1 pound/foot = 14.5939 N/m, 1 pound = 0.454 kg.

a. The uplift connection requirements are based on a 30 foot mean roof height located in Exposure B. For Exposures C and D and for other mean roof heights, multiply the above loads by the Adjustment Coefficients in Table R301.2(3).

b. The uplift connection requirements are based on the framing being spaced 24 inches on center. Multiply by 0.67 for framing spaced 16 inches on center and multiply by 0.5 for framing spaced 12 inches on center.

c. The uplift connection requirements include an allowance for 10 pounds of dead load.

d. The uplift connection requirements do not account for the effects of overhangs. The magnitude of the above loads shall be increased by adding the overhang loads found in the table. The overhang loads are also based on framing spaced 24 inches on center. The overhang loads given shall be multiplied by the overhang projection and added to the roof uplift value in the table.

e. The uplift connection requirements are based on wind loading on end zones as defined in Figure 6-2 of ASCE 7. Connection loads for connections located a distance of 20% of the least horizontal dimension of the building from the corner of the building are permitted to be reduced by multiplying the table connection value by 0.7 and multiplying the overhang load by 0.8.

f. For wall-to-wall and wall-to-foundation connections, the capacity of the uplift connector is permitted to be reduced by 100 pounds for each full wall above. (For example, if a 600-pound rated connector is used on the roof framing, a 500-pound rated connector is permitted at the next floor level down).

In utilizing Table R301.2(2) for component loads, the roof slope must be known. The determination of the component pressure is only necessary for the negative values shown in the table, as these values represent uplift pressures.

Code Text: *Load-bearing steel roof framing members shall comply with Figure R804.2(1) and the dimensional and minimum thickness requirements specified in Tables R804.2(1) and R804.2(2). The clear span of cold-formed steel ceiling joists shall not exceed the limits set forth in Tables R804.3.1(1) through R804.3.1(8). The horizontal projection of the rafter span, as shown in Figure R804.3, shall not exceed the limits set forth in Table R804.3.3(1).*

Discussion and Commentary: The conventional use of steel roof framing members is regulated in much the same manner as steel floor and wall framing systems. It is important that roof rafters and ceiling joists bear directly above load-bearing studs, with a maximum tolerance of $^3/_4$ inch between the centerline of the stud and the roof joist or rafter.

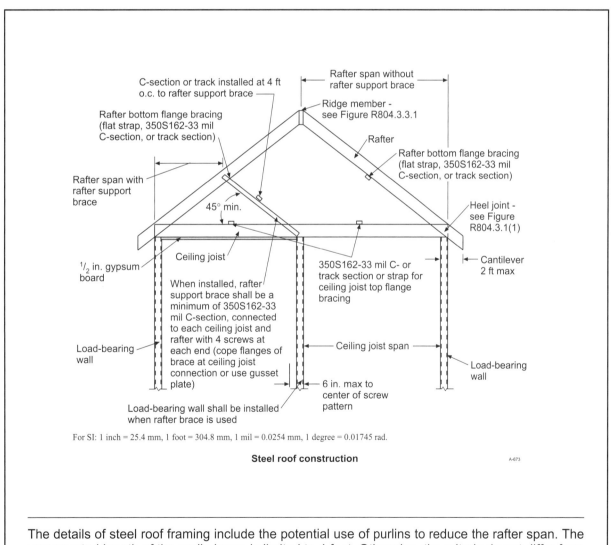

For SI: 1 inch = 25.4 mm, 1 foot = 304.8 mm, 1 mil = 0.0254 mm, 1 degree = 0.01745 rad.

Steel roof construction

A-673

The details of steel roof framing include the potential use of purlins to reduce the rafter span. The unsupported length of the purlin brace is limited to 4 feet. Otherwise, the criteria do not differ from those for light-frame wood construction.

Code Text: *Enclosed attics and enclosed rafter spaces formed where ceilings are applied directly to the underside of roof rafters shall have cross ventilation for each separate space by ventilating openings protected against the entrance of rain or snow. The total net free ventilating area shall not be less than* $^1/_{150}$ *of the area of the space ventilated except . . .* See two methods that will allow for a ventilating area reduction to $^1/_{300}$ of the area of the space ventilated. *Where eave or cornice vents are installed, insulation shall not block the free flow of air. A minimum of a 1-inch (25 mm) space shall be provided between the insulation and the roof sheathing and at the location of the vent.*

Discussion and Commentary: Large amounts of water vapor can migrate into the attic and condense on wood roof components. As the wetting and drying cycle continues, rotting and decay are possible. If the attic or enclosed rafter spaces are properly ventilated, water will not accumulate on the building components.

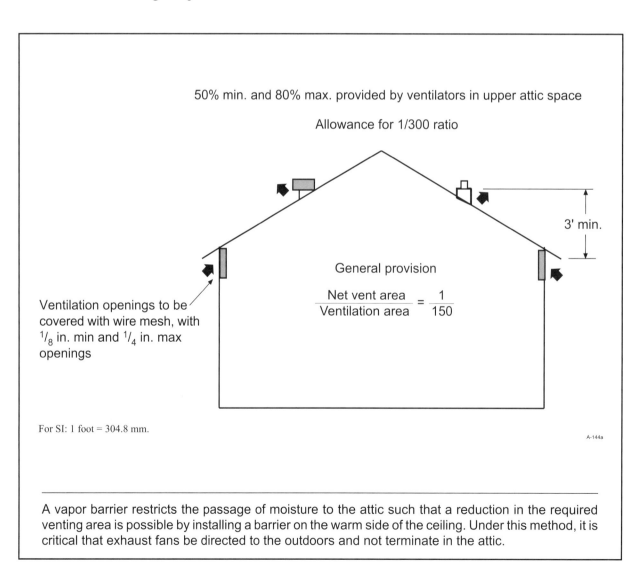

50% min. and 80% max. provided by ventilators in upper attic space

Allowance for 1/300 ratio

3' min.

General provision

$$\frac{\text{Net vent area}}{\text{Ventilation area}} = \frac{1}{150}$$

Ventilation openings to be covered with wire mesh, with $^1/_8$ in. min and $^1/_4$ in. max openings

For SI: 1 foot = 304.8 mm.

A-144a

A vapor barrier restricts the passage of moisture to the attic such that a reduction in the required venting area is possible by installing a barrier on the warm side of the ceiling. Under this method, it is critical that exhaust fans be directed to the outdoors and not terminate in the attic.

Code Text: *Unvented conditioned attic assemblies (spaces between the ceiling joists of the top story and the roof rafters) are permitted under the following conditions:* See four conditions that address 1) the prohibition of vapor retarders, 2) the application of air-impermeable insulation, 3) special criteria for warm humid locations, and 4) the installation of sufficient insulation.

Discussion and Commentary: In recent years, much research and experimentation has occurred regarding the concept of conditioned attic spaces. The allowance for such a practice, limited by a number of conditions set forth in the code, is based on the positive results that have occurred. Different requirements are in place to address the climatic region where the building is located.

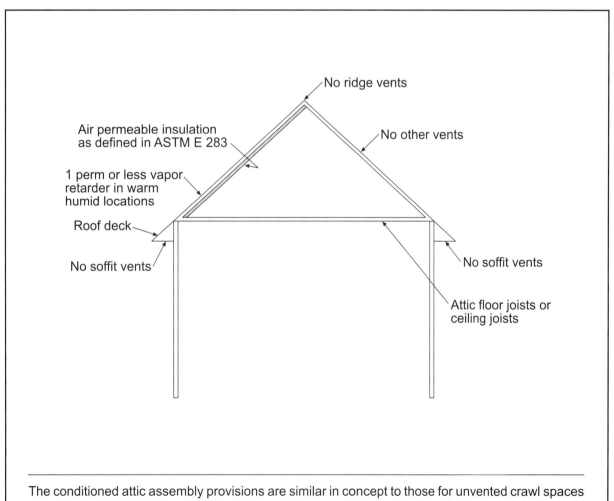

The conditioned attic assembly provisions are similar in concept to those for unvented crawl spaces as set forth in IRC Section R408.3. Both provisions are intended to simplify the code in relation to energy-efficiency requirements.

Code Text: *Buildings with combustible ceiling or roof construction shall have an attic access opening to attic areas that exceed 30 square feet (2.8 m²) and have a vertical height of 30 inches (762 mm) or more. The rough-framed opening shall not be less than 22 inches by 30 inches (559 mm by 762 mm) and shall be located in a hallway or other readily accessible location. A 30-inch (762 mm) minimum unobstructed headroom in the attic space shall be provided at some point above the access opening.*

Discussion and Commentary: During the life of a dwelling structure, it is quite probable that the attic will need to be accessed for a variety of reasons. Therefore, an adequately sized access opening must be provided with adequate headroom above the opening.

For SI: 1 inch = 25.4 mm.

Based on the typical center-to-center spacing of 24 inches for ceiling joists and roof trusses, the access opening is designed to occur between the members without any special modifications. A lip or frame supporting the access panel may encroach into the required 22-inch width.

Topic: Roof Covering Materials

Category: Roof Assemblies

Reference: IRC R902.1

Subject: Roof Classification

Code Text: *Roofs shall be covered with materials as set forth in Sections R904 and R905. Class A, B or C roofing shall be installed in areas designated by law as requiring their use or when the edge of the roof is less than 3 feet (914 mm) from a property line. Classes A, B and C roofing required to be listed by Section R902.1 shall be tested in accordance with UL 790 or ASTM E 108.*

Discussion and Commentary: The general scope of Chapter 9 includes the use of roof coverings as a weather-resistant material for protection of the building's interior. To a limited degree, the roof covering materials must also be resistant to external fire conditions. Because the primary consideration is that of fire spread from building to building, the requirements for roof covering effective against fire exposure are limited to those structures located in very close proximity to adjoining property lines.

Class A, B or C roof covering required

Less than 3'

For SI: 1 foot = 304.8 mm.

Although not specifically tested and classified, slate, clay and concrete roof tiles, exposed concrete roof decks, and metal sheets and shingles are considered equivalent to Class A roofing. Otherwise, a listed Class A, B or C roof covering must be used when the roof is located within 3 feet of the lot line.

Code Text: *Unless roofs are sloped to drain over roof edges, roof drains shall be installed at each low point of the roof. Where required for roof drainage, scuppers shall be placed level with the roof surface in a wall or parapet. The scupper shall be located as determined by the roof slope and contributing roof area.*

Discussion and Commentary: Although roofs on most buildings regulated by the IRC are distinctly pitched and simply drain roof water over the edge (often to a gutter and downspout system), occasionally a flat roof system is provided. In such situations, it is necessary to install roof drains as well as overflow drains. The overflow drains shall be installed with the inlet flow line located 2 inches above the low point of the roof. Overflow drains shall not be connected to roof drain lines and must discharge to a location approved by the building official. The *International Plumbing Code*® (IPC®), Section 1105, requires that the inlet area of roof drain strainers to be one and one-half times the area of the conductor or leader and, when they are the flat-surface type, two times the area of the conductor or leader.

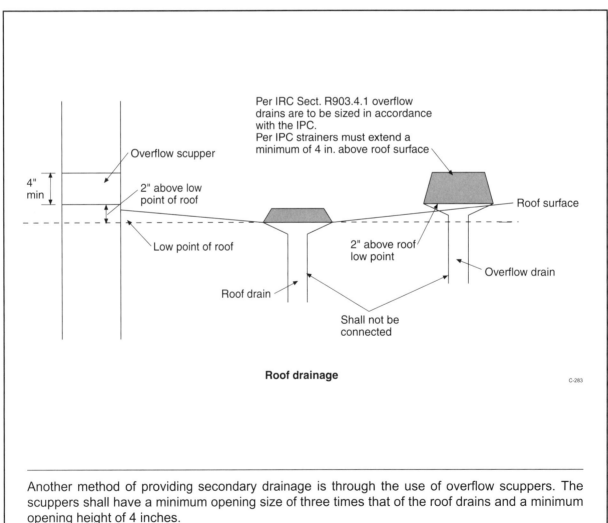

Per IRC Sect. R903.4.1 overflow drains are to be sized in accordance with the IPC.
Per IPC strainers must extend a minimum of 4 in. above roof surface

Overflow scupper

4" min

2" above low point of roof

Low point of roof

Roof surface

2" above roof low point

Roof drain

Overflow drain

Shall not be connected

Roof drainage

C-283

Another method of providing secondary drainage is through the use of overflow scuppers. The scuppers shall have a minimum opening size of three times that of the roof drains and a minimum opening height of 4 inches.

Code Text: *Asphalt shingles shall only be used on roof slopes of two units vertical in 12 units horizontal (2:12) or greater. For roof slopes from . . . 2:12 up to . . . 4:12, double underlayment application is required in accordance with Section R905.2.7. Fasteners for asphalt shingles shall be . . . of a length to penetrate through the roofing materials and a minimum of $^3/_4$ inch (19 mm) into the roof sheathing. Where the roof sheathing is less than $^3/_4$ inch (19 mm) thick, the fasteners shall penetrate through the sheathing. For normal application, asphalt shingles shall be secured to the roof with not less than four fasteners per strip shingle or two fasteners per individual shingle.*

Discussion and Commentary: A very common roofing material, asphalt shingles are composed of organic or glass felt coated with mineral granules.

Note: In areas where there has been a history of ice forming along the eaves causing a backup of water, special methods, rather than normal underlayment, shall extend up from eaves for enough to overlie a point 24 in. inside the wall line of the building.

source NRCA

For SI: 1 inch = 25.4 mm, °C = [(°F)-32/1.8].

Application of asphalt shingle on slopes between 2:12 and 4:12

A-674a

Where asphalt shingles are installed, they are regulated for sheathing, roof slope, underlayment, fasteners and attachment. As with other types of roof covering materials, they must also be installed in conformance with the manufacturer's installation instructions.

Code Text: *In areas where there has been a history of ice forming along the eaves causing a backup of water as designated in Table R301.2(1), an ice barrier that consists of at least two layers of underlayment cemented together or of a self-adhering polymer modified bitumen sheet, shall be used in lieu of normal underlayment and extend from the lowest edges of all roof surfaces to a point at least 24 inches (610 mm) inside the exterior wall line of the building.* See exception for detached accessory dwellings that contain no conditioned floor area.

Discussion and Commentary: Ice dams can form along a roof eave because rain and snow continually freezes and thaws, or frozen slush backs up in gutters. Therefore, the underlayment must be modified to prevent ice dams from forcing water under the roofing where it could cause damage to the residence's ceilings, walls and insulation. Beyond the 24-inch point, such special underlayment is deemed unnecessary as the interior of building provides adequate warmth to prevent ice dams from forming above the heated space.

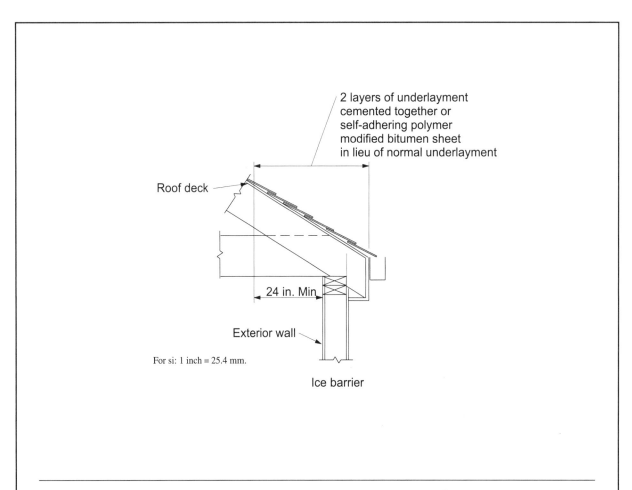

2 layers of underlayment cemented together or self-adhering polymer modified bitumen sheet in lieu of normal underlayment

Roof deck

24 in. Min.

Exterior wall

For si: 1 inch = 25.4 mm.

Ice barrier

The requirement for an ice barrier is not limited to the application of asphalt shingles. Similar requirements are applicable to roof coverings of mineral-surfaced roll roofing, slate and slate-type shingles, wood shingles and wood shakes.

Topic: Clay and Concrete Tile
Reference: IRC R905.3

Category: Roof Assemblies
Subject: Roof Covering Requirements

Code Text: *Clay and concrete tile shall be installed on roof slopes of two and one-half units vertical in 12 units horizontal ($2^1/_2$:12) or greater. For roof slopes from . . . $2^1/_2$:12 to . . . 4:12, double underlayment application is required in accordance with Section R905.3.3. Nails shall be . . . of sufficient length to penetrate the deck a minimum of $^3/_4$ inch (19.1 mm) or through the thickness of the deck, whichever is less.*

Discussion and Commentary: Tile has been used as a roofing material for many centuries. It not only provides for weather protection but is also resistant to exterior fire conditions. Both flat tile and roll tile, the two common configurations, can be interlocked through the use of ribs located along the tile edges.

Clay roll tile

Concrete flat tile

source NRCA

A-675

Much as they are for other roofing materials, special underlayment requirements are mandated for tile roofs having low slopes. In addition, the underlayment application is more highly regulated for structures located in areas subject to high winds.

Code Text: *Wood shingles shall be installed on slopes of three units vertical in 12 units horizontal (25-percent slope) or greater. Weather exposure for wood shingles shall not exceed those set in Table R905.7.5. Fasteners . . . shall be corrosion-resistant with a minimum penetration of $^1/_2$ inch (12.7 mm) into the sheathing. Wood shingles shall be attached to the roof with two fasteners per shingle, positioned no more than $^3/_4$ inch (19 mm) from each edge and no more than 1 inch (25 mm) above the exposure line.*

Discussion and Commentary: Wood shingles are wood roofing materials that are sawed in a manner to produce a uniform butt thickness. There are three different grades of wood shingles, with each grade available in three different lengths. The application method is specifically described so that following it will eliminate the potential for water intrusion.

Underlayment to comply with ASTM D 226, Type I

Roof deck (wood structural panel or wood plank)

No two joints in any three adjacent courses to be in direct alignment

Nailing $^3/_4$"

1"

Shingles must be doubled at eaves (starter course)

$1^1/_2$" min side lap

Joint or slots (shall be $^1/_4$ in. to $^3/_8$ in.)

Course line

Weather exposure limited by Table R905.7.5

For SI: 1 inch = 25.4 mm.

A-676

When fastening wood shingles to either spaced or solid sheathing, it is critical that care be taken so that the shingles are not split. Unless the sheathing is less than $^1/_2$-inch thick, the fasteners must penetrate into the sheathing a minimum of $^1/_2$ inch.

Topic: Wood Shakes

Reference: IRC R905.8

Category: Roof Assemblies

Subject: Roof Covering Requirements

Code Text: *Wood shakes shall only be used on slopes of three units vertical in 12 units horizontal (25-percent slope) or greater. Fasteners for wood shakes shall be corrosion-resistant, with a minimum penetration of $^1/_2$ inch (12.7 mm) into the sheathing. Wood shakes shall be attached to the roof with two fasteners per shake, positioned no more than 1 inch (25 mm) from each edge and no more than 2 inches (51 mm) above the exposure line. Shakes shall be interlaid with 18-inch-wide (457 mm) strips of not less than No. 30 felt shingled between each course in such a manner that no felt is exposed to the weather.*

Discussion and Commentary: Wood shakes are roofing materials that are split from logs, creating a more inconsistent surface finish than that of wood shingles. Shakes are either 18 inches or 24 inches long and graded as No. 1 or No. 2.

Weather exposure to be limited by Table R905.8.6

Underlayment to comply with ASTM D 226, Type I

Underlayment to comply with ASTM D 226, Type I

10"

10"

Slots $^1/_8$ in. to $^5/_8$ in. (naturally durable) $^1/_4$ in. to $^3/_8$ in. (preservative)

Offset slots $1^1/_2$ in. (min)

15 in. or 18 in. wood shake or wood shingle starter course

For SI: 1 inch = 25.4 mm.

A-677

A key element of a wood shake roof covering system is the application of interlayment installed shingle fashion between each course of shakes. By positioning the lower edge of the minimum No. 30 felt above the butt end of the shake, the felt extends twice the required weather exposure.

Quiz

Study Session 8
IRC Chapters 8 and 9

1. On a site having expansive soil, roof drainage water shall discharge a minimum of
   _____ from the foundation walls or to an approved drainage system.

   a. feet   (b) 5 feet

   c. feet   d. 10 feet

   Reference _901.3_____

2. Where the roof pitch is less than _3':12___, the structural members that support rafters
   and ceiling joists (such as ridge beams, hips, and valleys) shall be designed as beams.

   (a.) 3:12   b. 4:12

   c. 5:12   d. 6:12

   Reference _802.3_____

3. The maximum spacing of collar ties shall be ___4_____ on center.

   a. 16 inches   (b) 4 feet

   c. 5 feet   d. 8 feet

   Reference _802.3.1_____

4. Ends of ceiling joists shall be lapped a minimum of ___3"___ unless butted and toenailed to the supporting member.

    a. $1^1/_2$ inches

    b. 3 inches

    c. 4 inches

    d. 6 inches

    Reference ___802.3.2___

5. Where SPF #2 ceiling joists create an uninhabitable attic without storage and are spaced 24 inches on center, what is the maximum allowable span when 2-inch by 6-inch members are used?

    a. 11 feet, 2 inches

    b. 14 feet, 5 inches

    c. 14 feet, 9 inches

    d. 15 feet, 11 inches

    Reference ___Table 802.4 (1)___

6. A roof system is subjected to a 50 psf ground snow load and creates a dead load of 20 psf. The ceiling is not attached to the rafters. Assuming that rafter ties are provided at the top plate line, what is the maximum span of 2-inch by 8-inch Hem-Fir #1 rafters spaced at 24 inches on center?

    a. 9 feet, 9 inches

    b. 10 feet, 6 inches

    c. 11 feet, 6 inches

    d. 11 feet, 10 inches

    Reference ___Table R802.5.1 (4)___

7. Purlins shall be supported by braces having a maximum unbraced length of ___8___ and spaced at a maximum of ___4___ on center.

    a. 8 feet, 4 feet

    b. 8 feet, 6 feet

    c. 12 feet, 4 feet

    d. 12 feet, 6 feet

    Reference ___802.5.1___

8. The ends of a ceiling joist shall have a minimum of ___1½"___ bearing on wood and a minimum of ___1½"___ on metal.

    a. $1^1/_2$ inches, $1^1/_2$ inches

    b. $1^1/_2$ inches, 3 inches

    c. 3 inches, $1^1/_2$ inches

    d. 3 inches, 3 inches

    Reference ___802.6___

9. Notches in solid sawn lumber ceiling joists shall not be located in the middle _____ of the span and are limited in depth to _____ the depth of the joist.

   a. one-fourth, one-sixth
   b. one-fourth, one-third
   c. one-third, one-sixth
   d. one-third, one-third

   Reference _____

10. An opening in a roof system may be framed with a single header and single trimmer joists provided the header joist is located a maximum of _____ from the trimmer joist bearing.

    a. 3 feet
    b. 4 feet
    c. 6 feet
    d. 12 feet

    Reference _____

11. Where $^{15}/_{32}$-inch wood structural panels having a span rating of 32/16 are used as roof sheathing, what is the maximum span without edge support?

    a. 16 inches
    b. 24 inches
    c. 28 inches
    d. 32 inches

    Reference _____

12. A dwelling is located in an area having a 110 mph wind speed and Exposure B. The ground snow load is 20 psf. What is the maximum allowable rafter span for 800S162-54 33 ksi steel rafters installed at 24 inches on center with a roof slope of 12:12?

    a. 11 feet, 7 inches
    b. 13 feet, 9 inches
    c. 16 feet, 10 inches
    d. 22 feet, 7 inches

    Reference _____

13. In a steel-framed roof system, roof cantilevers shall not exceed _____ in horizontal projection.

    a. 12 inches
    b. 18 inches
    c. 24 inches
    d. 30 inches

    Reference _____

14. Openings provided for roof ventilation shall be covered with corrosion-resistant wire mesh with openings a minimum of __1/8__ and a maximum of __1/4__.

    a. $^1/_8$ inch, $^1/_4$ inch   b. $^1/_8$ inch, $^3/_8$ inch

    c. $^1/_4$ inch, $^3/_8$ inch   d. $^1/_4$ inch, $^1/_2$ inch

    Reference __806.1__

15. A required attic access opening shall have a minimum rough opening size of __22X30__.

    a. 18 inches by 24 inches   b. 18 inches by 30 inches

    c. 22 inches by 30 inches   d. 24 inches by 36 inches

    Reference __807.1__

16. Unless listed for lesser clearances, combustible insulation shall be separated a minimum of __3"__ from recessed lighting fixtures, fan motors and other heat-producing devices.

    a. 1 inch   b. 2 inches

    c. 3 inches   d. 6 inches

    Reference __808.1__

17. Class A, B or C roofing shall be installed unless the edge of the roof is a minimum of __902.1__ from the property line.

    a. 12 inches   b. 2 feet

    c. 3 feet   d. 5 feet

    Reference __902.1__.

18. In the attachment of asphalt shingles, special methods of fastening are required where the roof slope of _____ is exceeded.

    a. 12 units vertical in 12 units horizontal

    b. 16 units vertical in 12 units horizontal

    c. 18 units vertical in 12 units horizontal

    d. 20 units vertical in 12 units horizontal

    Reference __905.2.6__

19. Double underlayment is required for an asphalt shingle application where the roof has a maximum slope of _4:12_.

   a. 2:12   b. $2^1/_2$:12

   c. 3:12   d. 4:12

   Reference _905.2.7_

20. In an asphalt shingle roof application, a cricket need not be installed on the ridge side of any chimney with a maximum width of _30"_.

   a. 30 inches   b. 32 inches

   c. 36 inches   d. 42 inches

   Reference _905.2 & 3_

21. Nails used to attach concrete roof tiles to the roof deck shall penetrate the deck a minimum of _3/4"_ or through the thickness of the deck, whichever is less.

   a. $^1/_2$ inch   b. $^5/_8$ inch

   c. $^3/_4$ inch   d. $^7/_8$ inch

   Reference _905.3.6_

22. Wood roof shingles shall be laid with a minimum side lap of _$1^1/_2$"_ between joints in courses.

   a. $^3/_8$ inch   b. $^1/_2$ inch

   c. 1 inch   d. $1^1/_2$ inch

   Reference _905.7.5_

23. Where No. 2 wood shingles of naturally durable wood are installed on a roof having a 5:12 pitch, the maximum weather exposure for 16-inch shingles shall be _____.

   a. $3^1/_2$ inches   b. 4 inches

   c. $5^1/_2$ inches   d. $6^1/_2$ inches

   Reference _Table R905 7.5_

24. For preservative tapersawn shakes installed on a roof, the spacing between wood shakes shall be a minimum of _____ and a maximum of _____.

    a. $^1/_8$ inch, $^3/_8$ inch

    b. $^1/_8$ inch, $^5/_8$ inch

    c. $^1/_4$ inch, $^3/_8$ inch

    d. $^3/_8$ inch, $^5/_8$ inch

    Reference __905.8.6__

25. For a wood shake roof system, sheet metal roof valley flashing shall extend a minimum of __11"__ from the centerline of the valley in each direction.

    a. 4 inches

    b. 7 inches

    c. 11 inches

    d. 12 inches

    Reference __905.8.8__

26. A notch located in the end of a solid lumber ceiling joist shall have a maximum depth of __1/4"__.

    a. 1 inch

    b. 2 inches

    c. one-third the joist depth

    d. one-fourth the joist depth

    Reference __802.7.1__

27. In the framing of an opening in ceiling construction, approved hangers shall be used for the header joist to trimmer joist connections where the header joist span exceeds _____ feet.

    a. 4

    b. 6

    c. 8

    d. 12

    Reference __802.9__

28. Where eave or cornice vents are installed, a minimum __1"__ air space shall be provided between the insulation and the roof sheathing at the vent location.

    a. $^1/_2$-inch

    b. 1-inch

    c. 2-inch

    d. 3-inch

    Reference __806.3__

29. An access opening is required in attics of combustible construction where the attic space exceeds _____ square feet in area.

    a. 30
    b. 50

    c. 100
    d. 120

Reference R807.1

30. Overflow scuppers utilized for secondary roof drainage shall have a minimum size _____ of the roof drain.

    a. equal to that
    b. twice the size

    c. three times the size
    d. four times the size

Reference 903.4.1

31. A ridge board shall be a minimum of _____ -inch in nominal thickness and have a minimum depth equal to _____.

    a. 1, the cut end of the rafter

    b. 1, the nominal depth of the rafter

    c. 2, the actual depth of the rafter

    d. 2, the cut end of the rafter

Reference R802.3

32. What is the minimum total net free ventilating area required for a 1,500-square-foot attic space where 75 percent of the required ventilation area is provided by ventilators located in the upper portion of the attic space?

    a. 5 square feet
    b. 7.5 square feet

    c. 10 square feet
    d. 15 square feet

Reference 806.2

33. Where an unvented conditioned attic assembly is utilized for a residence located in a warm humid geographical location, the use of wood shakes on the roof would require a minimum continuous _____ -inch vented air space between the shakes and the roofing felt..

    a. $^1/_4$

    b. $^1/_2$

    c. $^3/_4$

    d. 1

Reference 804 # 3.2

34. Where required on an asphalt-shingled roof, an ice barrier shall extend from the lowest edges of the roof surfaces to a minimum of _____24_____ inches inside the exterior wall line of the building.

   a. 8
   b. 12
   c. 16
   d.) 24

   Reference _905.2.7.1_

35. A new roof covering of asphalt shingles cannot be installed without first removing existing roof coverings where the building is located in an area that has one or more hail days in a 20-year period that produce hail diameters exceeding _____ inch(es).

   a. $^3/_4$
   b. 1
   c.) $1^1/_2$
   d. $1^3/_4$

   Reference _907.3.1 Fig R903 S_

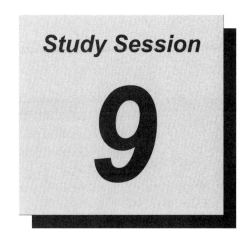

Study Session

9

2006 IRC Chapters 10 and 11
Chimneys/Fireplaces and Energy Efficiency

OBJECTIVE: To obtain an understanding of the provisions for chimneys and fireplaces, as well as the requirements for energy-efficient design and construction.

REFERENCE: Chapters 10 and 11, 2006 *International Building Code*

KEY POINTS:
- How must a masonry fireplace be supported and reinforced? 1001.1
- What is the minimum thickness of masonry fireplace walls? What is the maximum width of joints between firebricks? How are steel fireplace units regulated? 1001.5.1
- What is the minimum required depth of a firebox in a masonry fireplace? In a Rumsford fireplace? 1001.6 Exc.
- Of what material must a lintel be constructed? What is the minimum required bearing length at each end of the lintel? Where is the fireplace throat to be located?
- What construction materials are permitted for hearths and hearth extensions? What is the minimum thickness of a hearth? A hearth extension? What are the minimum dimensions of a hearth extension? 1001.10
- What is a masonry heater? How must it be supported and reinforced? 1002.3
- When is reinforcement mandated in the construction of a masonry chimney? How must such chimneys be supported? 1003.3
- Where are masonry chimneys prohibited from changing size or shape? How shall corbeling be done? Offsets? 1003.7
- At what minimum height is a masonry chimney required to terminate?
- What are the requirements for spark arrestors installed on masonry chimneys?
- What are the various methods and materials used as flue linings? How are multiple flues to be constructed? What methods are provided for determining the minimum required flue area? 1003.15.1 Options 1&2
- When is a chimney cleanout opening required? Where must it be located? 1003.17
- How far must combustible material be located from a masonry chimney located in the interior of a building? Located entirely outside a building? 1003.1T
- How is a factory-built chimney regulated?
- When is an exterior air supply required?

KEY POINTS:
(Cont'd)

- What is a climate zone and how is it determined? How does the climate zone relate to the energy efficiency requirements?

- When is the *International Energy Conservation Code*® (IECC®) applicable?

- How must insulation be labeled? When is a certificate of compliance permitted? How shall the insulation information be permanently identified in an attic area?

- How is the minimum required *R*-value for exterior walls regulated? Ceilings? Floors? Basement walls? Slab perimeter? How is the *R*-value determined?

- How is the performance of fenestration regulated? Replacement fenestration? Recessed light fixtures?

Code Text: *Footings for masonry fireplaces and their chimneys shall be constructed of concrete or solid masonry at least 12 inches (305 mm) thick and shall extend at least 6 inches (152 mm) beyond the face of the fireplace or foundation wall on all sides. Footings shall be founded on natural, undisturbed earth or engineered fill below frost depth. In areas not subjected to freezing, footings shall be at least 12 inches (305 mm) below finished grade.*

Discussion and Commentary: Because of the extensive weight of a masonry fireplace and chimney, the foundation is likely to support a greater load than the foundation for the surrounding construction. The minimum size mandated is expected to carry such a load. The relationship between the building foundation and the fireplace/chimney foundation must be considered in order to avoid detrimental settling.

For SI: 1 inch = 25.4 mm

Similar to other conventional foundation systems, the bottom of the footing must extend to below the frost line to eliminate the potential for movement due to freeze/thaw conditions. It is critical that the footing and foundation provide a solid base to avoid any possible settlement.

Topic: Firebox Walls

Reference: IRC R1001.5

Category: Chimneys and Fireplaces

Subject: Masonry Fireplaces

Code Text: *Masonry fireboxes shall be constructed of solid masonry units, hollow masonry units grouted solid, stone or concrete. When a lining of firebrick at least 2 inches (51 mm) in thickness or other approved lining is provided, the total minimum thickness of back and side walls shall each be 8 inches (203 mm) of solid masonry, including the lining. The width of joints between firebricks shall not be greater than $^1/_4$ inch (6 mm). When no lining is provided, the total minimum thickness of back and side walls shall be 10 inches (254 mm) of solid masonry.*

Discussion and Commentary: A variety of noncombustible materials is acceptable for the construction of masonry fireplaces. Where an approved lining is provided, the total masonry thickness is reduced from that needed where no lining is installed. Care should be taken to limit the joint width between firebricks in order to reduce the potential for joint deterioration.

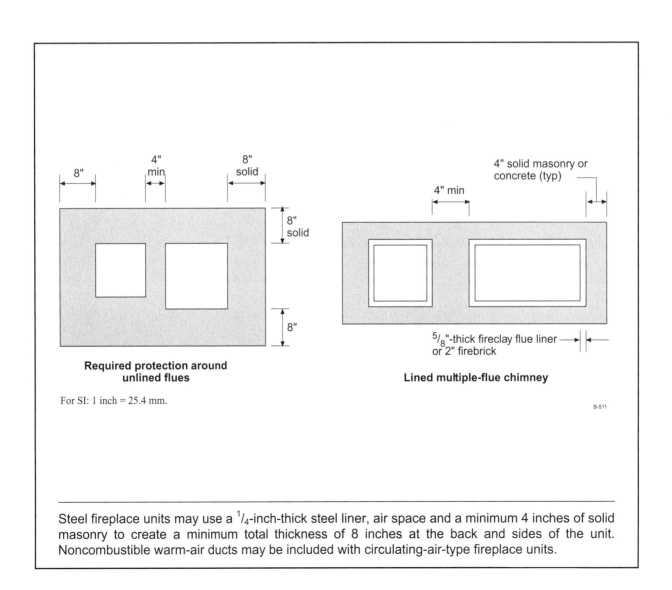

Required protection around unlined flues

Lined multiple-flue chimney

For SI: 1 inch = 25.4 mm.

B-511

Steel fireplace units may use a $^1/_4$-inch-thick steel liner, air space and a minimum 4 inches of solid masonry to create a minimum total thickness of 8 inches at the back and sides of the unit. Noncombustible warm-air ducts may be included with circulating-air-type fireplace units.

Code Text: *The firebox of a concrete or masonry fireplace shall have a minimum depth of 20 inches (508 mm). The throat shall not be less than 8 inches (203 mm) above the fireplace opening. The throat opening shall not be less than 4 inches (102 mm) in depth. The cross-sectional area of the passageway above the firebox, including the throat, damper and smoke chamber, shall not be less than the cross-sectional area of the flue.* See exception for Rumford fireplaces.

Discussion and Commentary: To provide for positive drafting, the size of the firebox is regulated, as is the location and depth of the throat opening. These criteria are based on many years of trial and error in the construction of fireplaces and chimneys.

For SI: 1 degree = 0.1745 rad.

B-513

The Rumford fireplace, designed by Count Rumford in the 18th century, was common worldwide until about 1850. Rumford fireplaces are tall and shallow to reflect more heat, and they have streamlined throats to eliminate turbulence and remove smoke with little loss of heated room air.

Code Text: *The minimum thickness of fireplace hearths shall be 4 inches (102 mm). The minimum thickness of hearth extensions shall be 2 inches (51 mm). See exception for permitted reduction in thickness. Hearth extensions shall extend at least 16 inches (406 mm) in front of and at least 8 inches (203 mm) beyond each side of the fireplace opening. Where the fireplace opening is 6 square feet (0.6 m^2) or larger, the hearth extension shall extend at least 20 inches (508 mm) in front of and at least 12 inches (305 mm) beyond each side of the fireplace opening.*

Discussion and Commentary: The hearth is the floor of the firebox, whereas the hearth extension is the projection of the hearth beyond the firebox opening. The purpose of the hearth extension is to provide a noncombustible area adjacent to the opening where sparks, embers and ashes from the firebox may fall without the risk of igniting of combustible flooring materials.

Plan

For SI: 1 inch = 25.4 mm, 1 square foot = 0.093 m^2.

B-512

Always noncombustible, a hearth extension must be a minimum of 2 inches thick unless the bottom of the firebox opening is located at least 8 inches above the top of the hearth extension. Under such conditions, the hearth extension need be only $^3/_8$-inch-thick material.

Code Text: *All wood beams, joists, studs and other combustible material shall have a clearance of not less than 2 inches (51 mm) from the front faces and sides of masonry fireplaces and not less than 4 inches (102 mm) from the back faces of masonry fireplaces.* See four exceptions for permitted reductions in clearance.

Discussion and Commentary: Radiant heat transfer through the materials used to construct a masonry fireplace and/or chimney necessitates a minimum separation between the masonry and combustible materials, such as wood floor, wall or roof framing. Radiant heat transfer, and to some degree direct flame impingement, are also of considerable concern at the firebox opening. Therefore, a minimum distance is mandated between the opening and any adjacent combustible material.

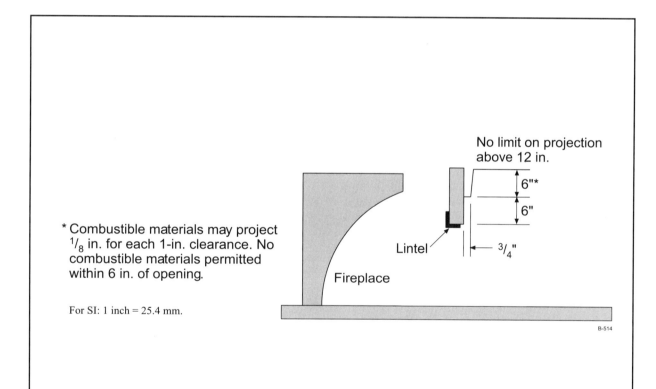

* Combustible materials may project $^1/_8$ in. for each 1-in. clearance. No combustible materials permitted within 6 in. of opening.

For SI: 1 inch = 25.4 mm.

No limit on projection above 12 in.

6"*

6"

Lintel

$^3/_4$"

Fireplace

B-514

Combustible woodwork, including trim or mantels, may be placed directly on the front of a masonry fireplace, provided such materials are not located within 6 inches of the fireplace opening. The projection of any combustible materials is strictly regulated where located within 12 inches of the opening, with a limitation of $^1/_8$ inch of projection for each 1-inch distance from the opening.

Code Text: *Masonry chimneys shall not be corbelled more than one-half of the chimney's wall thickness from a wall or foundation, nor shall a chimney be corbelled from a wall or foundation that is less than 12 inches (305 mm) thick unless it projects equally on each side of the wall* (exception for second story of two-story dwelling). *The projection of a single course shall not exceed one-half the unit height or one-third the unit bed depth, whichever is less.*

Discussion and Commentary: Corbeling occurs where masonry projects from a chimney surface in small increments as the chimney extends vertically. Two conditions are regulated: the maximum projection of a single course, and the limitations for the entire corbelled wall. A single course projection is limited to one-half of the individual unit height, and in no case more than one-third of the individual unit depth. If the projection exceeds these limitations, it is possible that the unit could fail due to its inability to carry the intended load. In addition, masonry walls are typically not permitted to be corbeled more than 50 percent of the chimney's wall thickness unless the projection occurs equally on each side of the wall. An unbalanced condition could potentially lead to cracking, or under extreme conditions, overturning of the wall.

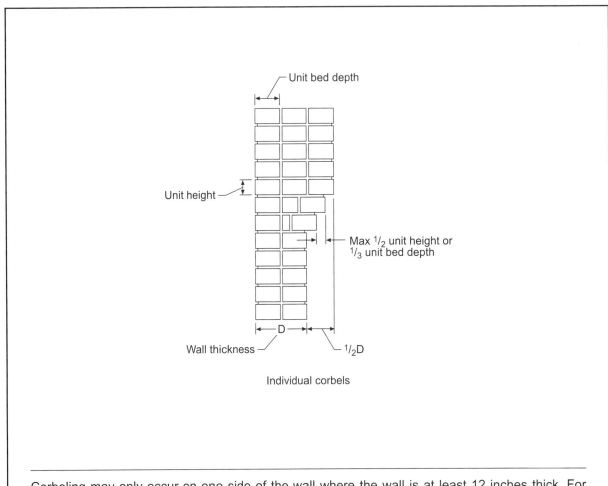

Individual corbels

Corbeling may only occur on one side of the wall where the wall is at least 12 inches thick. For thinner walls, the corbelling must project equally on both sides.

Code Text: *Chimneys shall extend at least 2 feet (610 mm) higher than any portion of a building within 10 feet (3048 mm), but shall not be less than 3 feet (914 mm) above the highest point where the chimney passes through the roof. Masonry chimney walls shall be constructed of solid masonry units or hollow masonry units grouted solid with not less than a 4-inch (102 mm) nominal thickness. Masonry chimneys shall be lined.*

Discussion and Commentary: A chimney must terminate some distance above a roof or other surface in order to provide for the necessary upward draft in the chimney. The minimum extensions mandated by the code, based on experience, were determined considering the possibility of downward drafts creating eddying, which neutralizes the required upward draft in the chimney as well as flue gas temperature.

Top of chimney must be at least 2 ft higher than peak or highest portion of roof within 10 ft horizontally

3 ft min height above roof where chimney penetrates

B-554c

For SI: 1 inch = 25.4 mm, 1 foot = 304.8 mm.

Changes in direction or dimensions of the chimney are prohibited in close proximity to structural components. Such changes create the potential for leaks or cracks in the chimney wall or flue lining, providing a point for ignition of combustible members located within any concealed areas.

Code Text: *Cleanout openings shall be provided within 6 inches (152 mm) of the base of each flue within every masonry chimney. See exception for chimney flues where cleaning is possible through the fireplace opening. The upper edge of the cleanout shall be located at least 6 inches (152 mm) below the lowest chimney inlet opening. The height of the opening shall be at least 6 inches (152 mm). The cleanout shall be provided with a noncombustible cover.*

Discussion and Commentary: The chimney interior can become coated with carbonaceous material that attacks the masonry and mortar, resulting in deposits that fall to the bottom of the chimney. Chimney cleaning also results in the accumulation of deposits at the chimney base. The deposits can be removed as necessary through the cleanout opening at the base of the chimney.

6 in. min

Cleanout opening, min 6 in. height

For SI: 1 inch = 25.4 mm.

B-554a

The opening for the chimney cleanout must be provided with a tight-fitting cover of cast iron or a similar material. Any space or gap created by a poor-fitting cover can create a hazardous condition or possibly result in a troublesome reduction in the intended draft effect.

Topic: Chimney Clearances	**Category:** Chimneys and Fireplaces
Reference: IRC R1003.18	**Subject:** Masonry Chimneys

Code Text: *Any portion of a masonry chimney located in the interior of the building or within the exterior wall of the building shall have a minimum air space clearance to combustibles of 2 inches (51 mm). Chimneys located entirely outside the exterior walls of the building, including chimneys that pass through the soffit or cornice, shall have a minimum air space clearance of 1 inch (25 mm). See exceptions for alternate chimney designs or systems.*

Discussion and Commentary: On account of the hot gases and products of combustion, the chimney walls become quite hot. As the heat is transferred through the masonry material, any combustible material in close proximity to the heated walls may reach the point of ignition. Thus, an air space of limited depth must be created around the chimney to help in the dissipation of heat.

The depth of the air space at a masonry chimney need not be as great where the chimney is located entirely outside of the exterior walls of the dwelling. The exterior surface will remain cooler because of its direct exposure to the ambient outdoor temperature.

Code Text: *Chimneys shall be provided with crickets when the dimension parallel to the ridgeline is greater than 30 inches (762 mm) and does not intersect the ridgeline. The intersection of the cricket and the chimney shall be flashed and counterflashed in the same manner as normal roof-chimney intersections. Crickets shall be constructed in compliance with Figure R1003.20 and Table R1003.20.*

Discussion and Commentary: The potential for water intrusion at chimney openings in roofs is of concern, particularly where the length of the chimney wall that runs perpendicular to the roof slope is sizable. By providing a cricket, positive water flow away from the roof opening can be ensured.

TABLE R1003.20
CRICKET DIMENSIONS

ROOF SLOPE	H
12 - 12	$^1/_2$ of W
8 - 12	$^1/_3$ of W
6 - 12	$^1/_4$ of W
4 - 12	$^1/_6$ of W
3 - 12	$^1/_8$ of W

Roof line

For SI: 1 inch = 25.4 mm.

Flue liner

1 in. min air space

W

B-510

The use of flashing and counterflashing at a roof opening such as a chimney is based on their ability to prevent moisture from breaching the weather protective membrane. Where corrosion-resistant metal is used as flashing, it must be minimum No. 26 galvanized sheet steel.

Topic: Exterior Air Intake

Reference: IRC R1006.2

Category: Chimneys and Fireplaces

Subject: Exterior Air Supply

Code Text: *The exterior air intake shall be capable of providing all combustion air from the exterior of the dwelling or from spaces within the dwelling ventilated with outside air such as non-mechanically ventilated crawl or attic spaces. The exterior air intake shall not be located within the garage or basement of the dwelling nor shall the air intake be located at an elevation higher than the firebox. The exterior air intake shall be covered with a corrosion-resistant screen of $^1/_4$-inch (6 mm) mesh.*

Discussion and Commentary: Unless it is possible to mechanically ventilate and control the indoor pressure of the room in which a fireplace is located so that the pressure remains neutral or positive, both factory-built and masonry fireplaces must have an exterior air supply. The presence of a complying exterior air intake allows for an adequate supply of combustion air to the fireplace.

The combustion (fresh) air intake

Factory-built fireplace installed in accordance with listing and manufacturer's instructions

B-515

Combustion air ducts are regulated for clearance, size and location. The outlet may be located in the back or sides of the firebox chamber, or as an alternative, located on or near the floor within a distance of 24 inches to the firebox opening.

Code Text: *Compliance shall be demonstrated by either meeting the requirements of the* International Energy Conservation Code® (IECC®) *or meeting the requirements of* Chapter 11. *Climate zones from Figure N1101.2 or Table N1101.2 shall be used in determining the applicable requirements from* Chapter 11. *Warm humid counties are listed in Table 1101.2.1.*

Discussion and Commentary: There are two permitted methods of addressing energy efficiency, each of which has a unique set of requirements. It is acceptable for the code user to evaluate both options and use the one that is more appropriate to the specific project. The IRC and the IECC approaches to energy efficiency compliance both utilize climate zones that are assigned to geographical areas of the United States based on multiple climate variables that address both cooling and heating considerations. The climate zone for the building under consideration is the basis for numerous requirements.

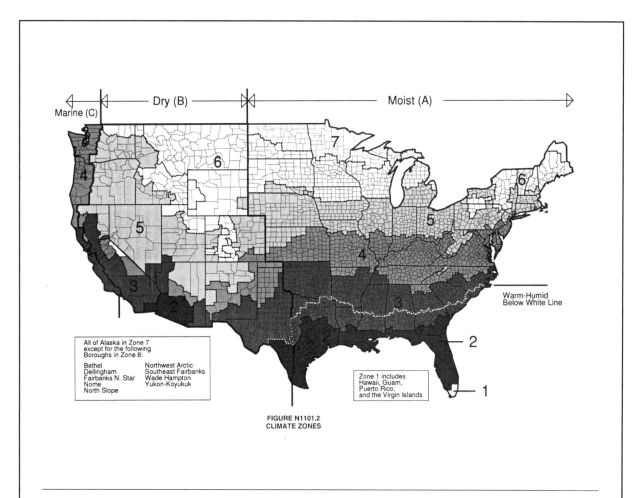

FIGURE N1101.2
CLIMATE ZONES

The eight climate zones have been divided into subcategories of marine, dry and moist to deal with varying levels of humidity. In addition, the southeastern portion of the country is designated as warm-humid, which results in specialized requirements.

Code Text: *An R-value identification mark shall be applied by the manufacturer to each piece of building thermal envelope insulation 12 inches (305 mm) or more wide. Alternately, the insulation installers shall provide a certification listing the type, manufacturer and R-value of insulation installed in each element of the building thermal envelope. U-factors of fenestration products (windows, door and skylights) shall be determined in accordance with NFRC 100 by an accredited, independent laboratory, and labeled and certified by the manufacturer. Products lacking such a labeled U-factor shall be assigned a default U-factor from Tables N1101.5(1) and N1101.5(2).*

Discussion and Commentary: The IRC requires certification of the installed density and *R*-value of blown-in or sprayed insulation in the wall cavities. Where installed in the attic, the certification must include both the initial installed thickness and the settled thickness, the coverage area and the number of bags of insulating material installed. Markers that indicate in 1-inch-high numbers the installed thickness of the insulation must be provided for every 300 square feet of attic area, attached to the trusses, rafters or joists.

TABLE N1101.5(1)
DEFAULT GLAZED FENESTRATION *U*-FACTORS

FRAME TYPE	SINGLE PANE	DOUBLE PANE	SKYLIGHT	
			Single	Double
Metal	1.2	0.8	2	1.3
Metal with thermal break	1.1	0.65	1.9	1.1
Nonmetal or metal clad	0.95	0.55	1.75	1.05
Glazed block	0.6			

TABLE N1101.5(2)
DEFAULT DOOR *U*-FACTORS

DOOR TYPE	*U*-FACTOR
Uninsulated metal	1.2
Insulated metal	0.6
Wood	0.5
Insulated, nonmetal edge, max 45% glazing, any glazing double pane	0.35

Fenestration includes skylights, roof windows, vertical windows, opaque doors, glazed doors, glass block and combination opaque/glazed doors. The U-factor for fenestration products shall be established in accordance with National Fenestration Rating Council (NFRC) 100.

Code Text: *A permanent certificate shall be posted on or in the electrical distribution panel. The certificate shall be completed by the builder or registered design professional. The certificate shall list the predominant R-values of insulation installed in or on ceiling/roof, walls, foundation (slab, basement wall, crawlspace wall and/or floor) and ducts outside conditioned spaces; U-factors for fenestration; and the solar heat gain coefficient (SHGC) of fenestration.*

Discussion and Commentary: The requirement mandating that energy-related information be provided on a certificate, and that that certificate be posted at the electrical panelboard, is intended to encourage an increased awareness of the energy-efficiency ratings for various building elements. The energy efficiency of the building as a system is a function of many elements considered as parts of a whole, and it is almost always impossible to have a proper identification and analysis of a building's energy efficiency once the structure is completed.

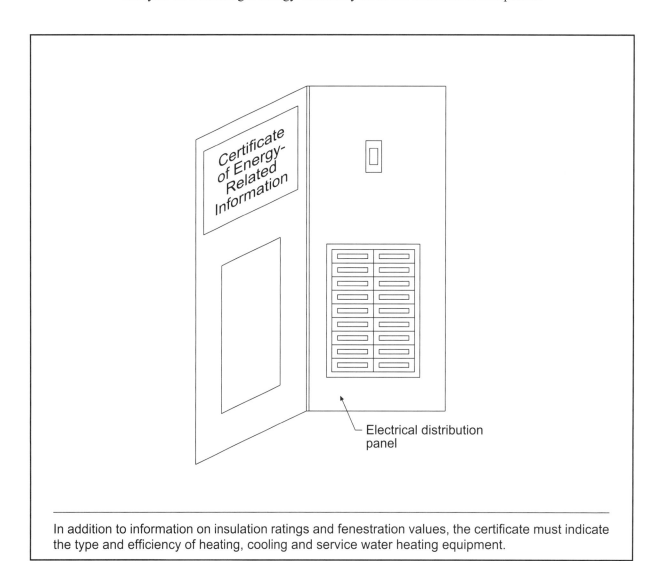

In addition to information on insulation ratings and fenestration values, the certificate must indicate the type and efficiency of heating, cooling and service water heating equipment.

Code Text: *The building thermal envelope shall meet the requirements of Table N1102.1 based on the climate zone specified in Table N1101.2.*

Discussion and Commentary: The insulation and fenestration requirements are established based upon two criteria: the type of building component and the climate zone in which the building is located. In addition to the maximum U-factor fenestration and skylight ratings, minimum insulation *R*-values are established for ceilings, wood frame walls, mass walls, floors, basement walls, slabs and crawl space walls.

TABLE N1102.1
INSULATION AND FENESTRATION REQUIREMENTS BY COMPONENT[a]

CLIMATE ZONE	FENESTRATION *U*-FACTOR	SKYLIGHT[b] *U*-FACTOR	GLAZED FENESTRATION SHGC	CEILING *R*-VALUE	WOOD FRAME WALL *R*-VALUE	MASS WALL *R*-VALUE	FLOOR *R*-VALUE	BASEMENT[c] WALL *R*-VALUE	SLAB[d] *R*-VALUE AND DEPTH	CRAWL SPACE WALL *R*-VALUE
1	1.2	0.75	0.40	30	13	3	13	0	0	0
2	0.75	0.75	0.40	30	13	4	13	0	0	0
3	0.65	0.65	0.40[e]	30	13	5	19	0	0	5/13
4 except Marine	0.40	0.60	NR	38	13	5	19	10/13	10, 2 ft	10/13
5 and Marine 4	0.35	0.60	NR	38	19 or 13 + 5[g]	13	30[f]	10/13	10, 2 ft	10/13
6	0.35	0.60	NR	49	19 or 13 + 5[g]	15	30[f]	10/13	10, 4 ft	10/13
7 and 8	0.35	0.60	NR	49	21	19	30[f]	10/13	10, 4 ft	10/13

a. R-values are minimums. U-factors and SHGC are maximums. R-19 insulation shall be permitted to be compressed into a 2 ′ 6 cavity.

b. The fenestration U-factor column excludes skylights. The solar heat gain coefficient (SHGC) column applies to all glazed fenestration.

c. The first R-value applies to continuous insulation, the second to framing cavity insulation; either insulation meets the requirement.

d. R-5 shall be added to the required slab edge R-values for heated slabs.

e. There are no solar heat gain coefficient (SHGC) requirements in the Marine Zone.

f. Or insulation sufficient to fill the framing cavity, R-19 minimum.

g. "13+5" means R-13 cavity insulation plus R-5 insulated sheathing. If structural sheathing covers 25% or less of the exterior, R-5 sheathing is not required where structural sheathing is used. If structural sheathing covers more than 25% of exterior, structural sheathing shall be supplemented with insulated sheathing of at least R-2.

A total UA (sum of U-factor times assembly area) method allows for comparing the entire building thermal envelope UA to the sum of individual U-factors. If the total UA is equivalent or less than the sum of the individual U-factors, then the building is considered in compliance.

Code Text: *The building thermal envelope shall be durably sealed to limit infiltration. The sealing methods between dissimilar materials shall allow for differential expansion and contraction. The following shall be caulked, gasketed, weatherstripped or otherwise sealed with an air barrier material, suitable film or solid material. 1) all joints, seams and penetrations, 2) site-built windows, doors and skylights, 3) openings between window and door assemblies and their respective jams and framing, 4) utility penetrations, 5) dropped ceilings or chases adjacent to the thermal envelope, 6) knee walls, 7) walls and ceiling separating the garage from conditioned spaces, 8) behind tubs and showers on exterior walls, 9) common walls between dwelling units, and 10) other sources of infiltration.*

Discussion and Commentary: Sealing the building envelope is critical to good thermal performance for the building. It will prevent warm, conditioned air from leaking out around doors, windows and other cracks during the heating season, thereby reducing the cost of heating the residence. During the hot summer months proper sealing will stop hot air from entering the residence, helping to reduce the air conditioning load.

Typical sources of air leakage in the home

B-516

Special attention must be paid during the construction process to ensure that any penetration in the building envelope is thoroughly sealed. This includes holes made for the plumbing, electrical and HVAC systems. In addition, all joints in airflow retarders must be taped for continuity.

Quiz

Study Session 9
IRC Chapters 10 and 11

1. The footing for a masonry chimney shall be a minimum of _12"_ thick.

   a   6 inches   b. 8 inches

   c. 10 inches   (d.) 12 inches

   Reference _1001.2_

2. A masonry chimney wall shall not change size or shape for a minimum of _6_ above or below where the chimney passes through floor, ceiling or roof components.

   (a.) 6 inches   b. 12 inches

   c. 24 inches   d. 30 inches

   Reference _1003.6_

3. In all cases, a masonry chimney shall extend a minimum of _36_ above the highest point where the chimney passes through the roof.

   a. 12 inches   b. 18 inches

   c. 24 inches   (d.) 36 inches

   Reference _1003.9_

4. A clay flue liner for a masonry chimney shall be carried vertically, with a maximum permitted slope of ___30°___ from the vertical.

   a. 15 degrees
   b. 22 1/2 degrees
   c. 30 degrees
   d. 45 degrees

   Reference ___1003.12___

5. When two or more flues are located in the same masonry chimney, they shall be separated by masonry wythes having a minimum thickness of ___4___.

   a. 2 inches
   b. 4 inches
   c. 6 inches
   d. 8 inches

   Reference ___1003.13___

6. Unless cleaning is possible through the fireplace opening, a cleanout opening shall be provided within ___6___ of the base of a flue in a masonry chimney.

   a. 6 inches
   b. 8 inches
   c. 12 inches
   d. 18 inches

   Reference ___1002.17___

7. Where a masonry chimney is constructed totally outside the exterior walls of a dwelling, a minimum air space clearance to combustibles of ___1___ is required.

   a. 0 inches; no clearance is required
   b. 1/2 inch
   c. 1 inch
   d. 2 inches

   Reference ___1003.18___

8. Foundations for masonry fireplaces and their chimneys shall extend a minimum of ___6___ beyond the face of the fireplace or support walls on all sides.

   a. 4 inches
   b. 6 inches
   c. 8 inches
   d. 12 inches

   Reference ___1001.2___

9. Where a 60-inch-wide masonry chimney and fireplace are located in Seismic Design Category D₁, the chimney shall be reinforced with ___6___ No. 4 continuous vertical reinforcing bars.

    a. 0; no reinforcement is required

    b. 2

    c. 4

    d. 6

    Reference 1001.8.1

10. What is the minimum required thickness of the back and side walls of a masonry fireplace constructed without a firebrick lining?

    a. 6 inches   b. 8 inches

    c. 10 inches   d. 12 inches

    Reference 1001.5

11. Where firebricks are used as lining for a masonry fireplace, the maximum joint spacing shall be ___¼___ between firebricks.

    a. $^1/_8$ inch   b. $^1/_4$ inch

    c. $^3/_8$ inch   d. $^1/_2$ inch

    Reference 1001.5

12. The damper or throat for a masonry fireplace shall be located a minimum of ___8___ above the lintel supporting masonry over the fireplace opening.

    a. 4 inches   b. 6 inches

    c. 8 inches   d. 12 inches

    Reference 1001.7

13. What is the minimum required thickness for the hearth of a masonry fireplace?

    a. $^3/_8$ inch   b. $^3/_4$ inch

    c. 2 inches   d. 4 inches

    Reference 1001.9.1

14. What is the minimum size of a hearth extension for a masonry fireplace having a 30-inch-high by 42-inch-wide fireplace opening?

    a. 16 inches by 42 inches   b. 16 inches by 58 inches

    c. 20 inches by 58 inches   (d.) 20 inches by 66 inches

Reference _1001.10_

15. A minimum $^3/_8$-inch-thick noncombustible hearth extension is permitted where the bottom of the firebox opening is a minimum of __8__ inches above the top of the hearth extension.

    a 6   b. (8)

    c. 12   d. 15

Reference _1001 9.2 EXP_

16. Wood studs shall have a minimum clearance of __4__ from the back face of a masonry fireplace.

    a 0 inches; no clearance is required

    b. 1 inch

    c. 2 inches

    (d.) 4 inches

Reference _1001.11_

17. Exposed combustible trim placed directly on the masonry fireplace front shall be located a minimum of __6__ from the fireplace opening.

    a. 1 inch   b. 2 inches

    c. 4 inches   (d.) 6 inches

Reference _1001.11 EXC. 4_

18. Combustible trim projecting a distance of 1 inch beyond the fireplace opening of a masonry fireplace shall be located a minimum of __12__ from the opening.

    a. 2 inches   b. 6 inches

    c. 8 inches   (d.) 12 inches

Reference _1001.11 EXC. 4_

19. Factory-built fireplaces shall be tested in accordance with __UL 127__ .

    a. ASTM C 199   b. ASTM C 315

    c. UL 103   (d.) UL 127

    Reference __1004.1__

20. Unless the room is mechanically ventilated and controlled so that the indoor pressure is neutral or positive, a combustion air passageway having a minimum size of __6__ square inches and a maximum size of __55__ square inches shall be provided as an exterior air supply for a factory-built or masonry fireplace.

    a. 6, 25   (b.) 6, 55

    c. 25, 55   d. 25, 75

    Reference __1006.4__

21. Unless a certification is provided by the insulation installers, an *R*-value identification mark shall be applied by the manufacturer to each piece of building thermal envelope insulation that is a minimum of __12__ inches wide.

    (a.) 12   b. 15

    c. 18   d. 22

    Reference __N1101.4__

22. Exposed insulation applied to the exterior of basement walls shall be protected with a covering that extends a minimum of __6__ inches below grade.

    a. (6)   b. 12

    c. 18   d. 24

    Reference __N1101.6.1__

23. In Climate Zone 2, what is the minimum *R*-value required for a wood-frame exterior wall?

    a. 11   b. (13)

    c. 15   d. 19

    Reference __N1102.1 Table N1102.1__

24. Other than site-built elements, windows, skylights and sliding glass doors shall have a maximum air infiltration rate of _____ cfm per square foot.

    a. 0.05   b. 0.1

    c. 0.3   d. 1.0

    Reference N1102.4.2

25. Unless located completely inside the building thermal envelope, supply and return air ducts shall be insulated to a minimum of _R-8_.

    a. R-6   b. R-8

    c. R-11   d. R-13

    Reference N1103.2.1

26. Concrete foundations for masonry chimneys shall extend a minimum of _6_ inches beyond each side of the exterior dimensions of the chimney.

    a. 3   b. 6

    c. 8   d. 12

    Reference 1003.2

27. The net free area of a spark arrestor for a masonry chimney shall be a minimum of _4_ time(s) the net free area of the outlet of the chimney flue it serves.

    a. one   b. two

    c. three   d. four

    Reference 1003.9.1 #1

28. Where a 54-inch-wide masonry chimney is constructed above a roof having a slope of 8:12, a required cricket shall have a minimum height of _18_ inches.

    a. 9   b. 13$\frac{1}{2}$

    c. 18   d. 27

    Reference 1003.20 T. 1003.20

29. Except for Rumford fireplaces, the firebox of a masonry fireplace shall have a minimum depth of _20_ inches.

    a. 12   b. 18

    (c.) 20   d. 24

    Reference _1001 6_

30. In general, a minimum clearance of _36_ inches is required between the outside surface of a masonry heater and combustible materials.

    a. 6   b. 12

    c. 18   (d.) 36

    Reference _1002.5_

31. Where a steel fireplace unit incorporating a steel firebox lining is installed with solid masonry to form a masonry fireplace, it shall be constructed with minimum ___1/4"___ -inch steel.

    a. $^1/_8$ b. $^3/_{16}$

    (c) $^1/_4$ d. $^5/_{16}$

    Reference _1001.5.1_

32. A Rumsford fireplace shall have a minimum fireplace depth of _14_ inches where the fireplace opening is 42 inches in width.

    a. 12   (b.) 14

    c. 15   d. 20

    Reference _1001.6 tbl_

33. Where the hearth extension for a masonry fireplace is located at the same level as the bottom of the firebox opening, the minimum required thickness of the hearth extension is ___2___ inches.

    a. $1^1/_2$ (b.) 2

    c. 3   d. 4

    Reference _1001.9.2_

34. Unlisted combustion air ducts installed to serve a factory-built or masonry fireplace shall be installed with a minimum _____1_____ -inch clearance to combustibles for those portions of the duct within _____5_____ feet of the duct outlet.

   a.) 1, 5   b. 1, 10

   c. 2, 5   d. 2, 10

   Reference ___1006.3___

35. The thickness of blown-in cellulose roof/ceiling insulation shall be provided on markers installed throughout the attic space at a rate of one marker for every _____30_____ square feet maximum.

   a. 100   b. 200

   c.) 300   d. 500

   Reference ___N1101 4.1___

2006 IRC Chapters 13 and 14

General Mechanical System Requirements and Heating/Cooling Equipment

OBJECTIVE: To gain an understanding of the general provisions for the installation of mechanical appliances, equipment and systems, including requirements regulating heating and cooling equipment.

REFERENCE: Chapters 13 and 14, 2006 *International Residential Code*

KEY POINTS:
1303.1
- What information must appear on the factory-applied nameplate affixed to an appliance?

M 1305.1
- How much working space is needed in front of the control side of an appliance in order to service the appliance? How much of a reduction is allowed in front of a room heater? M 1305.1

M 1305.1.1
- Where a central furnace is installed in an alcove or compartment, what is the minimum total width of the enclosed area? What is the minimum working clearance required along the sides of the furnace? Along the back? On the top? How much front clearance is mandated where the firebox is open to the atmosphere?

M 1305.1.2
- How large a passageway must be provided to an appliance installed in a compartment, alcove, basement or similar space? How must access to an attic appliance be addressed? M 1305.1.3 For appliances installed in under-floor spaces? M 1305.1.4

6" M 1305.14.1
- What minimum clearance is required between the ground and the appliance where the appliance is suspended from the floor? How far must the excavation extend beyond the appliance enclosure? – M1305.14.2

M 1306.1
- What is the minimum clearance needed between an appliance and unprotected combustible construction? How may such clearances be reduced? What is the minimum required clearance for solid fuel appliances? M1306.2 M1306.2.1

M 1307.2
- In what seismic design categories must water heaters be anchored or strapped to resist horizontal displacement due to earthquake forces? What prescriptive method is set forth?

M 1307.3
- How high above a floor in a garage is an appliance with an ignition source required to be located?

- What special provisions apply to hydrogen-generating and refueling operations?

M1308 ?
- How shall piping used as a portion of a mechanical system be protected where installed in concealed locations through holes and notches in framing members?

KEY POINTS: M1406.1 • How are radiant heating systems required to be installed? Duct heaters M1408.3
(Cont'd) • What minimum clearance from a wall is mandated for a floor furnace? What is the minimum clearance from doors, draperies or similar combustible objects? What is the minimum required clearance from the furnace register to a combustible projecting object above?

M1409.2 #2 — • Where are vented wall furnaces prohibited from being located? What clearances are mandated between the furnace air inlet or outlet and the swing of a door?

M1410.2 • How must vented room heaters be installed? What minimum clearance is required on all sides of the appliance?

M1411.3.1 • How is condensate from refrigeration cooling equipment to be addressed? What methods are described for an auxiliary or secondary drain system? What is the minimum drain pipe size?

M1414.1 • What limitations are mandated for the use of fireplace stoves?

Code Text: *Central furnaces within compartments or alcoves shall have a minimum working space clearance of 3 inches (76 mm) along the sides, back and top with a total width of the enclosing space being at least 12 inches (305 mm) wider than the furnace. Furnaces having a firebox open to the atmosphere shall have at least a 6-inch (152 mm) working space along the front combustion chamber side.* See exception for replacement appliances.

Discussion and Commentary: A minimum amount of clearance is desired around a central furnace to allow for access to the unit, provide space to facilitate the removal of the furnace, create an air space for ventilating and cooling the appliance, and protect any surrounding combustible materials. The increased clearance at the front protects against flame roll-out and provides free movement of combustion air.

24 in. min opening

For SI: 1 inch = 25.4 mm.

B-517

The established clearances are applicable to new installations even though lesser clearances are permitted by the manufacturer. Replacement appliances installed in existing compartments may have lesser clearances if in conformance with the manufacturer's instructions.

Code Text: *Attics containing appliances requiring access shall be provided with an opening and a clear and unobstructed passageway large enough to allow removal of the largest appliance, but not less than 30 inches (762 mm) high and 22 inches (559 mm) wide and not more than 20 feet (6096 mm) in length when measured along the centerline of the passageway from the opening to the appliance.* See exceptions to 1) eliminate the passageway and level service space where appliance can be serviced and removed through the required opening, and 2) permit passageway length of 50 feet maximum where minimum 6-foot by 22-inch unobstructed passageway is provided.

Discussion and Commentary: By placing the mechanical equipment in the attic, it is possible to better utilize the building's floor area. However, it is important that such equipment be accessible for service or removal. In addition to a properly sized access opening, a continuous passageway of solid flooring must be provided where needed to reach the appliance. A minimum 30-inch by 30-inch service area is required in front of any appliance access.

For SI: 1 inch = 25.4 mm.

A light fixture is mandated to illuminate the passageway and the appliance's location. The switch controlling the fixture must be located at the entry to the passageway. In addition, a receptacle outlet shall be located at or near the appliance for use during maintenance or repair.

Code Text: *Underfloor spaces containing appliances requiring access shall be provided with an unobstructed passageway large enough to remove the largest appliance, but not less than 30 inches (762 mm) high and 22 inches (559 mm) wide, nor more than 20 feet (6096 mm) in length when measured along the centerline of the passageway from the opening to the appliance. If the depth of the passageway or the service space exceeds 12 inches (305 mm) below the adjoining grade, the walls shall be lined with concrete or masonry extending 4 inches (102 mm) above the adjoining grade in accordance with Chapter 4* (Foundations). See exceptions to 1) eliminate the passageway and level service space where appliance can be serviced and removed through the required opening, and 2) permit passageway length of 50 feet maximum where minimum 6-foot by 22-inch unobstructed passageway is provided.

Discussion and Commentary: The requirements for access to an underfloor furnace are similar to those for furnaces located in attics and other concealed areas. A switched light fixture and receptacle outlet are mandated to allow for safe and convenient appliance repair and maintenance.

Adequate clearance must be provided for an underfloor furnace. If the appliance is suspended from the floor, it must be at least 6 inches above the ground. Where the underfloor space is excavated, clearances of 12 inches on the sides and 30 inches on the control side are required.

Code Text: *Appliances shall be installed with the clearances from unprotected combustible materials as indicated on the appliance label and in the manufacturer's installation instructions.*

Discussion and Commentary: The clearances established by the manufacturer's installation instructions are assumed to be from unprotected combustible construction. These clearances create the necessary airspace for reducing the conduction of heat that might ignite adjacent combustible materials. In addition, convection cooling provided by the airspace assists in the proper operation of the appliance. It is important to note that the provisions and guidelines are not intended to apply to the installation of unlisted appliances.

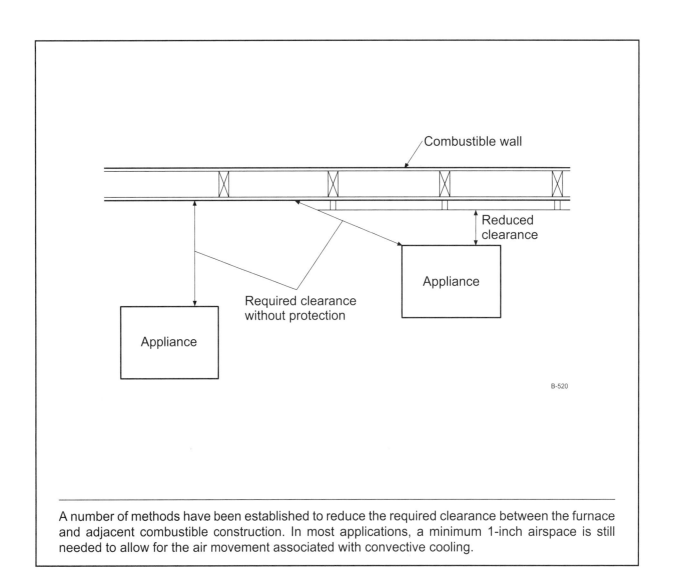

A number of methods have been established to reduce the required clearance between the furnace and adjacent combustible construction. In most applications, a minimum 1-inch airspace is still needed to allow for the air movement associated with convective cooling.

Topic: Clearance Reduction

Category: Mechanical Systems

Reference: IRC M1306.2

Subject: Clearances from Combustible Construction

Code Text: *Reduction of clearances shall be in accordance with the appliance manufacturer's instructions and Table M1306.2. Forms of protection with ventilated air space shall conform to the following requirements: 1) not less than 1-inch (25 mm) air space shall be provided between the protection and combustible wall surface; 2) air circulation shall be provided by having edges of the wall protection open at least 1 inch (25 mm); 3) if the wall protection is mounted on a single flat wall away from corners, air circulation shall be provided by having the bottom and top edges, or the side and top edges open at least 1 inch (25 mm); and 4) wall protection covering two walls in a corner shall be open at the bottom and top edges at least 1 inch (25 mm).*

Discussion and Commentary: The allowances for reduced clearances require that a 1-inch air space be maintained to allow the air circulation needed to keep the temperature rise within acceptable limits.

TABLE M1306.2
REDUCTION OF CLEARANCES WITH SPECIFIED FORMS OF PROTECTION[a, b, c, d, e, f, g, h, i, j, k]

TYPE OF PROTECTION APPLIED TO AND COVERING ALL SURFACES OF COMBUSTIBLE MATERIAL WITHIN THE DISTANCE SPECIFIED AS THE REQUIRED CLEARANCE WITH NO PROTECTION [See Figures M1306.1 and M1306.2]	WHERE THE REQUIRED CLEARANCE WITH NO PROTECTION FROM APPLIANCE, VENT CONNECTOR, OR SINGLE WALL METAL PIPE IS:									
	36 inches		18 inches		12 inches		9 inches		6 inches	
	Allowable clearances with specified protection (inches)[b]									
	Use column 1 for clearances above an appliance or horizontal connector. Use column 2 for clearances from an appliance, vertical connector and single-wall metal pipe.									
	Above column 1	Sides and rear column 2	Above column 1	Sides and rear column 2	Above column 1	Sides and rear column 2	Above column 1	Sides and rear column 2	Above column 1	Sides and rear column 2
3$^1/_2$-inch thick masonry wall without ventilated air space	—	24	—	12	—	9	—	6	—	5
$^1/_2$-in. insulation board over 1-inch glass fiber or mineral wool batts	24	18	12	9	9	6	6	5	4	3
24 gage sheet metal over 1-inch glass fiber or mineral wool batts reinforced with wire on rear face with ventilated air space	18	12	9	6	6	4	5	3	3	3
3$^1/_2$-inch thick masonry wall with ventilated air space	—	12	—	6	—	6	—	6	—	6
24 gage sheet metal with ventilated air space	18	12	9	6	6	4	5	3	3	2
$^1/_2$-inch thick insulation board with ventilated air space	18	12	9	6	6	4	5	3	3	3
24 gage sheet metal with ventilated air space over 24 gage sheet metal with ventilated air space	18	12	9	6	6	4	5	3	3	3
1-inch glass fiber or mineral wool batts sandwiched between two sheets 24 gage sheet metal with ventilated air space.	18	12	9	6	6	4	5	3	3	3

For SI: 1 inch = 25.4 mm, 1 pound per cubic foot = 16.019 kg/m³, °C = [(°F)-32/1.8], 1 Btu/(h × ft² × °F/in.) = 0.001442299 (W/cm² × °C/cm).

a. Reduction of clearances from combustible materials shall not interfere with combustion air, draft hood clearance and relief, and accessibility of servicing.

b. Clearances shall be measured from the surface of the heat producing appliance or equipment to the outer surface of the combustible material or combustible assembly.

c. Spacers and ties shall be of noncombustible material. No spacer or tie shall be used directly opposite appliance or connector.

d. Where all clearance reduction systems use a ventilated air space, adequate provision for air circulation shall be provided as described. (See Figures M1306.1 and M1306.2.)

e. There shall be at least 1 inch between clearance reduction systems and combustible walls and ceilings for reduction systems using ventilated air space.

f. If a wall protector is mounted on a single flat wall away from corners, adequate air circulation shall be permitted to be provided by leaving only the bottom and top edges or only the side and top edges open with at least a 1-inch air gap.

g. Mineral wool and glass fiber batts (blanket or board) shall have a minimum density of 8 pounds per cubic foot and a minimum melting point of 1,500°F.

h. Insulation material used as part of a clearance reduction system shall have a thermal conductivity of 1.0 Btu inch per square foot per hour °F or less. Insulation board shall be formed of noncombustible material.

i. There shall be at least 1 inch between the appliance and the protector. In no case shall the clearance between the appliance and the combustible surface be reduced below that allowed in this table.

j. All clearances and thicknesses are minimum; larger clearances and thicknesses are acceptable.

k. Listed single-wall connectors shall be permitted to be installed in accordance with the terms of their listing and the manufacturer's instructions.

Because solid-fuel-burning appliances can produce radiant high-intensity heat and have wide variations in heat output, there are restrictions on the use of Table M1306.2 for reducing clearances to combustible construction. A minimum of 12 inches must always be maintained.

Code Text: *Appliances designed to be fixed in position shall be fastened or anchored in an approved manner. In Seismic Design Categories D₁ and D₂, water heaters shall be anchored or strapped to resist horizontal displacement caused by earthquake motion. Strapping shall be at points within the upper one-third and lower one-third of the appliance's vertical dimensions. At the lower point, the strapping shall maintain a minimum distance of 4 inches (102 mm) above the controls.*

Discussion and Commentary: In those areas determined to have a high earthquake risk, it is important that any water heater be fastened in place to avoid damage. The strapping should be adequately attached to the studs or other framing members. It must be located an adequate distance from the control panel to allow access and use.

Wall sheathing

Located in upper ¹/₃ of height

Approved strapping

Located in lower ¹/₃ of height and min of 4 in. above controls

Flexible gas pipe connection

For SI: 1 inch = 25.4 mm.

B-521

Large water heaters weigh several hundred pounds; if they are not secured, the water, gas and electrical connections may be damaged or severed in an earthquake, creating a hazardous situation. In addition to strapping, approved flexible connectors should be used.

Topic: Elevation of Ignition Source
Reference: IRC M1307.3, M1307.3.1

Category: Mechanical Systems
Subject: Appliance Installation

Code Text: *Appliances having an ignition source shall be elevated such that the source of ignition is not less than 18 inches (457 mm) above the floor in garages. For the purpose of Section M1307.3, rooms or spaces that are not part of the living space of a dwelling unit and that communicate with a private garage through openings shall be considered to be part of the garage. Appliances located in a garage or carport shall be protected from impact by automobiles.*

Discussion and Commentary: Gasoline leakage or spillage in a garage is a possible danger. Gasoline fumes will evaporate from liquid puddles at the floor level. Any potential ignition source must be elevated to keep open flame, spark-producing elements and heating elements above the gasoline fume level.

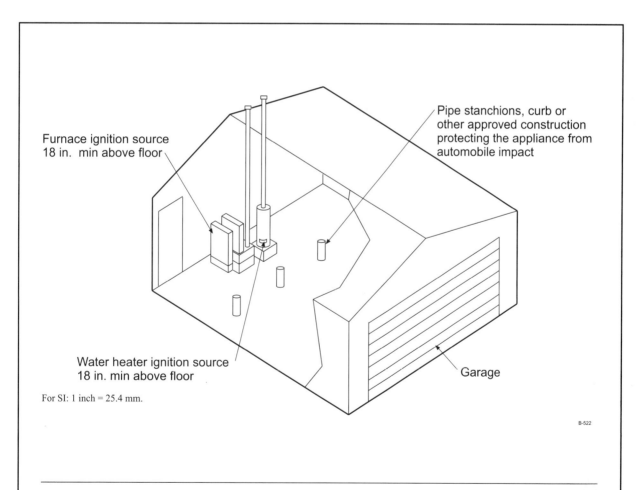

Furnace ignition source 18 in. min above floor

Pipe stanchions, curb or other approved construction protecting the appliance from automobile impact

Water heater ignition source 18 in. min above floor

Garage

For SI: 1 inch = 25.4 mm.

B-522

Appliances or equipment exposed to impact by vehicles could create a hazardous situation if their fuel or power connections were loosened or broken. Unless located out of the path of potential impact, the appliances must be protected by a barrier such as curb wheel stops or pipe stanchions.

Code Text: *In concealed locations where piping, other than cast-iron or galvanized steel, is installed through holes or notches in studs, joists, rafters or similar members less than 1.5 inches (38 mm) from the nearest edge of the member, the pipe shall be protected by shield plates. Protective shield plates shall be a minimum of 0.062-inch-thick (1.6 mm) steel, shall cover the area of the pipe where the member is notched or bored, and shall extend a minimum of 2 inches (51 mm) above sole plates and below top plates.*

Discussion and Commentary: It is critical that piping passing through joists, studs, rafters and other structural members be protected from damage, typically caused by wallboard fasteners and/or wood structural panels. Therefore, steel shield plates are to be utilized where the distance from the member surface to the notch or bored hole exceeds a specified distance.

For SI: 1 inch = 25.4 mm

Acceptable methods of drilling, notching or otherwise altering structural framing members for mechanical piping components are set forth in the provisions of Chapter 5 for floor framing, Chapter 6 for wall framing and Chapter 8 for roof-ceiling framing.

Code Text: *Floor registers of floor furnaces shall be installed not less than 6 inches (152 mm) from a wall. Wall registers of floor furnaces shall be installed not less than 6 inches (152 mm) from the adjoining wall at inside corners. The furnace register shall be located not less than 12 inches (305 mm) from doors in any position, draperies or similar combustible objects. The furnace register shall be located at least 5 feet (1524 mm) below any projecting combustible materials. The floor furnace burner assembly shall not project into an occupied under-floor area. The floor furnace shall not be installed in concrete floor construction built on grade. The floor furnace shall not be installed where a door can swing within 12 inches (305 mm) of the grille opening.*

Discussion and Commentary: Flat floor furnaces and wall register floor furnaces are designed to be built directly into a wood-framed floor. However, there are clearances necessary because of the heat generated in the combustion chamber and heat coming from the register.

For SI: 1 inch = 25.4 mm.

B-523

Floor furnaces are vented appliances and are installed in an opening in the floor. Such units supply heat to the room by gravity convection and direct radiation and typically serve as the sole source of space heating. They are sometimes used in cottages and small homes.

Code Text: *Floor furnaces shall be installed not closer than 6 inches (152 mm) to the ground. Clearance may be reduced to 2 inches (51 mm), provided that the lower 6 inches (152 mm) of the furnace is sealed to prevent water entry. Where excavation is required for a floor furnace installation, the excavation shall extend 30 inches (762 mm) beyond the control side of the floor furnace and 12 inches (305 mm) beyond the remaining sides.*

Discussion and Commentary: Floor furnaces must have supports independent of the floor register. A 6-inch clearance must also be maintained from the ground to protect against moisture, rust and corrosion of the furnace casing. This may be reduced to 2 inches if the unit is sealed against flooding. Floor furnaces are not permitted to be supported on the ground.

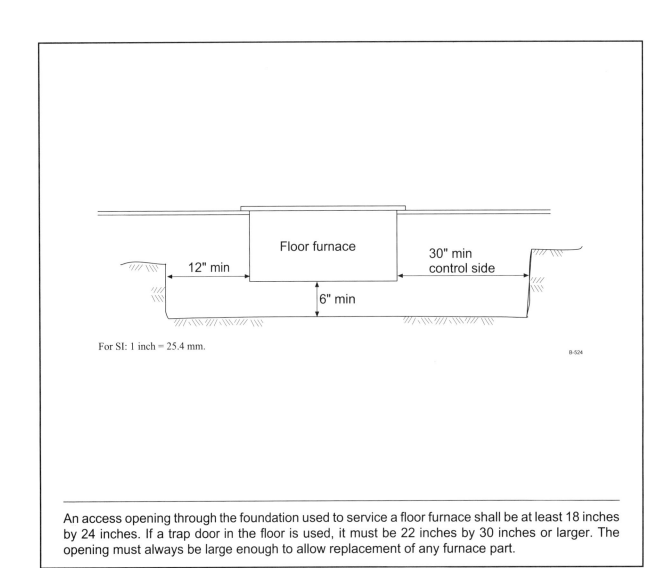

For SI: 1 inch = 25.4 mm.

B-524

An access opening through the foundation used to service a floor furnace shall be at least 18 inches by 24 inches. If a trap door in the floor is used, it must be 22 inches by 30 inches or larger. The opening must always be large enough to allow replacement of any furnace part.

Code Text: *Vented wall furnaces shall be located where they will not cause a fire hazard to walls, floors, combustible furnishings or doors. Vented wall furnaces installed between bathrooms and adjoining rooms shall not circulate air from bathrooms to other parts of the building. Vented wall furnaces shall not be located where a door can swing within 12 inches (305 mm) of the furnace air inlet or outlet measured at right angles to the opening. Doorstops or door closers shall not be installed to obtain this clearance.*

Discussion and Commentary: Provided with the necessary internal clearances, wall furnaces are designed for installation in combustible stud cavities. To allow for necessary air movement and reduce the potential for the ignition of combustible materials, adequate clearances are mandated.

Unlike forced-air furnaces, wall furnaces are not designed for the distribution of hot air through ductwork. Wall furnaces may be of the gravity-type or possibly constructed with small blowers. The attachment of ducts to either type of unit could potentially cause overheating.

Code Text: *Condensate from all cooling coils or evaporators shall be conveyed from the drain pan outlet to an approved place of disposal. Condensate shall not discharge into a street, alley or other areas so as to cause a nuisance. In addition to the requirements of Section M1411.3, a secondary drain or auxiliary drain pan shall be required for each cooling or evaporator coil where damage to any building components will occur as a result of overflow from the equipment drain pan or stoppage in the condensate drain piping. Such piping shall maintain a minimum horizontal slope in the direction of discharge of not less than $^1/_8$ unit vertical in 12 units horizontal (1-percent slope). Drain piping shall be a minimum of $^3/_4$-inch (19 mm) nominal pipe size.*

Discussion and Commentary: The condensate created by cooling coils and evaporators can be a nuisance at best, a hazard at worst. Condensation can damage or ruin the appearance of building finish material, and over time undetected water damage can lead to possible structural failure.

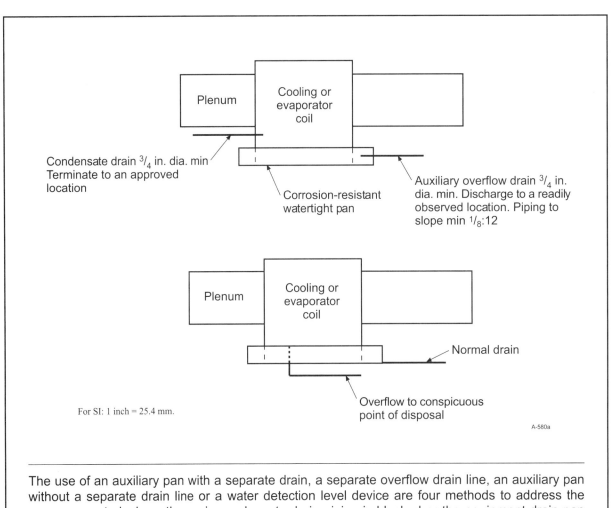

For SI: 1 inch = 25.4 mm.

A-580a

The use of an auxiliary pan with a separate drain, a separate overflow drain line, an auxiliary pan without a separate drain line or a water detection level device are four methods to address the concern created where the main condensate drain piping is blocked or the equipment drain pan overflows.

Code Text: *Foundations and supports for outdoor mechanical systems shall be raised at least 3 inches (76 mm) above the finished grade, and shall also conform to the manufacturer's installation instructions.*

Discussion and Commentary: All mechanical units installed outdoors are subject to corrosion and other environmental effects. The installation of such units on the ground could subject them to additional and unnecessary effects of drainage water and possible ponded water. Therefore, all mechanical systems must be raised to a minimum height of 3 inches above the surrounding finished grade.

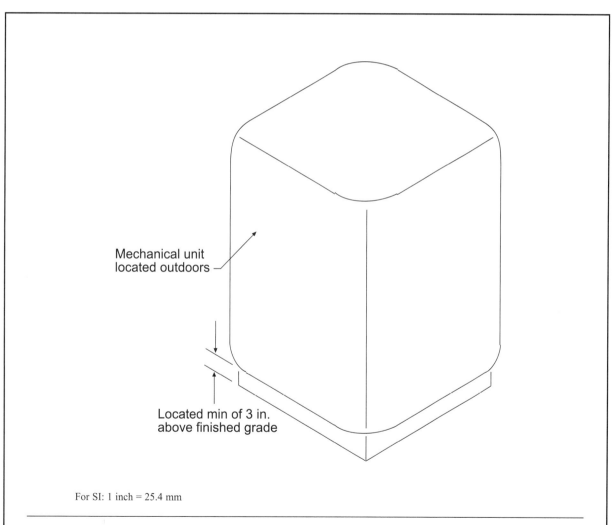

Mechanical unit located outdoors

Located min of 3 in. above finished grade

For SI: 1 inch = 25.4 mm

The protection afforded by raising the mechanical unit a small distance above the ground surface is sometimes necessary because of the unit's production of defrost water, whereas the presence of storm surface drainage is also often responsible for damage to any unit placed outdoors.

Code Text: *Fireplace stoves shall be listed, labeled and installed in accordance with the terms of the listing. Fireplace stoves shall be tested in accordance with UL 737. Hearth extensions for fireplace stoves shall be installed in accordance with the listing of the fireplace stove. The supporting structure for a hearth extension for a fireplace stove shall be at the same level as the supporting structure for the fireplace unit. The hearth extension shall be readily distinguishable from the surrounding floor area.*

Discussion and Commentary: A fireplace stove is a free-standing, chimney-connected solid fuel-burning heater designed to be operated with the fire chamber doors in either the open or closed position. A fireplace stove, including its hearth extension, must be installed in strict accordance with the listing.

Fireplace stove
- Listed, labeled and installed per UL 737

Heath extension
- Supporting structure at same level as structure for fireplace unit
- Readily distinguishable from surrounding floor area

Because a fireplace stove can be operated with the chamber doors in the open position, it is important that a hearth extension be provided to protect the floor in front of the opening from sparks and embers, as well as shield the surrounding floor surface from radiant heat.

Study Session 10
IRC Chapters 13 and 14

1. In order for someone to service an appliance other than a room heater, a minimum working space of ___30___ inches deep by 30 inches wide shall be provided in front of the control side of the appliance.

   a. 24
   b. 30
   c. 36
   d. 48

   Reference _M1305.1_

2. For the purpose of providing working space, a central furnace located within a closet or similar compartment shall be provided with a minimum of ___3___ clearance along the sides, back and top of the furnace, with the total width of the enclosing space a minimum of _3" 12"_ wider than the furnace.

   a. 3 inches, 6 inches
   b. 3 inches, 12 inches
   c. 6 inches, 12 inches
   d. 6 inches, 30 inches

   Reference _M1305.1.1_

3. A minimum ___6___ working space shall be provided along the front combustion chamber side of a central furnace having a firebox open to the atmosphere.

   a. 6-inch
   b. 12-inch
   c. 24-inch
   d. 30-inch

   Reference _M1305.1.1_

4. Where an appliance is installed in a compartment accessed by a passageway, the passageway shall have a minimum unobstructed width of ___24___, but not less than necessary to allow removal of the largest appliance in the space.

   a. 24 inches   b. 30 inches
   c. 36 inches   d. 42 inches

   Reference N1305.1.2

5. Where an appliance is installed in an attic, the clear access opening shall be a minimum of ___22" x 30'___, but in no case less than necessary to allow removal of the largest appliance.

   a. 20-inches by 30-inches   b. 22-inches by 30-inches
   c. 30-inches by 30-inches   d. 36-inches by 36-inches

   Reference M1305.1.3

6. Appliances suspended from the floor shall have a minimum clearance of ___6___ from the ground.

   a. 3 inches   b. 4 inches
   c. 6 inches   d. 8 inches

   Reference M1305.1.4.1

7. Excavations for appliance installations shall extend a minimum depth of _____ below the appliance with a minimum clearance of ___12___ on all sides except the control side.

   a. 3 inches, 6 inches   b. 6 inches, 6 inches
   c. 6 inches, 12 inches   d. 12 inches, 12 inches

   Reference M1305.1.4.2

8. Where ventilated air space is used to reduce the required clearance between an appliance and unprotected combustible materials, what is the minimum air space that is typically required?

   a. $^1/_2$ inch   b. 1 inch
   c. 2 inches   d. 3 inches

   Reference M1306.2

9. An appliance requires a minimum clearance without protection of 12 inches above the appliance to combustible material. If $^1/_2$-inch-thick insulation board with a ventilated air space is used as protection, the minimum clearance may be reduced to ___6___.

   a. 4 inches   (b) 6 inches

   c. 9 inches   d. 12 inches

   Reference Table M1306.2

10. A solid-fuel-burning appliance requires a minimum clearance without protection of 18 inches from the sides and back to combustible material. If 24 gage sheet metal with a ventilated air space is used as protection, the minimum clearance may be reduced to ___12___.

    a. 4 inches   b. 6 inches

    c. 9 inches   (d) 12 inches

    Reference M1306.2.1 & Table M1306.2

11. Where an appliance with an ignition source is located in the garage, the source of ignition shall be located a minimum of ___18___ above the garage floor.

    a. 12 inches   (b) 18 inches

    c. 36 inches   d. 48 inches

    Reference M1307.3

12. A combustion air opening for a central furnace shall be unobstructed for a minimum distance of ___6___ in front of the opening.

    a. 3 inches   (b) 6 inches

    c. 12 inches   d. 30 inches

    Reference M1402.3

13. The unobstructed total area of the outside and return air ducts or openings to a heat pump shall be a minimum of ___6___ square inches per 1,000 Btu/h output rating or as indicated by the conditions of the listing of the heat pump.

    (a) 6   b. 10

    c. 12.5   d. 20

    Reference M1403.1

14. In order to permit the free drainage of defrost water, the supports or foundation for the outdoor unit of a heat pump shall be raised a minimum of ___3___ above the ground.

    a. 2 inches   b. 3 inches

    c. 4 inches   d. 6 inches

    Reference ___M1403.2___

15. What is the maximum permitted operating temperature for a radiant heating system used on a gypsum assembly?

    a. 90°F   b. 110°F

    c. 125°F   d. 140°F

    Reference ___M1406.5___

16. Unless listed and labeled for closer installation, a duct heater shall be located a minimum of ___48___ from a heat pump or air conditioner.

    a. 18 inches   b. 24 inches

    c. 36 inches   d. 48 inches

    Reference ___M1407.3___

17. The furnace register of a vented floor furnace shall be located a minimum of _____ from doors in any position, draperies and other similar combustible objects.

    a. 3 inches   b. 6 inches

    c. 12 inches   d. 60 inches

    Reference ___M1408.3 #3___

18. The access opening through the foundation wall to a vented floor furnace shall have a minimum size of ___18 x 24___ .

    a. 18-inches by 24-inches   b. 20-inches by 30-inches

    c. 22-inches by 30-inches   d. 30-inches by 30-inches

    Reference ___M1408.4___

19. Where the lower 6 inches of a vented floor furnace is sealed to prevent water entry, a minimum ___2___ clearance shall be provided between the furnace and the ground.

    (a.) 2-inch   b. 3-inch

    c. 4-inch   d. 6-inch

    Reference _M1408.5 # 3_

20. A minimum clearance of ___12___ shall be provided between the inlet or outlet of a vented wall furnace and a door at any point in its swing, measured at a right angle to the opening.

    a. 6 inches   (b.) 12 inches

    c. 18 inches   d. 36 inches

    Reference _M1409.2 # 2_

21. Unless listed and labeled for such use, cooling coils of refrigeration cooling equipment shall not be located _upstream_ of heat exchangers.

    a. within 12 inches   b. within 36 inches

    (c.) upstream   d.) downstream

    Reference _M1411.2_

22. Drain piping for refrigeration cooling equipment shall be a minimum of _3/4"_ nominal pipe size.

    a. $^1/_2$ inch   (b.) $^3/_4$ inch

    c. 1 inch   d. $1^1/_2$ inch

    Reference _M1411.3.1_

23. Where an auxiliary drain pan with a separate drain is installed under the cooling coils where condensate will occur, the minimum depth of the pan shall be _1 1/2"_.

    a. 1 inch   (b.) $1^1/_2$ inches

    c. 2 inches   d. 3 inches

    Reference _M1411.3.1 # 1_

24. Evaporative coolers shall be installed on a level platform or base located a minimum of _____3_____ above the adjoining ground.

a. 3 inches   b. 4 inches

c. 6 inches   d. 12 inches

Reference _M1413.1_

25. The supporting structure for a hearth extension for a fireplace stove shall be located _same level_ the supporting structure for the fireplace unit.

a. at least 6 inches below   b. at least 3 inches below

c. at the same level as   d. at least 3 inches above

Reference _M1414.2_

26. Where an appliance is located in an underfloor space, a minimum _____ level service space shall be provided at the front or service side of the appliance.

a. 30-inch by 30-inch   b. 36-inch by 30-inch

c. 36-inch by 36-inch   d. 42-inch by 36-inch

Reference _M1305.1.4_

27. The required lower strapping for a water heater in Seismic Design Categories D_1 and D_2 shall be located a minimum of _4_ inches above the controls.

a. 3   b. 4

c. 6   d. 8

Reference _M1307.2_

28. A private garage containing a hydrogen-generating appliance shall be limited to a maximum of _850_ square feet in area.

a. 720   b. 800

c. 850   d. 1,000

Reference _M1307.4.1_

29. Where piping other than cast-iron or galvanized steel is installed through bored holes of studs within a concealed wall space, steel shield plates are not required where the holes are a minimum of _____ inch(es) from the nearest edge of the stud.

    a. $^3/_4$ b. $1^1/_4$

    c. $1^3/_8$ d. $1^1/_2$

Reference ᴍ 1308.2

30. Unless listed for use on combustible floors without floor protection, a floor-mounted vented room heater shall be located so that noncombustible flooring materials extend a minimum of __18__ inches beyond the appliance on all sides.

    a. 12   b. 18

    c. 24   d. 36

Reference M1410.2

31. Where access to an appliance located in an attic is provided by an unobstructed passageway at least 6 feet high and 22 inches wide, the passageway shall have a maximum length of __50__ .

    a. 10 feet   b. 20 feet

    c. 50 feet   d. an unlimited distance

Reference M1305.1.3 #2

32. The two permanent outdoor openings required to be provided in a private garage containing a hydrogen-generating appliance shall have a minimum free area of __1/2__ per 1,000 cubic feet of garage volume.

    a. $^1/_2$ square foot

    b. 1 square foot

    c. 2 square feet

    d. 5 square feet

Reference M1307.4.1.1

33. Foundations and supports for outdoor mechanical systems shall be raised a minimum of __3__ inches above the finished grade.

    a. 2   b. 3

    c. 4   d. 6

Reference M1305.3

34. The register of a vented floor furnace shall be located a minimum of _____ below any projecting combustible materials.

    a. 12 inches

    b. 2 feet

    c. 3 feet

    ⓓ 5 feet

    Reference M1408.3# 7

35. Secondary drain piping from a cooling coil or evaporator shall maintain a minimum horizontal slope in the direction of discharge of ___1/8___ unit vertical in 12 units horizontal.

    a. $^1/_{12}$

    b.) $^1/_8$

    c. $^1/_4$

    d. $^1/_2$

    Reference M1411.3.1

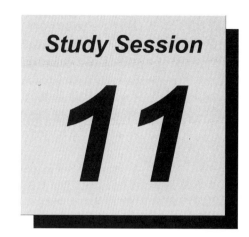

Study Session

11

2006 IRC Chapters 15, 16 and 19

Exhaust Systems, Duct Systems and Special Fuel-Burning Equipment

OBJECTIVE: To gain an understanding of the general provisions for the installation of exhaust systems, duct systems and special fuel-burning equipment, such as ranges, ovens and sauna heaters.

REFERENCE: Chapters 15, 16 and 19, 2006 *International Residential Code*

KEY POINTS:
M1502.2
- Where must mechanical exhaust air terminate? What locations are strictly prohibited? M1502.2
M1502.6
- What are the installation requirements for clothes dryer exhaust systems? What is the maximum length of a flexible transition duct? Where is such a duct permitted?
M1502.c
- What is the maximum permitted length of a clothes dryer exhaust duct? How does the presence of one or more bends affect the allowable length? M1502.c
M1503.1
- How is a range hood required to be ducted? What are the physical characteristics of the duct? What materials are permitted for duct construction? M1503.2
M 1505.1
- How must a domestic open-top broiler unit be ducted? What is the required clearance between the overhead exhaust hood and combustible construction? The minimum clearance between the cooking surface and the hood?
M1507.3 & Table
- What are the minimum exhaust rates for ventilating kitchens? Bathrooms?
M1601.1.1 #1
- What is the maximum discharge air temperature permitted for equipment connected to an above-ground duct system? The maximum temperature when gypsum products are used? #5 The maximum flame spread? #6
M1601.1.1 #7
- Under what conditions can stud cavities be used as return air ducts? 7-7.1 7.2 7.3 7.4
M1601.1.2
- What are the limitations on the installation of an underground duct system? What is the maximum duct temperature permitted for underground ducts? What is the minimum size of the access opening to an under-floor plenum? M1601.4.4
M1601.2.1 #1
- What is the maximum permitted flame spread for duct insulation coverings and linings? The maximum permitted smoke-developed index? M1601.2.1 #1
M1601.2.1 #2
- How should external duct insulation be identified? At what intervals must the identification be provided? M1601.2.1 #3
M1601.2.2
- What is the maximum permitted length of vibration isolators?
M1601.3.2
- What means of support are addressed for metal ducts? Nonmetallic ducts? M1601.3.2

KEY POINTS:
(Cont'd)

- How are underfloor plenums to be formed? How shall the plenum connect to the furnace? What degree of access to the plenum is required. *M1601.4.2* *M1601.4.3* *M1601.4.4*
- Which sources of outside or return air are prohibited? *M1602.2* *#s 1-5*
- What is the minimum vertical clearance required between the cook-top of a range and any unprotected combustible material? Under what conditions may the clearance be reduced? *M1901.1*
- How must sauna heaters be installed and protected from contact? What is the temperature limit for the heater's thermostat? *M1902.1 & M1902.2* *M1902.4*

Code Text: *The maximum length of a clothes dryer exhaust duct shall not exceed 25 feet (7620 mm) from the dryer location to the wall or roof termination. The maximum length of the duct shall be reduced 2.5 feet (762 mm) for each 45-degree (0.8 rad) bend and 5 feet (1524 mm) for each 90-degree (1.6 rad) bend. The maximum length of the exhaust duct does not include the transition duct.* See exceptions for 1) installations where the make and model of the dryer is known and the manufacturer's installation instructions provide a maximum length, and 2) where large-radius bends are installed and the maximum length is provided by engineering calculation.

Discussion and Commentary: Clothes dryer exhausts present a different problem than other exhaust systems because the exhaust air is laden with moisture and lint. The air must be vented to the outside and not discharged into an attic or crawl space because wood structural members could be adversely affected. Exhaust outlets must be equipped with a backdraft damper to prevent cold air, rain, snow, rodents and vermin from entering the vent.

Minimum duct size as recommended by manufacturer shall terminate on the outside of the building

Backdraft damper

Screens not permitted

Clothes Dryer

- Maximum duct length of 25 ft
- *Subtract 5 ft for each 90° bend (2.5 ft for 45°)
- Transition duct not included in length

For SI: 1 foot = 304.8 mm, 1 degree = 0.1745 rad.

B-526

The length restrictions placed on the vent ensure that the dryer exhaust blower will be able to drive sufficient air volume to carry off the moisture. Only when the make and model of the dryer is known, or where an engineered calculation is provided, can the length possibly be increased.

Code Text: *Domestic open-top broiler units shall be provided with a metal exhaust hood, not less than 28 gage, with a clearance of not less than 0.25 inch (6 mm) between the hood and the underside of combustible materials or cabinets. A clearance of at least 24 inches (610 mm) shall be maintained between the cooking surface and the combustible material or cabinet.*

Discussion and Commentary: A minimal clearance is needed between the hood and any cabinets or similar items to ensure that the surface material of the hood does not come in direct contact with the combustible material. The vertical clearance protects overhead cabinets from ignition should a fire initiate on the cooktop. It is important that the length of the hood be at least that of the boiler unit to allow for complete coverage.

For SI: 1 inch = 25.4 mm.

B-527

A boiler unit exhaust hood must discharge to the exterior and be provided with a backdraft damper. A range hood must also exhaust directly to the outdoors unless a listed and labeled ductless range hood is installed in accordance with the manufacturer's instructions.

Code Text: *Where toilet rooms and bathrooms are mechanically ventilated, the ventilation equipment shall be installed in accordance with Section M1507. Exhaust air from bathrooms and toilet rooms shall not be recirculated within a residence or to another dwelling unit and shall be exhausted directly to the outdoors. Ventilation systems shall be designed to have the capacity to exhaust the minimum air flow rate determined in accordance with Table M1507.3.*

Discussion and Commentary: The mechanical ventilation rates for bathrooms and water closet compartments are also set forth in Section R303.3 as an exception to the general requirement for natural ventilation. The fundamental provisions require such spaces to be provided with natural ventilation at a minimum rate of $1\frac{1}{2}$ square feet of exterior openings.

TABLE M1507.3
MINIMUM REQUIRED EXHAUST RATES FOR
ONE- AND TWO-FAMILY DWELLINGS

AREA TO BE VENTILATED	VENTILATION RATES
Kitchens	100 cfm intermittent or 25 cfm continuous
Bathrooms—Toilet Rooms	Mechanical exhaust capacity of 50 cfm intermittent or 20 cfm continuous

For SI: 1 cubic foot per minute = 0.0004719 m³/s.

Kitchen ventilation can be addressed by two different methods, as is the case for bathroom ventilation. Where continuous kitchen ventilation is provided, a minimum rate of 25 cfm is required. If an intermittent exhaust system is utilized, it must be rated for at least 100 cfm.

Code Text: *Equipment connected to duct systems shall be designed to limit discharge air temperature to a maximum of 250°F (121°C). Factory-made air ducts shall be constructed of Class 0 or Class 1 materials as designated in Table M1601.1.1(1). Minimum thicknesses of metal duct material shall be as listed in Table M1601.1.1(2).*

Discussion and Commentary: The temperature limit for discharge air is based on the possible use of nonmetallic ducts. Higher air temperatures could cause their rapid deterioration. The flame spread is strictly limited to 25 for Class 1 ducts and 0 for Class 0 ducts. Where metal ducts are used, they shall be of substantial construction to ensure that their structural integrity is maintained. Ducts fabricated from lighter gage materials will experience an unacceptable amount of distortion in use.

TABLE M1601.1.1(1)
CLASSIFICATION OF FACTORY-MADE AIR DUCTS

DUCT CLASS	MAXIMUM FLAME-SPREAD RATING
0	0
1	25

TABLE M1601.1.1(2)
GAGES OF METAL DUCTS AND PLENUMS USED FOR HEATING OR COOLING

TYPE OF DUCT	SIZE (inches)	MINIMUM THICKNESS (inch)	EQUIVALENT GALVANIZED SHEET GAGE	APPROXIMATE ALUMINUM B & S GAGE
Round ducts and enclosed rectangular ducts	14 or less	0.013	30	26
	over 14	0.016	28	24
Exposed rectangular ducts	14 or less	0.016	28	24
	over 14	0.019	26	22

For SI: 1 inch = 25.4 mm.

Gypsum products may be used in the fabrication of return air ducts or plenums, provided the air temperature is limited and condensation on the duct surfaces is not a concern. Excessive temperatures and high humidity are both potential causes of damage to the gypsum materials.

Code Text: *Stud wall cavities and the spaces between solid floor joists to be utilized as air plenums shall comply with the following conditions: 7.1) such cavities or spaces shall not be utilized as a plenum for supply air; 7.2) such cavities or spaces shall not be part of a required fire-resistance-rated assembly; 7.3) stud wall cavities shall not convey air from more than one floor level; and 7.4) stud wall cavities and joist space plenums shall be isolated from adjacent concealed spaces by tight-fitting fire blocking in accordance with Section R602.8.*

Discussion and Commentary: The use of stud spaces and panned joist spaces as return air ducts is acceptable since the air temperature is such not to cause a hazard or deterioration of the construction materials. Isolation of the return air plenum from other unused spaces is necessary so that debris will not be drawn into the furnace.

Wall studs

Fireblock above

Grille

Fireblocking material on bottom of joists. Air between floor joists used as return air duct

Gypsum nailed over both sides of studs

Bottom plate cutaway opening sawed in floor

Blocking of 2 in. lumber forms end of cold-air duct between joists

For SI: 1 inch = 25.4 mm, 1 degree = 0.1745 rad.

B-528

Often, walls or floor-ceiling assemblies such as dwelling unit separations or exterior walls adjacent to property lines must be of fire-resistant-rated construction. Such assemblies must not be used as air plenums.

Code Text: *Underground duct systems shall be constructed of approved concrete, clay, metal or plastic. The maximum duct temperature for plastic ducts shall not be greater than 150°F (66°C). Metal ducts shall be protected from corrosion in an approved manner or shall be completely encased in concrete not less than 2 inches (51 mm) thick. Nonmetallic ducts shall be installed in accordance with the manufacturer's installation instructions.*

Discussion and Commentary: Ducts fabricated from the listed materials should not collapse when installed underground or within concrete. Ducts installed in or beneath concrete should be properly sealed and secured before the concrete pour. Otherwise, leakage of concrete into the ducts could block the flow of air when the system is placed in operation. For the same reason, underground ducts should be secured to avoid damage during backfilling.

Note:
1.   Min of 2 in. concrete encasement around metal duct.
2.   Approved material designed for this use and installed in accordance with manufacturer's instructions.

For SI: 1 inch = 25.4 mm.   B-529

Where plastic pipe is used in an underground duct system, the duct temperature is limited to avoid the possibility of deterioration or distortion of the plastic materials. To ensure the adequacy of plastic ducts, specific material and structural criteria are referenced in several ASTM standards.

Code Text: *Duct coverings and linings, including adhesives where used, shall have a flame-spread index not higher than 25, and a smoke-developed index not over 50 when tested in accordance with ASTM E 84. External duct insulation and factory-insulated flexible ducts shall be legibly printed or identified at intervals not greater than 36 inches (914 mm) with the name of the manufacturer; the thermal resistance R-value at the specified installed thickness; and the flame spread and smoke-developed indexes of the composite materials.*

Discussion and Commentary: Ducts that pass through nonconditioned areas should be insulated to prevent excessive heat loss or heat gain in cooling installations. Insulation also prevents the cool surfaces of air-conditioning ducts from condensing moisture from the surrounding unconditioned air.

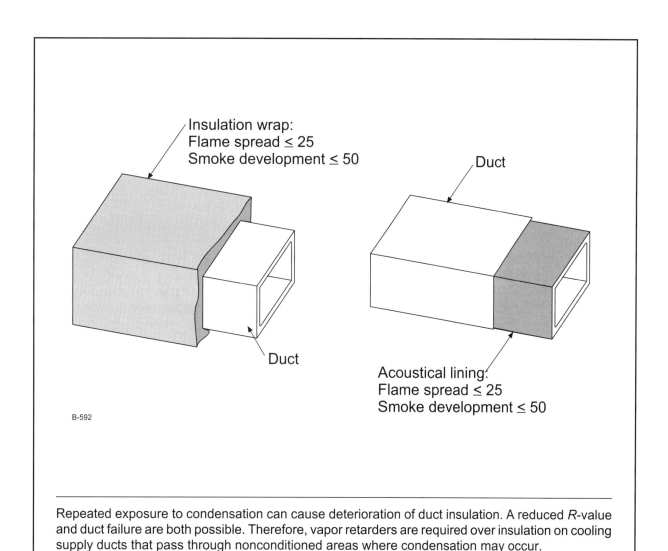

Insulation wrap:
Flame spread ≤ 25
Smoke development ≤ 50

Duct

Duct

Acoustical lining:
Flame spread ≤ 25
Smoke development ≤ 50

B-592

Repeated exposure to condensation can cause deterioration of duct insulation. A reduced R-value and duct failure are both possible. Therefore, vapor retarders are required over insulation on cooling supply ducts that pass through nonconditioned areas where condensation may occur.

Code Text: *Joints of duct systems shall be made substantially airtight by means of tapes, mastics, gasketing or other approved closure systems. Duct connections to flanges of air distribution system equipment or sheet metal fittings shall be mechanically fastened. Crimp joints for round ducts shall have a contact lap of at least $1^1/_2$ inches (38 mm) and shall be mechanically fastened by means of at least three sheet metal screws or rivets equally spaced around the joint. Metal ducts shall be supported by $^1/_2$-inch (13 mm) wide 18-gage metal straps or 12-gage galvanized wire at intervals not exceeding 10 feet (3048 mm) or other approved means.*

Discussion and Commentary: Duct leakage should be minimized to prevent energy loss. Adequate duct support is necessary to maintain alignment and limit deflection.

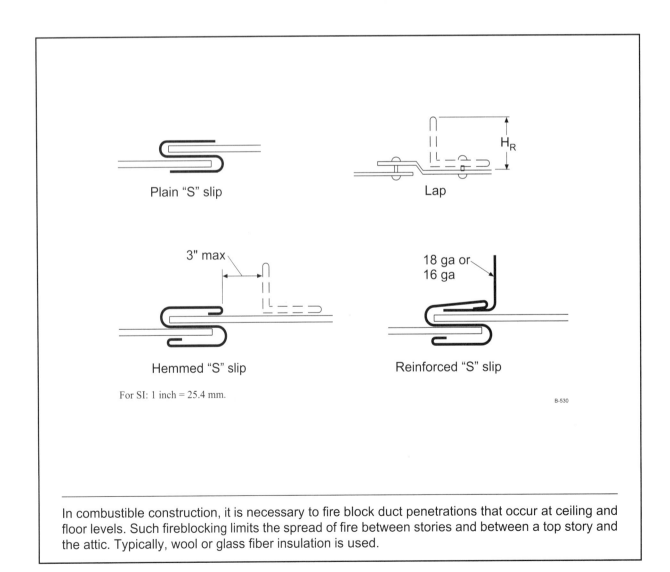

Plain "S" slip

Lap

H_R

3" max

Hemmed "S" slip

18 ga or 16 ga

Reinforced "S" slip

For SI: 1 inch = 25.4 mm.

B-530

In combustible construction, it is necessary to fire block duct penetrations that occur at ceiling and floor levels. Such fireblocking limits the spread of fire between stories and between a top story and the attic. Typically, wool or glass fiber insulation is used.

Code Text: *Fuel gas lines and plumbing waste cleanouts shall not be located within an* under-floor space used as a supply plenum. *The space shall be cleaned of loose combustible materials and scrap, and shall be tightly enclosed. The ground surface of the space shall be covered with a moisture barrier having a minimum thickness of 4 mils (0.1 mm). The under-floor space, including the sidewall insulation, shall be formed by materials having flame-spread ratings not greater than 200 when tested in accordance with ASTM E 84.*

Discussion and Commentary: A counterflow furnace is typically used to discharge conditioned air to the underfloor plenum. Floor registers are installed in openings cut through the floor for delivery of the conditioned air to the space above. Access to the plenum area must be provided by a minimum 18-inch by 24-inch (457 mm by 610 mm) opening.

Return air

Furnace

18 in. x 24 in. access

Floor diffuser

6 in. min extension below joists

Under-floor plenum space

Vapor barrier with a minimum thickness of 4 mils

For SI: 1 inch = 25.4 mm.

B-531

To limit the temperature of the under-floor space, the furnace must be provided with controls that
1) start the air-circulating fan when the air in the furnace bonnet reaches at least 150°F, and
2) limit the outlet air temperature to 200°F.

Topic: Required Area
Reference: IRC M1602

Category: Duct Systems
Subject: Return Air

Code Text: *Return air shall be taken from inside the dwelling. Dilution of return air with outdoor air shall be permitted.*

Discussion and Commentary: Recirculation of air contaminated with objectionable odors, fumes or flammable vapors is prohibited because of potential health and safety hazards. Therefore, return air ducts must not serve kitchens, bathrooms, closets, attached garages or other dwelling units. Also, outside air intakes must be located carefully to preclude objectionable or contaminated air from being drawn into the air distribution system. Air intakes above the roof must be located a minimum of 10 feet laterally from a plumbing vent or fuel-burning appliance vent, unless the source of noxious fumes is more than 3 feet higher than the outside air intake. In that case, a lesser lateral separation is acceptable.

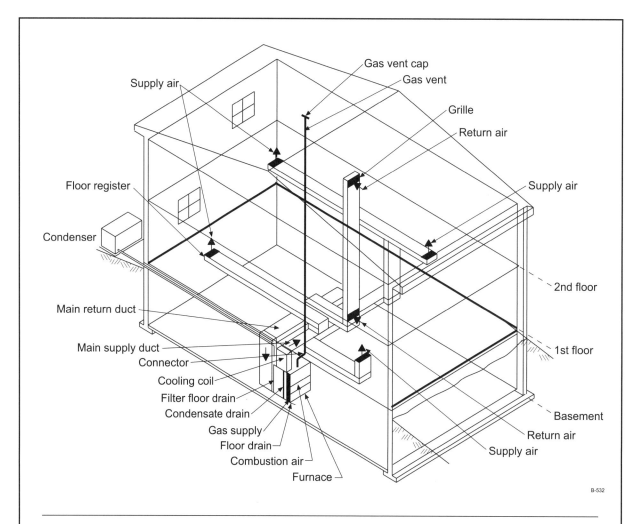

To prevent the intrusion of animals and insects, it is necessary to provide a mesh screen having openings no greater than $1/2$ inch. However, in order to maintain the necessary openings free and clear from obstruction, a minimum dimension of $1/4$ inch is mandated.

2006 IRC Study Companion

Code Text: *Freestanding or built-in ranges shall have a vertical clearance above the cooking top of not less than 30 inches (762 mm) to unprotected combustible material. Reduced clearances are permitted in accordance with the listing and labeling of the range hoods or appliances.*

Discussion and Commentary: Vertical clearance to combustibles is important because cooking range tops produce considerable heat. Additionally, overheated utensils, particulary those containing grease, may be the source of flame. An adequate clear space above the range also ensures that hoods, shelves and other potential obstructions do not encroach into the working space needed for cooking operations. It is also important that the installation of any household cooking appliance not interfere with combustion air.

Min 30 in. clearance, cooking top to underside of combustible material

Range

For SI: 1 inch = 25.4 mm.

B-533

Although vented or unvented range hoods are often installed over ranges, the code does not require hoods to be installed. Where installed however, the hoods and exhaust methods must be in compliance with the code and the manufacturer's installation instructions.

Topic: Locations and Protection; Controls   **Category:** Special Fuel-burning Equipment
Reference: IRC M1902.1, M1902.4   **Subject:** Sauna Heaters

Code Text: *Sauna heaters shall be protected from accidental contact by persons with a guard of material having a low thermal conductivity, such as wood. The guard shall have no substantial effect on the transfer of heat from the heater to the room. Sauna heaters shall be equipped with a thermostat that will limit room temperature to not greater than 194°F (90°C). Where the thermostat is not an integral part of the heater, the heat-sensing element shall be located within 6 inches (152 mm) of the ceiling.*

Discussion and Commentary: In a small space such as a sauna room, it is quite possible that contact with the heating element will occur unless the element is adequately guarded. The guard must be of a type that allows the heated air to circulate throughout the room, as concentrated heat and a combustible guard could create a fire hazard.

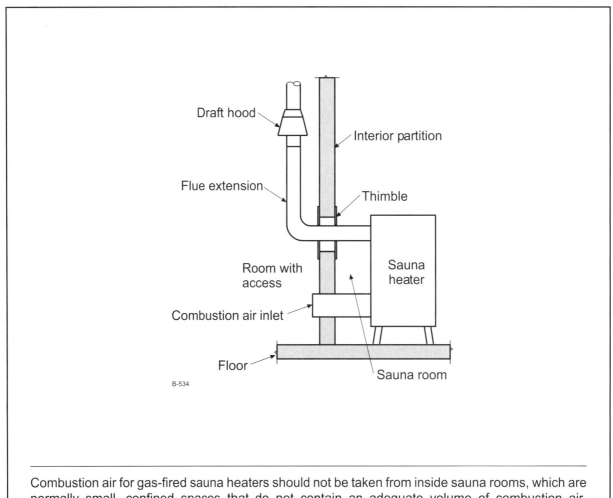

Combustion air for gas-fired sauna heaters should not be taken from inside sauna rooms, which are normally small, confined spaces that do not contain an adequate volume of combustion air. Combustion air and venting shall be in compliance with Chapters 17 and 18.

Code Text: *Exhaust ducts shall terminate on the outside of the building. Exhaust duct terminations shall be in accordance with the dryer manufacturer's installation instructions. Exhaust ducts shall terminate not less than 3 feet (914 mm) in any direction from openings into buildings. Exhaust duct terminations shall be equipped with a backdraft damper. Screens shall not be installed at the duct termination.*

Discussion and Commentary: Dryer exhaust contains high levels of moisture and highly combustible clothes fibers; therefore, they must terminate at the building's exterior. To avoid any hazard created by the exhaust termination, there must be a clearance of at least 3 feet to any other openings into the building. This would include operable windows, doors and air intakes.

For SI: 1 foot = 304.8 mm

The provisions regulating clothes dryer exhaust systems are not applicable to listed and labeled condensing (ductless) clothes dryers. This type of dryer is electrically heated and recirculates the same air through the clothes drum.

Study Session 11
IRC Chapters 15, 16 and 19

1. Exhaust ducts for clothes dryers shall not _____.

   a. extend to the exterior of the dwelling

   (b.) be connected with sheet-metal screws

   c. be equipped with a backdraft damper

   d. be connected to flexible transition ducts

   Reference M1502.5

2. Unless modified by the manufacturer's installation instructions, what is the maximum permitted length of a clothes dryer exhaust duct when measured from the dryer location to the wall or roof termination?

   a. 18 feet   b. 20 feet

   (c) 25 feet   d. 30 feet

   Reference M1502.6

3. Unless modified by the manufacturer's installation instructions or determined by engineering calculation, what is the maximum permitted length of a clothes dryer exhaust duct that has one 90-degree bend and two 45-degree bends?

   a. 25 feet   b. 20 feet

   c. $17^1/_2$ feet   (d.) 15 feet

   Reference M1502.6

4. Which of the following requirements is not applicable to a duct serving a range hood?

    a. It shall have a smooth surface.

    b. It shall be airtight.

    c. It shall be equipped with a backdraft damper.

    d. It may terminate in a crawl space.

    Reference M1503.1

5. When serving a domestic kitchen cooking appliance equipped with a down-draft exhaust system, a schedule 40 PVC exhaust duct shall extend a maximum of ___1"___ above grade outside the building.

    a. 1 inch   b. 6 inches

    c. 12 inches   d. 36 inches

    Reference 1503.2 Exc 4

6. A domestic open-top broiler unit shall be provided with a minimum clearance of ___1/2"___ between the hood and the underside of combustible material or cabinets.

    a. 0 inches; no clearance is required

    b. ¹/₄ inch

    c. ¹/₂ inch

    d. 1 inch

    Reference M1505.1

7. For a domestic open-top boiler unit, a minimum of ___24"___ shall be maintained between the cooking surface and combustible material or cabinet.

    a. 24 inches   b. 28 inches

    c. 30 inches   d. 32 inches

    Reference M1505.1

8. Heating equipment connected to an above-ground duct system shall be designed to limit discharge air temperature to a maximum of ___250°___.

    a. 125°F   b. 200°F

    c. 225°F   d. 250°F

    Reference M1601.1.1 #1

9. When serving heating, cooling or ventilation equipment, above-ground factory-made duct systems shall be constructed of materials classified as _0 or 1_.

    a. Class 0 only   b. Class 1 only

    (c) Class 0 or Class 1   d. Class 1 or Class 2

Reference _M1601.1.1 #2_

10. A 16-inch exposed rectangular duct that is a portion of an above-ground duct system shall have a minimum equivalent galvanized sheet gage of _28_.

    a. No. 22   b. No. 26

    (c) No. 28   d. No. 30

Reference _Table M1601.1 (2)_

11. Gypsum board may be used in the construction of a return air plenum in an above-ground duct system, provided the maximum air temperature is _125°_ and the exposed surfaces are not subject to condensation.

    (a) 125°F   b. 150°F

    c. 180°F   d. 210°F

Reference _M1601.1.1 # 5_

12. What is the maximum flame spread index permitted for materials used to construct above-ground duct systems?

    a. 25   b. 75

    (c) 200   d. 450

Reference _M1601.1.1 # 6_

13. Where used in an underground duct system, the maximum permitted duct temperature for plastic ducts shall be _150°_.

    a. 125°F   (b) 150°F

    c. 180°F   d 210°F

Reference _M1601.1.2_

14. When protected by concrete, metal ducts installed in an underground duct system shall be encased with a minimum thickness of ___2.___.

   a. 1 inch   (b.) 2 inches

   c. 3 inches   d. 4 inches

   Reference ___M1601.1.2___

15. Duct coverings and linings for factory-made ducts shall have a maximum flame spread index of ___25___ and a maximum smoke-developed index of ___50___.

   (a) 25, 50   b. 25, 200

   c. 50, 200   d. 50, 450

   Reference ___M1601.2.1 #1___

16. Factory-insulated flexible factory-made ducts shall be identified at maximum intervals of ___3 FT___ with the name of the manufacturer, the thermal resistance *R*-value, the flame spread index and the smoke-developed index.

   (a.) 3 feet   b. 5 feet

   c. 6 feet   d. 10 feet

   Reference ___M1601 2.1 #3___

17. Where vibration isolators are installed between mechanical equipment and metal ducts, they shall have a maximum length of ___10___.

   a. 6 inches   (b) 10 inches

   c. 12 inches   d. 24 inches

   Reference ___M1601.2.2___

18. Where metal ducts are supported by $^{1}/_{2}$-inch wide metal straps, the straps shall be minimum ___18___ and located a maximum of ___10___ on center.

   a. 12-gage, 6 feet   b. 18-gage, 6 feet

   (c.) 18-gage, 10 feet   d. 22-gage, 10 feet

   Reference ___M1601.3.2___

19. An under-floor space used as a supply plenum shall be formed by materials having a maximum flame spread rating of _200_.

    a. 25   b. 50

    c. 75   (d) 200

Reference _M1601.4.2_

20. Where the under-floor space is used as a supply plenum, a duct shall extend from the furnace supply outlet to a minimum of _6_ inch(es) below the combustible framing.

    a. 1   b. 3

    (c.) 6   d. 8

Reference _M1601.4.3_

21. Unless the discharge outlet is a minimum of _____ above the outside air inlet, outside air for a forced-air heating system shall be taken a minimum of _____ from an appliance vent outlet, a vent opening from a plumbing drainage system, or the discharge outlet of an exhaust fan.

    a. 2 feet, 6 feet   b. 2 feet, 10 feet

    c. 3 feet, 6 feet   (d) 3 feet, 10 feet

Reference _M1602.2 #1_

22. Which one of the following spaces is acceptable as a source for return air for a forced-air heating system?

    (a) sleeping room   b. bathroom

    c. kitchen   d. closet

Reference _M1602.2 #4_

23. Outdoor air inlets for a return air system shall be covered with screens having a minimum opening size of _1/4_ and a maximum opening size of _1/2_.

    (a) $^{1}/_{4}$ inch, $^{1}/_{2}$ inch   b. $^{1}/_{4}$ inch, $^{3}/_{4}$ inch

    c. $^{3}/_{8}$ inch, $^{1}/_{2}$ inch   d. $^{3}/_{8}$ inch, $^{3}/_{4}$ inch

Reference _M1602.3_

24. Built-in or freestanding ranges shall have a minimum vertical clearance of _30_ above the cooking top to an unprotected combustible construction.

   a. 24 inches   (b.) 30 inches

   c. 32 inches   d. 36 inches

   Reference _M 1901.1_

25. Sauna heaters shall be equipped with a thermostat that will limit room temperature to a maximum of _194°_.

   a. 120°F   b. 140°F

   c. 172°F   (d.) 194°F

   Reference _M 1902.4_

26. A mechanically ventilated kitchen area shall be provided with a minimum ventilation rate of _25_ cfm where continuous ventilation is utilized.

   a. 20   (b.) 25

   c. 50   d. 100

   Reference _TM1507.3_

27. A bathroom ventilated by an intermittent mechanical means shall have a minimum mechanical exhaust capacity of _50_ cfm.

   a. 20   b. 25

   (c.) 50   d. 100

   Reference _T M1507.3_

28. The access through the floor to an under-floor plenum shall provide a minimum opening size of _18x24_

   (a.) 18 inches by 24 inches   b. 20 inches by 24 inches

   c. 22 inches by 30 inches   d. 24 inches by 30 inches

   Reference _M1601.4.4_

29. Where an under-floor space is used as a supply plenum, the furnace shall be equipped with an approved automatic control that limits the outlet air temperature to a maximum of __200__ °F.

   a. 150   b. 200

   c. 225   d. 250

   Reference M1601.4.5

30. Where the thermostat used to limit sauna room temperature is not an integral part of the sauna heater, the heat-sensing element shall be located a maximum of __6__ inches below the ceiling.

   a. 3   b. 4

   c. 6   d. 8

   Reference M1902.4

31. A clothes dryer exhaust duct shall terminate outside of the building a minimum of __3__ feet in any direction from openings into the building.

   a. 2   b. 3

   c. 5   d. 10

   Reference M1502.2

32. A flexible transition duct for a clothes dryer exhaust system is limited to a maximum length of __8__ feet.

   a. 3   b. 5

   c. 6   d. 8

   Reference M1502.4

33. Where used in an above-ground duct system, factory-made air ducts constructed of Class 0 materials shall have a maximum flame spread rating of __0__.

   a. 0   b. 25

   c. 75   d. 200

   Reference M1601.1.1(1)

34. Crimp joints for round metal ducts shall have a minimum contact lap of _1¹/₂"_ inch(es).

    a. $^{3}/_{4}$ b. 1

    c. $1^{1}/_{2}$ d. 2

Reference _M1603 1_

35. Unless in conformance with the criteria for underground duct systems, ducts shall be installed with a minimum of ____4____ inches separation from earth.

    a. 3   b. 4

    c. 6   d. 12

Reference _M1601.3.6_

Study Session

12

2006 IRC Chapters 17, 18, 20, 21, 22 and 23

Combustion Air, Chimneys and Vents, Boilers/Water Heaters, Hydronic Piping, Special Piping and Storage Systems, and Solar Systems

OBJECTIVE: To provide an understanding of the general provisions for combustion air, chimneys, vents, boilers, water heaters, hydronic piping, special piping and storage systems, and solar systems.

REFERENCE: Chapters 17, 18, 20, 21, 22 and 23, 2006 *International Residential Code*

KEY POINTS:
M1701.1.1
- How shall combustion air be obtained in buildings of unusually tight construction? Under what condition may combustion air be taken from inside a building through infiltration? 1701.1.1

M1701.3
- Where are volume dampers prohibited?

M1701.4
- Which rooms in a dwelling unit are prohibited as sources for combustion air? # 1-3

M1701.5
- What is considered the net free area of a metal louver when not specified by the manufacturer? A wood louver? M1701.5

M1702.2
2
- What is considered a confined space? How many combustion air openings are required in such a space? Where are they to be located? What is the minimum area of each opening when communicating with the outdoors by vertical ducts? Horizontal ducts? M1703.2.1

M1703.2.1
- What is the minimum allowable cross-sectional dimension of a rectangular combustion air duct? 3"

M1703.3 # 1-4
- Under what conditions may combustion air be taken from the attic? From an under-floor area? M1703.4

M1703.5
- What are the minimum and maximum opening sizes permitted for screens used to cover outside combustion air openings? Not Less Than 1/4 and no greater Than 1/2"

M1701.11
- What venting method is to be used where two or more listed appliances are connected to a common natural draft venting system? # 1-3

Table M1803.2
- What is the minimum thickness for a single-wall metal pipe connector? 0.019

M1804.2.3
- How far above the highest connected appliance outlet must a natural draft appliance vent terminate? How far above the bottom of a wall vent must a natural draft gas vent terminate?

M1805.2
- Where must a chimney connector enter a masonry chimney? How is the size of a chimney flue that is connected to more than one appliance calculated? M1805.3

KEY POINTS:
(Cont'd)

M2001.3

- Where are valves required for a boiler? What type of gauges are mandated? What is a boiler low-water cutoff? M2002.5 *(M2002.2 & M2002.3)*

M2003.1.1
- What requirements apply to expansion tanks used with hot water boilers?

M2005.2
- Where is the installation of fuel-fired water heaters prohibited? In what manner can access be provided to a water heater in an under-floor or attic area? M2005.2.1

M2101.1
- What are the general provisions related to hydronic piping systems? How must floor heating systems be tested? M2103.3

M2203.1
- How are special piping and storage systems to be installed?

M2301.2
- What special conditions apply to the installation of a solar energy system?

Topic: Prohibited Sources

Category: Combustion Air

Reference: IRC M1701.4

Subject: General Provisions

Code Text: *Combustion air ducts and openings shall not connect appliance enclosures with space in which the operation of a fan may adversely affect the flow of combustion air. Combustion air shall not be obtained from an area in which flammable vapors present a hazard. Fuel-fired appliances shall not obtain combustion air from any of the following rooms or spaces: 1) sleeping rooms, 2) bathrooms, or 3) toilet rooms. See two exceptions allowing specific appliances in such rooms.*

Discussion and Commentary: Exhaust fans, as just one example, can produce negative pressures in a space, thereby reducing the availability of combustion air. Air that could have served as combustion air could be consumed instead as makeup air for an exhaust system. Openings to the outdoors and infiltration might not be able to provide the required combustion air if also providing makeup air for exhaust fans or appliances such as clothes dryers.

14"

8"

Free area considered as 84 sq in. (75% of 112)

Metal Louver

14 in. × 8 in. = 112 sq in. of gross area

14"

8"

Free area considered as 28 sq in. (25% of 112)

Wood Louver

For SI: 1 inch = 25.4 mm, 1 square inch = 645 mm².

C-279

In calculating the free area of combustion air openings fitted with louvers, metal louvers obstruct only approximately 25 percent of the opening, wood louvers 75 percent. Prefabricated louvers may use the manufacturer's "free area" designation rather than the percentage method.

Code Text: *Where the space in which the appliance is located does not meet the criteria specified in Section M1702.1 (unconfined space), two permanent openings to adjacent spaces shall be provided so that the combined volume of all spaces meets the criterion. One opening shall be within 12 inches (305 mm) of the top and one within 12 inches (305 mm) of the bottom of the space. Each opening shall have a free area equal to a minimum of 1 square inch per 1,000 Btu/h (2.20 mm²/kW) input rating of all appliances installed within the space, but not less than 100 square inches (64 415 mm²).*

Discussion and Commentary: If the volume of the space in which the fuel-burning equipment is installed is greater than 50 cubic feet per 1,000 Btu/h of the aggregate input rating of the equipment, then infiltration of air into a residence normally provides sufficient air for combustion, ventilation and draft hood dilution.

One of the required openings is within 12 in. of the ceiling

Two permanent openings 1 sq in. per 1,000 Btu/h for each opening (Min area for each opening is 100 sq in.)

Furnace

Confined space

One of the required openings is within 12 in. of the floor

For SI: 1 inch = 25.4 mm, 1 square inch = 645 mm², 1 Btu/h = 0.2931 W

B-535

A dwelling built with unusually tight construction (plastic film vapor retarders, heavy caulking and sealing, etc.) may have less than 0.5 air changes per hour and will need to have air supplied from the outdoors or spaces freely communicating with the outdoors.

Code Text: *Where the space in which fuel-burning appliances are located does not meet the criterion for indoor air specified in Section M1702, outside combustion air shall be provided through openings or ducts. One opening shall be within 12 inches (305 mm) of the top of the enclosure, and one within 12 inches (305 mm) of the bottom of the enclosure. Where communicating with the outdoors by means of vertical ducts, each opening shall have a free area of at least 1 square inch per 4,000 Btu/h (550 mm²/kW) of total input rating of all appliances in the space.*

Discussion and Commentary: With two openings located as required, the warm air rises and is vented to the outdoors through the upper duct. The lower duct draws in colder, heavier air from the outdoors, which is heated by the appliances and rises toward the ceiling. Thus, the two openings create a natural draft that provides a constant source of combustion air.

Confined space

Top duct is within 12 in. of ceiling. Bottom duct is within 12 in. of floor.

Fuel-fired appliance

Outdoor air ducts 1 sq in. per 2,000 Btu/h each

For SI: 1 inch = 25.4 mm, 1 square inch = 645 mm², 1 Btu/h = 0.2931 W

B-536

Horizontal ducts must have a minimum opening area of 1 square inch per 2,000 Btu/h of total input ratings. Larger openings are mandated for horizontal ducts than for vertical ducts because the draft is somewhat less and the resistance is greater.

Code Text: For *combustion air obtained from an attic area, 1) the attic ventilation shall be sufficient to provide the required volume of combustion air; 2) the combustion air opening shall be provided with a metal sleeve extending from the appliance enclosure to at least 6 inches (152 mm) above the top of the ceiling joists and ceiling insulation; 3) an inlet air duct within an outlet air duct shall be an acceptable means of supplying attic combustion air to an appliance room provided that the inlet duct extends at least 12 inches (305 mm) above the top of the outlet duct in the attic space; and 4) the end of ducts that terminate in an attic shall not be screened.*

Discussion and Commentary: A primary concern when taking combustion air from an attic space is that insulation is usually present. The minimum 6-inch extension of the metal sleeve should ensure against insulation from entering the sleeve, while the prohibition against screening eliminates the potential for the screen to be blocked by loose insulation.

Note: Attic should be sufficiently vented for combustion air to be taken from attic.

For SI: 1 inch = 25.4 mm.

B-538

Where a ventilated crawl space is to be used to obtain combustion air, the free area of the openings between the crawl space and the outdoors must be at least twice the area required for the combustion air openings.

Code Text: *Manually operated dampers shall not be installed except in connectors or chimneys serving solid-fuel-burning appliances. Automatically operated dampers shall conform to UL 17* (Vent or Chimney Connector Dampers for Oil-fired Appliances) *and be installed in accordance with the terms of their listing and label. The installation shall prevent firing of the burner when the damper is not opened to a safe position.*

Discussion and Commentary: A damper installed in a solid fuel-burning appliance is usually opened manually during the process of starting a fire. If the damper is not opened, smoke will spill back into the room, alerting the user. For gas or liquid fuel-fired appliances, however, the user may not be aware of a partially or completely closed damper, and a hazardous condition could develop.

Automatic flue dampers and automatic vent dampers

B-550

Because damper failure can result in a hazardous condition, automatic dampers must be listed and labeled, and installed by a qualified installer so that the manufacturer's installation instructions and the terms of the listing are followed.

Code Text: *Vent and chimney connectors shall be installed in accordance with the manufacturer's installation instructions and within the space where the appliance is located. A chimney connector or vent connector shall not pass through any floor or ceiling. The horizontal run of an uninsulated connector to a natural draft chimney shall not exceed 75 percent of the height of the vertical portion of the chimney above the connector. The horizontal run of a listed connector to a natural draft chimney shall not exceed 100 percent of the height of the vertical portion of the chimney above the connector.*

Discussion and Commentary: A chimney connector is simply a pipe that connects a fuel-burning appliance to a chimney. A vent connector serves a similar purpose. The ideal chimney or vent configuration is a vertical system, even though it is not always practical. The connector length is limited because of flow resistance of the connector and heat loss through the connector wall.

TABLE M1803.3.4
CHIMNEY AND VENT CONNECTOR CLEARANCES
TO COMBUSTIBLE MATERIALS[a]

TYPE OF CONNECTOR	MINIMUM CLEARANCE (inches)
Single–wall metal pipe connectors:	
Oil and solid-fuel appliances	18
Oil appliances listed for use with Type L vents	9
Type L vent piping connectors:	
Oil and solid-fuel appliances	9
Oil appliances listed for use with Type L vents	3[b]

For SI: 1 inch = 25.4 mm.

a. These minimum clearances apply to unlisted single-wall chimney and vent connectors. Reduction of required clearances is permitted as in Table M1306.2.

b. When listed Type L vent piping is used, the clearance shall be in accordance with the vent listing.

Connectors must be provided with minimum clearances to combustibles. The clearances established in Table M1803.3.4 may be reduced if the combustible material is protected by one of the methods established in Table M1306.2.

Code Text: *Vents passing through a roof shall extend through flashing and terminate in accordance with the manufacturer's installation requirements. Decorative shrouds shall not be installed at the termination of vents except where such shrouds are listed and labeled for use with the specific venting system and are installed in accordance with the manufacturer's installation instructions.*

Discussion and Commentary: The vent system must be physically protected to prevent damage to the vent and to prevent combustibles from coming into contact with or being placed too close to the vents. This protection is usually provided by enclosing the vent in chases, shafts or cavities during the building construction. Physical protection is not required in the room or space where the vent originates (at the appliance connection) and is not required in such locations as attics that are not occupied or used for storage.

Vent termination through a roof   C-278

Direct-vent appliances, often referenced as "sealed combustion" appliances, have an enclosed combustion chamber and a conduit for bringing all combustion air directly from the outdoors. Rather than prescribe specific provisions, the code relies on the manufacturer's installation requirements when addressing vent terminals.

Code Text: *Vents for natural draft appliances shall terminate at least 5 feet (1524 mm) above the highest connected appliance outlet, and natural draft gas vents serving wall furnaces shall terminate at an elevation at least 12 feet (3658 mm) above the bottom of the furnace. Type L vents shall terminate with a listed and labeled cap in accordance with the vent manufacturer's installation instructions not less than 2 feet (610 mm) above the roof and not less than 2 feet (610 mm) above any portion of the building within 10 feet (3048 mm).*

Discussion and Commentary: Natural draft appliances must terminate at a high enough point to ensure that the venting system will perform properly under all conditions. Type L vents are used for oil-burning appliances and are designed for higher temperature flue gases than those encountered in gas-fired appliances.

For SI: 1 foot = 304.8 mm.

Type BW gas vent

B-539

Unless the vent is an integral part of a listed and labeled appliance, an individual vent for a single appliance must have a cross-sectional area equivalent to or greater than that of the connector to the appliance, with a minimum area of 7 square inches.

Code Text: *In addition to the requirements of the IRC, the installation of boilers shall conform to the manufacturer's instructions. The manufacturer's rating data, the nameplate and operating instructions of a permanent type shall be attached to the boiler. Boilers shall have all controls set, adjusted and tested by the installer. Boilers shall be equipped with pressure-relief valves with minimum rated capacities for the equipment served. Pressure-relief valves shall be set at the maximum rating of the boiler. Discharge shall be piped to drains by gravity to within 18 inches (457 mm) of the floor or to an open receptor.*

Discussion and Commentary: A boiler is defined as a self-contained appliance from which hot water is circulated for heating purposes and then returned to the appliance. It operates at a maximum water pressure of 160 psig and at a maximum water temperature of 250°F. To ensure a safe and proper installation, boilers must be constructed to the appropriate standards and installed in accordance with the manufacturer's instructions.

The installer must provide a complete set of boiler operating instructions for use by the owner. It is wise to attach such instructions directly to the boiler for immediate access. In additon, a complete control diagram must be furnished and located with the operating instructions.

Code Text: *Hot water boilers shall be provided with expansion tanks. Nonpressurized expansion tanks shall be securely fastened to the structure or boiler and supported to carry twice the weight of the tank filled with water. Pressurized expansion tanks shall be consistent with the volume and capacity of the system. Tanks shall be capable of withstanding a hydrostatic test pressure of two and one-half times the allowable working pressure of the system.*

Discussion and Commentary: Inactive hydronic heating systems are filled with water at room temperature. When the boiler is fired, the temperature of the water increases, causing the volume of the water to increase. Without an expansion tank, hydrostatic pressure would rapidly build within the boiler and piping system, activating the pressure-relief valve. An expansion tank provides space for the water to expand as it is heated and keeps the water pressure in the normal operating range while the boiler is operating.

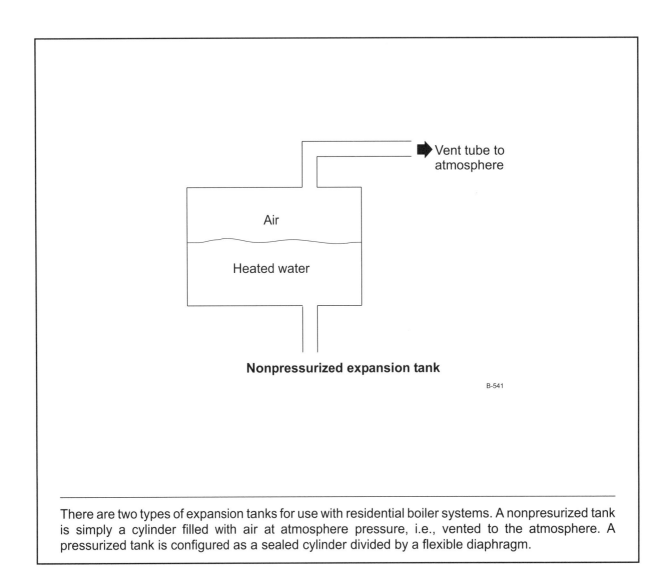

There are two types of expansion tanks for use with residential boiler systems. A nonpresurized tank is simply a cylinder filled with air at atmosphere pressure, i.e., vented to the atmosphere. A pressurized tank is configured as a sealed cylinder divided by a flexible diaphragm.

Code Text: *Hydronic piping shall conform to Table M2101.1. Approved piping, valves, fittings and connections shall be installed in accordance with the manufacturer's installation instructions. The potable water system shall be protected from backflow in accordance with the provisions listed in Section P2902. Piping shall be installed so that piping, connections and equipment shall not be subjected to excessive strains or stresses. Provisions shall be made to compensate for expansion, contraction, shrinkage and structural settlement.*

Discussion and Commentary: A variety of piping materials is acceptable for use in hydronic piping systems. Because of its high resistance to corrosion and strength, crosslinked polyethylene (PEX) tubing is the only type of material that should be used underground. Plastic piping is only allowed in hydronic systems where the maximum water temperature will not exceed 180°F. Higher temperatures can weaken the pipe and may lead to bursting.

TABLE M2101.1
HYDRONIC PIPING MATERIALS

MATERIAL	USE CODE[a]	STANDARD[b]	JOINTS	NOTES
Brass pipe	1	ASTM B 43	Brazed, welded, threaded, mechanical and flanged fittings	
Brass tubing	1	ASTM B 135	Brazed, soldered and mechanical fittings	
Chlorinated poly (vinyl chloride) (CPVC) pipe and tubing	1, 2, 3	ASTM D 2846	Solvent cement joints, compression joints and threaded adapters	
Copper pipe	1	ASTM B 42, B 302	Brazed, soldered and mechanical fittings threaded, welded and flanged	
Copper tubing (type K, L or M)	1, 2	ASTM B 75, B 88, B 251, B 306	Brazed, soldered and flared mechanical fittings	Joints embedded in concrete
Cross-linked polyethylene (PEX)	1, 2, 3	ASTM F 876, F 877	(See PEX fittings)	Install in accordance with manufacturer's instructions.
Cross-linked polyethylene/aluminum/cross-linked polyethylene-(PEX-AL-PEX) pressure pipe	1, 2	ASTM F 1281 or CAN/ CSA B137.10	Mechanical, crimp/insert	Install in accordance with manufacturer's instructions.
PEX Fittings		ASTM F 1807 ASTM F 1960 ASTM F 2098	Copper-crimp/insert fittings, cold expansion fittings, stainless steel clamp, insert fittings	Install in accordance with manufacturer's instructions
Plastic fittings PEX		ASTM F 1807		
Polybutylene (PB) pipe and tubing	1, 2, 3	ASTM D 3309	Heat-fusion, crimp/insert and compression	Joints in concrete shall be heat-fused.
Polyethylene (PE) pipe, tubing and fittings (for ground source heat pump loop systems)	1, 2, 4	ASTM D 2513; ASTM D 3350; ASTM D 2513; ASTM D 3035; ASTM D 2447; ASTM D 2683; ASTM F 1055; ASTM D 2837; ASTM D 3350; ASTM D 1693	Heat-fusion	
Polyproplylene (PP)	1, 2, 3	ISO 15874 ASTM F 2389	Heat-fusion joints, mechanical fittings, threaded adapters, compression joints	
Soldering fluxes	1	ASTM B 813	Copper tube joints	
Steel pipe	1, 2	ASTM A 53; A 106	Brazed, welded, threaded, flanged and mechanical fittings	Joints in concrete shall be welded. Galvanized pipe shall not be welded or brazed.
Steel tubing	1	ASTM A 254	Mechanical fittings, welded	

For SI: °C = [(°F)-32]/1.8.
a. Use code:
    1. Above ground.
    2. Embedded in radiant systems.
    3. Temperatures below 180°F only.
    4. Low temperature (below 130°F) applications only.
b. Standards as listed in Chapter 43.

Pipe openings through concrete or masonry must be sleeved to compensate for expansion of the piping due to hot water. As the flow of water causes the pipe to move and expand when the boiler cycles, sufficient clearance is necessary to avoid stress damage to the piping.

Code Text: *Supply tanks shall be listed and labeled and shall conform to UL 58 for underground tanks and UL 80 for inside tanks. The maximum amount of fuel oil stored above ground or inside of a building shall be 660 gallons (2498 L). Supply tanks (located within a building) larger than 10 gallons (38 L) shall be placed not less than 5 feet (1524 mm) from any fire or flame either within or external to any fuel-burning appliance. Tanks installed outside above ground shall be a minimum of 5 feet (1524 mm) from an adjoining property line. Such tanks shall be suitably protected from the weather and from physical damage.*

Discussion and Commentary: The cross connection of two supply tanks on the same level is permitted if the total capacity of the system does not exceed 660 gallons. This procedure, which uses gravity flow from one tank to the other, provides an option for installation purposes.

Underground tanks must be at least 1 foot from the wall of a basement, pit, or property line to ensure that settlement of surrounding materials does not place a stress on the tank or piping. Tanks must also be covered with at least 12 inches of earth for physical protection.

Code Text: *Solar energy collectors, controls, dampers, fans, blowers and pumps shall be accessible for inspection, maintenance, repair and replacement. Where mounted on or above the roof coverings, the collectors and supporting structure shall be constructed of noncombustible materials or fire-retardant-treated wood equivalent to that required for the roof construction. Roof and wall penetrations shall be flashed and sealed to prevent entry of water, rodents and insects in accordance with Chapter 9.*

Discussion and Commentary: Developed due to an emphasis placed on conservation of our natural resources, solar energy systems allow for an alternative method in providing space heating or cooling, domestic water heating, and pool heating. The effectiveness of these systems depends on the size of the solar array, the energy requirements of the dwelling and the availability of sunlight.

VACUUM BREAKER
AIR VENT
TEMPERATURE SENSOR
COLLECTORS
FLOW
COLLECTOR LOOP
DIFFERENTIAL THERMOSTAT
THERMOMETER
LEVEL GAUGE
TEMPERATURE SENSOR
WATER LEVEL
FILL VALVE
VALVES
TEMPERATURE AND PRESSURE-RELIEF VALVE
HOUSE WATER LOOP
MIXING VALVE
VALVE
AUXILIARY HEATING ELEMENTS
FLOW
DRAIN
DRAIN
STORAGE TANKS
VALVES
HEAT EXCHANGER COILS
PUMP
THERMOMETER
B-543

Some water heaters and boilers are designed to operate with "solar assistance." These units heat water by ordinary methods and have additional connections for solar preheating. Preheated water reduces the amount of energy that is consumed with ordinary heating methods.

Study Session 12
IRC Chapters 17, 18, 20, 21, 22 and 23

1. In a building of ordinary tightness, the combustion air for fuel-burning appliances may be obtained from infiltration where the volume of the room or space is a minimum _____50_____ cubic feet per 1,000 Btu/h.

   a. 20   b. 25

   c. 40   d. 50

   Reference M1701.1.1

2. In general, combustion air is permitted to be taken from which one of the following spaces?

   a. sleeping room   b. toilet room

   c. kitchen   d. bathroom

   Reference M1701.4

3. When determining the free area for combustion air openings, a free area of _75%_ of the gross area is to be used for metal louvers unless the free area is specified by the louver manufacturer.

   a. 25 percent   b. 40 percent

   c. 60 percent   d. 75 percent

   Reference M1701.5

4. The gross area of a wood louver used to cover a combustion air opening is 240 square inches. For the purpose of determining combustion air, what is the net free area?

   (a.) 60 square inches   b. 96 square inches

   c. 144 square inches   d. 180 square inches

   Reference <u>M 1701.5</u>

5. Where normal infiltration is not adequate to provide the necessary combustion air, a minimum of two permanent openings shall be provided, with the openings located a maximum of ___12___ from the top and bottom of the space.

   a. 4 inches   b. 6 inches

   c. 8 inches   (d.) 12 inches

   Reference <u>M1702.2</u>

6. The two combustion air openings required to serve a confined space shall each have a minimum free area of ___100___ square inches.

   a. 60   b. 80

   (c.) 100   d. 120

   Reference <u>M 1702.2</u>

7. Where all air is taken from inside the building, the net free area for each of two required combustion air openings shall be a minimum of 1 square inch per ___1000___ Btu/h input rating of all appliances in a confined space.

   (a.) 1,000   b. 2,000

   c. 3,000   d. 4,000

   Reference <u>M 1702.2</u>

8. Where combustion air is taken from the outdoors by means of vertical ducts, each opening shall have a free area of at least 1 square inch per ___4,000___ Btu/h of total input rating of all appliances in the space.

   a. 1,000   b. 2,000

   c. 3,000   (d.) 4,000

   Reference <u>M1703.2.1</u>

9. Where combustion air is taken from the outdoors by means of horizontal ducts, each opening shall have a free area of at least 1 square inch per _2,000_ Btu/h of total input rating of all appliances in the space.

   a. 1,000　　　　　　　　　　ⓑ 2,000

   c. 3,000　　　　　　　　　　d. 4,000

   Reference M 1703.2.1

10. Where combustion air is taken from the outdoors, what is the minimum cross-sectional dimension permitted for a rectangular combustion air duct?

    ⓐ 3 inches　　　　　　　　b 4 inches

    c. 6 inches　　　　　　　　d. 8 inches

    Reference M1703.2.1

11. Outside combustion air openings shall be covered with a corrosion-resistant screen or equivalent protection having a minimum opening size of _1/4_ inch and a maximum opening size of _1/2_ inch.

    a. $^1/_8$, $^1/_4$　　　　　　　　ⓑ $^1/_4$, $^1/_2$

    c. $^1/_2$, $^3/_4$　　　　　　　　d. $^1/_2$, 1

    Reference M1703.5

12. Where passing through a wall or partition, a single-wall metal pipe used as a vent connector to connect fuel-burning appliances to a vent shall be guarded by a ventilated metal thimble a minimum of _4_ inches larger in diameter than the vent connector.

    a. 2　　　　　　　　　　　　b. 3

    ⓒ 4　　　　　　　　　　　　d. 6

    Reference M1803.3.1 #2

13. For other than direct vent appliances, the vent termination for a mechanical draft system shall be mounted a minimum of _3_ feet horizontally from an oil tank or gas meter.

    ⓐ 3　　　　　　　　　　　　b. 4

    c. 15　　　　　　　　　　　d. 10

    Reference M1804.2.6 #5

14. Where combustion air is taken from the attic, the combustion air opening shall be provided with a metal sleeve that extends from the appliance enclosure to a minimum height of ___6___ above the top of the ceiling joists and insulation.

    a. 3 inches   b. 4 inches

    c. 6 inches   d. 12 inches

    Reference M1703.3 #2

15. Where combustion air is taken from under-floor areas, the required free opening area to the outside shall be a minimum of ___twice___ the required combustion air opening.

    a. one-half   b. 60 percent

    c. twice   d. three times

    Reference M1703.4

16. Where an 8-inch-diameter single wall metal pipe is used as a vent connector, it shall have a minimum galvanized sheet metal gage of ___24___.

    a. No. 22   b. No. 24

    c. No. 26   d. No. 28

    Reference Table M1803.2

17. Vent and chimney connectors shall have a minimum slope of ___1/4___ rise per foot of run.

    a. $^1/_8$ inch   b. $^1/_4$ inch

    c. $^3/_8$ inch   d. $^1/_2$ inch

    Reference M1803.3

18. The horizontal run of a listed vent connector to a natural draft chimney shall be a maximum of ___100%___ of the height of the vertical portion of the chimney above the connector.

    a. 50 percent   b. 75 percent

    c. 100 percent   d. 125 percent

    Reference M1803.2

19. Unless modified by Table M1306.2, what is the minimum required clearance between an unlisted Type L vent piping connector serving a solid fuel appliance and any combustible materials?

    a. 3 inches   b. 6 inches

    c. 9 inches   d. 18 inches

Reference _Table M 1803.3.4_

20. A natural gas draft vent serving a wall furnace shall terminate a minimum height of _12"_ above the bottom of the furnace.

    a. 5 feet   b. 8 feet

    c. 10 feet   d. 12 feet

Reference _M1504.2.3_

21. Chimney flues connected to more than one appliance shall be sized based on the area of the largest connector plus _50_ of the areas of additional chimney connectors.

    a. 50 percent   b. 75 percent

    c. 100 percent   d. 125 percent

Reference _M1805.3_

22. The pressurized expansion tank for a boiler shall be capable of withstanding a minimum hydrostatic test pressure _2½_ the allowable working pressure of the system.

    a. equal to   b. twice

    c. two and one-half times   d. three times

Reference _M2003.1.1_

23. Hydronic piping to be embedded in concrete shall be tested by applying a minimum hydrostatic pressure of _100_ for a minimum time period of _30_.

    a. 50 psi, 15 minutes   b. 50 psi, 30 minutes

    c. 100 psi, 15 minutes   d. 100 psi, 30 minutes

Reference _M2103.3_

24. Fill piping for oil tanks shall terminate outside of a building a minimum distance of ___2 ft___ from any building opening at the same or lower level.

    a. 12 inches   (b.) 2 feet

    c. 5 feet   d. 10 feet

    Reference ___M2203 .3___

25. Solar energy systems shall be equipped with means to limit the maximum water temperature of the system fluid entering or exchanging heat with any pressurized vessel inside the dwelling to a maximum of ___180°___.

    a. 120°F   b. 150°F

    (c.) 180°F   d. 210°F

    Reference ___M2301. 2q___

26. Fuel-fired water heaters shall not be installed in a ___closet___

    a. bathroom   b. bedroom

    c. kitchen   (d.) storage closet

    Reference ___M2005. 2___

27. PEX tubing used for hydronic piping shall be supported at maximum intervals of _____ when installed in a horizontal position.

    (a.) 2 feet, 8 inches   b. 4 feet

    c. 6 feet, 6 inches   d. 8 feet

    Reference ___Table M2101.9___

28. Hydronic piping shall be tested hydrostatically at a minimum pressure of ___100___ psi for a minimum of ___15___ minutes.

    a. 5, 15   b. 10, 30

    c. 60, 30   (d.) 100, 15

    Reference ___M2101 .10___

29. A fuel oil storage tank installed inside of a building shall have a maximum capacity of
    _600_ gallons.

    a. 330   b. 660

    c. 1,000   d. 2,000

    Reference _M2201.2_

30. Pressure at the fuel oil supply inlet to an appliance shall be a maximum of _3_ psi.

    a. 3   b. 5

    c. 10   d. 15

    Reference _M2204.3_

31. Where an inlet air duct within an outlet air duct is used to obtain combustion air from an
    attic area, the inlet duct shall extend a minimum of _12_ inches above the top of
    the outlet duct.

    a. 3   b. 4

    c. 6   d. 12

    Reference _M1703.3 #3_

32. Vents for natural draft appliances shall terminate a minimum of _5_ feet above
    the highest connected appliance outlet.

    a. 3   b. 5

    c. 10   d. 12

    Reference _M1804.23_

33. Discharge piping beyond the pressure relief valve of a boiler shall terminate a maximum
    of _18_ inch(es) above the floor or to an open receptor.

    a. 1   b. 6

    c. 12   d. 18

    Reference _M2002.4_

34. Fuel oil tanks installed outside above ground shall be located a minimum of ___S___ feet from an adjoining property line.

    (a.) 5   b. 10

    c. 15   d. 20

    Reference ___M2201.2.2___

35. Vent piping for a fuel oil tank shall have a minimum pipe size of ___1¼___ inch/es.

    a. $^3/_4$ b. 1

    (c.) $1^1/_4$ d. $1^1/_2$

    Reference ___M2203.2___

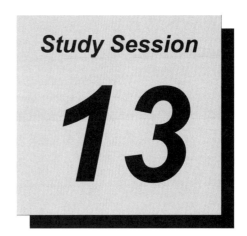

Study Session

13

2006 IRC Chapter 24
Fuel Gas

OBJECTIVE: To provide an understanding of the general provisions for fuel gas piping systems, fuel gas utilization equipment and related accessories, venting systems and combustion air configurations.

REFERENCE: Chapter 24, 2006 *International Residential Code*

KEY POINTS:
G2406.2 (303.3)

• In what locations are the installation of fuel-fired appliances prohibited? What concerns are applicable to outdoor installations? G2406.3 (303.6)

G2407.5

• How shall combustion, ventilation and dilution air be provided in buildings of unusually tight construction? G2407.1 (304.1)

• How shall openings be provided for a confined space where all combustion air is taken from inside the building? From outside the building? From both inside and outside the

G2407.7 — building? In what manner are combustion air ducts regulated? G 2407 11 (304.11)

• Where must equipment and appliances having an ignition source be located in a private garage? G2408.2 (305.3)

T G2409 2 • How may the required clearances between combustible materials and chimneys; vents; fuel gas appliances, devices, and equipment; and kitchen exhaust equipment be reduced?

G2412.5(401.5) • What identifying mark is to be used on exposed gas piping? What type of pipe material is exempt from the labeling requirement? Same (Steel piping)

G2413.2 (402.2) • How is the maximum gas demand determined? How is the gas piping to be sized? G2413.3 (402.3)

• What materials are acceptable for use as gas piping? When are materials allowed to be used again? G2412.2 (403.2)

G2414.7 (403.7) • When is a protective coating mandated for gas piping? How are field threading operations

G2414 9.2 (403.9.2) regulated? What methods are acceptable for pipe joints? G2414.10.1 (403.12.1)

G2415.1 —> • Where is gas piping prohibited? What are the limitations for gas piping in concealed

G2415.3(404.3) —> locations? How must such piping be protected where installed through a foundation or basement wall? G2415.4 (404.4)

× G2415.5(404.5) —> • What methods of protection against physical damage are identified for gas piping? How

G2415.7(404.7) ⇒ should pipe be protected against corrosion? What is the minimum required burial depth? G2415.9 (404.9)

G2416.2 (405.2) ⇒ • How shall changes in direction be made for metallic pipe? For plastic pipe? G2416.3 (405.3)

KEY POINTS:
(Cont'd)

- What are the general conditions for the inspection and testing of fuel gas piping?
- Where must gas shutoff valves be located? Where must appliance shutoff valves be installed? Equipment shutoff valves for a fireplace?
- What is the maximum permitted length for an appliance fuel connector serving a range or domestic clothes dryer? Serving other appliances?
- At what maximum intervals must fuel gas piping be supported?
- What methods are available for terminating a gas vent? Where are decorative shrouds permitted?
- Where must chimneys serving gas-fired appliances terminate? At what minimum height must gas vents terminate above a roof? Where must a Type B or Type L gas vent terminate? A Type B-W vent? A single-wall metal pipe?
- How are vented wall furnaces to be located? Floor furnaces? Unit heaters?
- What are prohibited sources for outside or return air for a forced-air heating system?

Code Text: *Makeup air requirements for the operation of exhaust fans, kitchen ventilation systems, clothes dryers and fireplaces shall be considered in determining the adequacy of a space to provide combustion air requirements.*

Discussion and Commentary: The introduction of makeup air is critical to the proper operation of all fuel-burning appliances located in areas subject to the effects of exhaust systems. Too little makeup air will cause excessive negative pressures to develop, not only reducing the exhaust airflow but also interfering with appliance venting and potentially causing combustion byproducts to discharge into the building.

For SI: 1 cubic foot per minute = 0.0004719 m³/s

DEPLETION OF COMBUSTION AIR

Natural draft appliances also compete among themselves for combustion air. The appliance venting system that produces the strongest draft, such as a solid fuel appliance or a fireplace, can cause combustion air shortages for the appliance venting systems that produce a weaker draft.

Code Text: *In concealed locations, where piping other than black or galvanized steel is installed through holes or notches in wood studs, joists, rafters or similar members less than 1.5 inches (38 mm) from the nearest edge of the member, the pipe shall be protected by shield plates. Shield plates shall be a minimum of $^1/_{16}$-inch-thick (1.6 mm) steel, shall cover the area of the pipe where the member is notched or bored, and shall extend a minimum of 4 inches (102 mm) above sole plates, below top plates and to each side of a stud, joist or rafter.*

Discussion and Commentary: The provision is intended to minimize the possibility that nails or screws will be driven into the gas pipe or tube. Because nails and screws sometimes miss the stud, rafter, joist, or sole or top plates, the shield plates must extend beyond the stud itself.

For SI: 1 inch = 25.4 mm.

Pipe protection

C-277

Piping must not be installed in any solid concrete or masonry floor construction. The potential for pipe damage due to slab settlement, cracking or the corrosive action of the floor material makes it imperative that the gas piping be installed in channels or casings, or similarly protected.

Code Text: *Piping shall be supported with pipe hooks, metal pipe straps, bands, brackets or hangers suitable for the size of piping, of adequate strength and quality, and located at intervals so as to prevent or damp out excessive vibration. Piping shall be anchored to prevent undue strains on connected equipment and shall not be supported by other piping. Supports, hangers, and anchors shall be installed so as not to interfere with the free expansion and contraction of the piping between anchors.*

Discussion and Commentary: Suspended gas piping must be supported properly by hangers and straps to prevent straining or bending, which could cause leaks and create a hazardous condition. The supporting material should be compatible with the piping material to avoid electrolytic action.

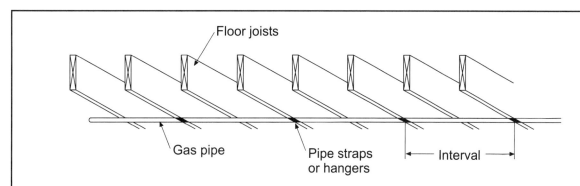

Gas pipe supports

B-544

TABLE G2424.1
SUPPORT OF PIPING

STEEL PIPE, NOMINAL SIZE OF PIPE (inches)	SPACING OF SUPPORTS (feet)	NOMINAL SIZE OF TUBING SMOOTH-WALL (inch O.D.)	SPACING OF SUPPORTS (feet)
$^1/_2$	6	$^1/_2$	4
$^3/_4$ or 1	8	$^5/_8$ or $^3/_4$	6
$1^1/_4$ or larger (horizontal)	10	$^7/_8$ or 1 (horizontal)	8
$1^1/_4$ or larger (vertical)	Every floor level	1 or Larger (vertical)	Every floor level

For SI: 1 inch = 25.4 mm, 1 foot = 304.8 mm.

Buried piping should be laid on a solid bed. This could be 4 to 6 inches of sand, gravel or tamped earth. The backfill should not contain any large rocks or frozen chunks of earth that might impose loads that could damage the pipe.

Code Text: *Where a sediment trap is not incorporated as a part of the gas utilization equipment, a sediment trap shall be installed downstream of the equipment shut-off valve as close to the inlet of the equipment as practical. The sediment trap shall be either a tee fitting with a capped nipple in the bottom opening of the run of the tee or other device approved as an effective sediment trap. Illuminating appliances, ranges, clothes dryers and outdoor grills need not be so equipped.*

Discussion and Commentary: Sediment traps cause the gas flow to change direction 90 degrees at the sediment collection point, thus causing solid or liquid contaminants to drop out of the gas flow. Sediment traps prevent debris from entering the gas controls where such contaminants could cause hazardous appliance malfunction.

Appliance connection

Sediment trap

B-545

Most appliance and equipment manufacturers mandate the installation of a sediment trap. The code requires compliance with manufacturers' instructions and listings, so it is consistent that the installation of a trap is a basic code provision.

Code Text: *Each appliance shall be provided with a shutoff valve separate from the appliance. The shutoff valve shall be located in the same room as the appliance, not further than 6 feet (1829 mm) from the appliance, and shall be installed upstream from the union, connector or quick disconnect device it serves. Such shutoff valves shall be provided with access.* See exception for vented decorative appliances.

Discussion and Commentary: The fuel gas connection to every gas appliance must be provided with an individual gas shutoff valve to permit maintenance, repair, replacement or temporary disconnection. The shutoff valve must be adjacent to the appliance, conspicuously located, and within reach to permit it to be easily located and operated in the event of an emergency. For example, a gas cock serving an appliance on the first floor could not be located in the basement.

1
Lubricated flat-head
plug cock

2
Lubricated square-head
plug cock

3
Lubricated lever-handle
plug cock

Typical gas shutoff valves

B-546

Valves are more likely to leak than piping and fittings, and therefore valve locations are restricted. Locating fuel gas valves in concealed spaces unduly restricts access to the valve in the event of an emergency or servicing the equipment to which the valve is attached.

Code Text: *Connectors shall have an overall length not to exceed 3 feet (914 mm), except for range and domestic clothes dryer connectors, which shall not exceed 6 feet (1829 mm) in length. Only one connector shall be used for each appliance.* See exception for rigid metallic piping. *Connectors shall not be concealed within, or extended through, walls, floors, partitions, ceilings or appliance housings.* See exception for fireplace inserts. *A shutoff valve not less than the nominal size of the connector shall be installed ahead of the connector.*

Discussion and Commentary: Flexible (semirigid) appliance fuel connectors are primarily for use with cooking ranges and clothes dryers where the gas connection is located behind the appliance and some degree of flexibility is necessary to facilitate the hook-up. Longer lengths for range and dryer installations allow the connector to be coiled or otherwise arranged to allow limited movement of the appliance without stressing the connector. Rigid connectors are not practical for such movable appliances.

Gas pipe

Shutoff valve

3'* max

*6 ft for ranges and clothes dryers

For SI: 1 foot = 304.8 mm.

B-547

Flexible connectors, usually constructed of brass or stainless steel, are labeled with tags of metal rings placed over the tubing. The manufacturer's installation instructions must be followed to protect the connector from damage and to prevent leakage.

Code Text: *The type of venting system to be used shall be in accordance with Table G2427.4. Plastic piping used for venting appliances listed for use with such venting materials shall be approved. Special gas vents shall be listed and installed in accordance with the special gas vent manufacturer's installation instructions.*

Discussion and Commentary: Various types of vents are listed in Table G2427.4 with the corresponding types of appliances that can be served by the vent.

TABLE G2427.4
TYPE OF VENTING SYSTEM TO BE USED

APPLIANCES	TYPE OF VENTING SYSTEM
Listed Category I appliances Listed appliances equipped with draft hood Appliances listed for use with Type B gas vent	Type B gas vent (Section G2427.6) Chimney (Section G2427.5) Single-wall metal pipe (Section G2427.7) Listed chimney lining system for gas venting (Section G2427.5.2) Special gas vent listed for these appliances (Section G2427.4.2)
Listed vented wall furnaces	Type B-W gas vent (Sections G2427.6, G2436)
Category II appliances	As specified or furnished by manufacturers of listed appliances (Sections G2427.4.1, G2427.4.2)
Category III appliances	As specified or furnished by manufacturers of listed appliances (Sections G2427.4.1, G2427.4.2)
Category IV appliances	As specified or furnished by manufacturers of listed appliances (Sections G2427.4.1, G2427.4.2)
Unlisted appliances	Chimney (Section G2427.5)
Decorative appliances in vented fireplaces	Chimney
Direct-vent appliances	See Section G2427.2.1
Appliances with integral vent	See Section G2427.2.2

The type and size of the vent must be as dictated by the manufacturer's installation instructions for the appliance. The design and installation instructions prevented by the vent manufacturer must also be consulted when designing a vent system for any particular application.

Code Text: *Factory-built chimneys shall be installed in accordance with the manufacturers' instructions. Factory-built chimneys used to vent appliances that operate at positive vent pressure shall be listed for such application.*

Discussion and Commentary: All prefabricated chimney systems must bear the label of an approved agency. The label states information such as the type of appliance the chimney was tested for use with, a reference to the manufacturer's installation instructions and the minimum required clearances to combustibles. The manufacturer's instructions contain sizing criteria and the requirements for every aspect of a factory-built chimney installation, which include component assembly, clearances to combustibles, support, terminations, connections, protection from damage and fireblocking.

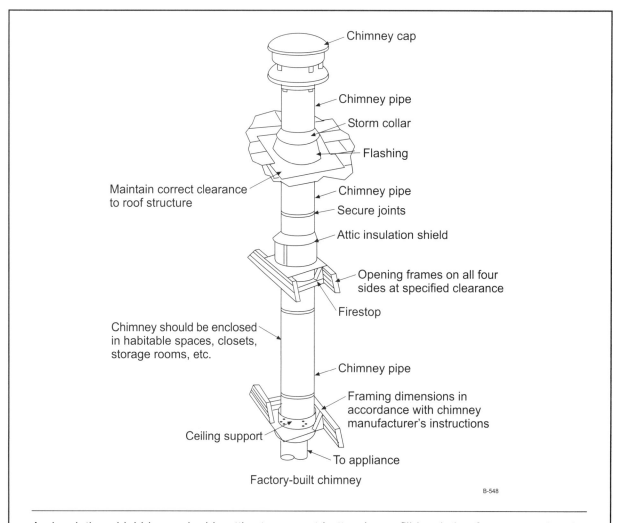

Chimney cap
Chimney pipe
Storm collar
Flashing
Maintain correct clearance to roof structure
Chimney pipe
Secure joints
Attic insulation shield
Opening frames on all four sides at specified clearance
Firestop
Chimney should be enclosed in habitable spaces, closets, storage rooms, etc.
Chimney pipe
Framing dimensions in accordance with chimney manufacturer's instructions
Ceiling support
To appliance
Factory-built chimney

B-548

An insulation shield is required in attics to prevent batt or loose-fill insulation from contacting the chimney. The attic shield maintains the airspace clearance between the chimney and the insulation material to reduce the chimney temperatures transmited from the chimney surfaces.

Code Text: *Gas vents that are 12 inches (305 mm) or less in size and located not less than 8 feet (2438 mm) from a vertical wall or similar obstruction shall terminate above the roof in accordance with Figure G2427.6.4. Gas vents that are over 12 inches (305 mm) in size or are located less than 8 feet (2438 mm) from a vertical wall or similar obstruction shall terminate not less than 2 feet (610 mm) above the highest point where they pass through the roof and not less than 2 feet (610 mm) above any portion of a building within 10 feet (3048 mm) horizontally.* See other requirements for direct-vent systems, appliances with integral vents, and appliances utilizing mechanical draft systems.

Discussion and Commentary: Figure G2427.6.4 indicates the need for a greater vent height above the roof as the roof approaches being a vertical surface. The greater the roof pitch, the greater effect of the wind striking the roof surface. However, the height should be limited as much as possible because of several concerns. Vent piping exposed to the outdoors encourages condensation in cold weather, could require guy wires or braces depending on height and is considered aesthetically unattractive.

Roof pitch	H (min) ft	m
Flat to 6/12	1.0	0.30
6/12 to 7/12	1.25	0.38
Over 7/12 to 8/12	1.5	0.46
Over 8/12 to 9/12	2.0	0.61
Over 9/12 to 10/12	2.5	0.76
Over 10/12 to 11/12	3.25	0.99
Over 11/12 to 12/12	4.0	1.22
Over 12/12 to 14/12	5.0	1.52
Over 14/12 to 16/12	6.0	1.83
Over 16/12 to 18/12	7.0	2.13
Over 18/12 to 20/12	7.5	2.27
Over 20/12 to 21/12	8.0	2.44

Gas vent termination

B-549

It is a common misapplication for code users to apply chimney termination height requirements to vents, thereby causing vents to extend above roofs much higher than required in many cases. For roof pitches up to a 6/12 pitch, only a 1-foot extension is necessary.

Code Text: *Where two or more gas appliances are vented through a common vent connector or vent manifold, the common vent connector or vent manifold shall be located at the highest level consistent with available headroom and the required clearance to combustible materials and shall be sized in accordance with Section G2428 or other approved engineering methods.*

Discussion and Commentary: A manifold is defined as a vent or connector that is a lateral (horizontal) extension of the lower end of a common vent. Manifolds must be installed as high as possible to provide for the most appliance connector rise, which is always beneficial for venting performance. Connector rise takes advantage of the energy of hot gases discharging directly from the appliance and develops flow velocity and draft in addition to that produced by the common vent.

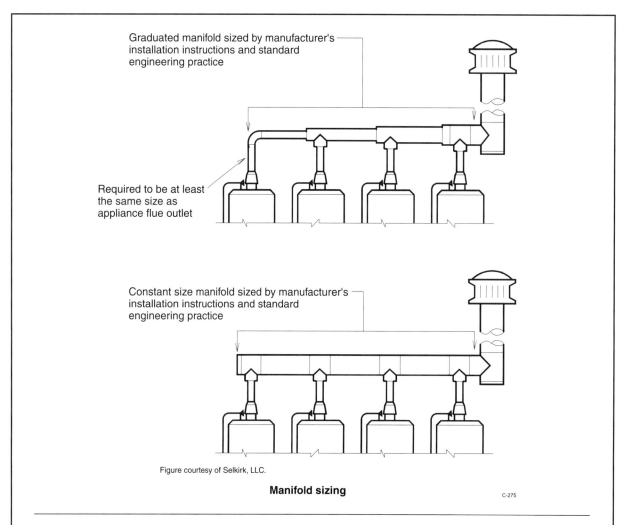

Graduated manifold sized by manufacturer's installation instructions and standard engineering practice

Required to be at least the same size as appliance flue outlet

Constant size manifold sized by manufacturer's installation instructions and standard engineering practice

Figure courtesy of Selkirk, LLC.

Manifold sizing

C-275

An option is available where there are only two draft-hood-equipped appliances. This alternative method is intended for use only with the alternative sizing method of Section G2427.6.9.1 and is not compatible with Section G2428.

Code Text: *An automatically operated vent damper shall be of a listed type.*

Discussion and Commentary: The installation of an automatic damper must be in strict accordance with the manufacturer's installation instructions. A malfunctioning or improperly installed flue damper could cause the appliance to malfunction and discharge the products of combustion directly into the building interior. Because automatic damper failure could result in a hazardous condition, automatic dampers are required to be listed and labeled. Automatic flue dampers are energy-saving devices designed to close off or restrict an appliance flue passageway when the appliance is not operating and is in its "off" cycle. Such devices save energy by trapping residual heat in a heat exchanger after the burners shut off and by preventing the escape of conditioned room air up the vent. Thus, the appliance efficiency can be boosted, and building air infiltration can be reduced.

Automatic flue dampers and automatic vent dampers

B-550

Manual dampers can only be used with solid fuel-burning appliances, in that the user will be aware of a closed damper due to the smoke. In the case of a gas-fired appliance, the user may not know of a partially or completely closed damper, and a hazardous condition could develop.

Topic: Flame Safeguard Device

Reference: IRC G2432.2

Category: Fuel Gas

Subject: Decorative Appliances for Fireplaces

Code Text: *Decorative appliances for installation in approved solid fuel-burning fireplaces, with the exception of those tested in accordance with ANSI Z21.84, shall utilize a direct ignition device, an ignitor or a pilot flame to ignite the fuel at the main burner, and shall be equipped with a flame safeguard device. The flame safeguard device shall automatically shut off the fuel supply to a main burner or group of burners when the means of ignition of such burners becomes inoperative.*

Discussion and Commentary: The typical flame safeguard device used with gas log set appliances is a combination manual control valve, pressure regulator, pilot feed and magnetic pilot safety mechanism with a thermocouple generator. If the pilot flame is extinguished, the drop in thermocouple output voltage will cause the control valve to "lock out" in the closed position, thereby preventing the flow of gas to the main burner and the pilot burner.

Gas log with label and flame safeguard device

Gas log installation

B-551

Gas-fired decorative log sets and log lighters are examples of accessory appliances designed for installation in solid fuel-burning fireplaces. Gas log sets provide some radiant heat; however, their primary function is to create an aesthetically pleasant simulation of a wood log fire.

Topic: Support and Clearance	**Category:** Fuel Gas
Reference: IRC G2444.2, G2444.4	**Subject:** Unit Heaters

Code Text: *Suspended-type unit heaters shall be supported by elements that are designed and constructed to accommodate the weight and dynamic loads. Hangers and brackets shall be of noncombustible material. Suspended-type unit heaters shall be installed with clearances to combustible materials of not less than 18 inches (457 mm) at the sides, 12 inches (305 mm) at the bottom and 6 inches (152 mm) above the top where the unit heater has an internal draft hood or 1 inch (25 mm) above the top of the sloping side of the vertical draft hood.*

Discussion and Commentary: As with all suspended fuel-fired appliances, a support failure could result in a fire, explosion or injury to building occupants. The supports themselves must be properly designed. Equally important are the structural members to which the supports are attached, such as rafters, beams, joists and purlins. All brackets, pipes, rods, angle iron, structural members and fasteners must be designed for the dead and dynamic loads.

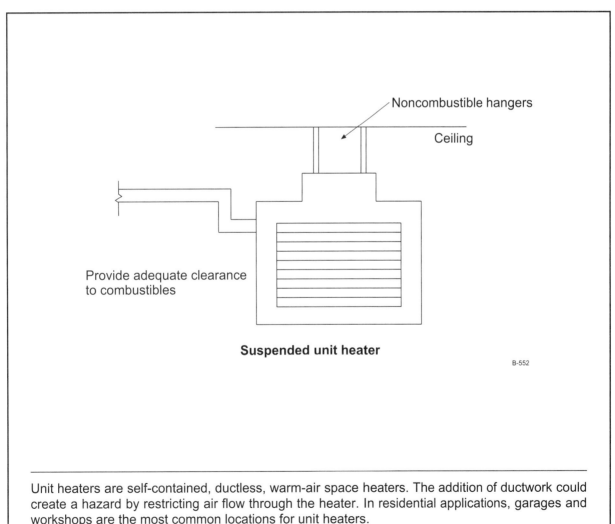

Noncombustible hangers

Ceiling

Provide adequate clearance to combustibles

Suspended unit heater

B-552

Unit heaters are self-contained, ductless, warm-air space heaters. The addition of ductwork could create a hazard by restricting air flow through the heater. In residential applications, garages and workshops are the most common locations for unit heaters.

Code Text: *Unvented room heaters shall be equipped with an oxygen-depletion-sensitive safety shutoff system. The system shall shut off the gas supply to the main and pilot burners when the oxygen in the surrounding atmosphere is depleted to the percent concentration specified by the manufacturer, but not lower than 18 percent. The system shall not incorporate field adjustment means capable of changing the set point at which the system acts to shut off the gas supply to the room heater.*

Discussion and Commentary: The oxygen depletion sensor is basically a pilot burner that is extremely sensitive to the oxygen content in the combustion air. If the oxygen content (approximately 80 percent) drops to a predetermined level, the pilot flame will destabilize and extinguish or become incapable of sufficiently heating the thermocouple or thermopile generator, resulting in main gas control valve shutdown and lockout.

B-930

Unvented room heaters are limited to a maximum input rating of 40,000 Btu/h. When determining the maximum unvented room heater rating allowed, the aggregate input rating of all unvented appliances in a room is limited to 20 Btu/h per cubic foot of volume of the room.

Quiz

Study Session 13
IRC Chapter 24

1. In general, fuel-fired appliances are permitted to be located in which one of the following spaces?

   a. sleeping rooms
   b. storage closets
   c. bathrooms
   d. garages ✎(circled)

   Reference M2406.2

2. Where located in a hazardous location, fuel-gas equipment and appliances having an ignition source shall be elevated such that the ignition source is located a minimum of ___18___ above the floor.

   a. 6 inches
   b. 12 inches
   c. 18 inches ✎(circled)
   d. 30 inches

   Reference G2408.2

3. Unless protected from motor vehicle damage, fuel-gas appliances located in private garages shall be installed with a minimum clearance of ___6ft___ above the floor.

   a. 60 inches
   b. 72 inches ✎(circled)
   c. 80 inches
   d. 84 inches

   Reference G2408.3

4. Assume that a fuel-gas appliance requires a minimum clearance without protection of 36 inches above the appliance to combustible material. If $^1/_2$-inch thick insulation board over 1-inch glass fiber batts is used as protection, the minimum clearance may be reduced to _24"_ .

a.	9 inches	b.	12 inches
c.	18 inches	(d.)	24 inches

Reference _Table 62+04.2_

5. Assume that a fuel-gas appliance requires a minimum clearance without protection of 9 inches from the sides and back to combustible material. If 0.024 sheet metal with a ventilated air space is used as protection, the minimum clearance may be reduced to _3"_ .

a.	2 inches	(b.)	3 inches
c.	4 inches	d.	6 inches

Reference _Table 62404.2_

6. For other than steel pipe, exposed gas piping shall be identified by a yellow label spaced at maximum intervals of _5 ft"_ .

(a.)	5 feet	b.	6 feet
c.	8 feet	d.	10 feet

Reference _62412.5_

7. Unless special conditions are met, the maximum design operating pressure for gas piping systems located inside buildings shall be _5 psig_ .

(a.)	5 psig	b.	10 psig
c.	15 psig	d.	20 psig

Reference _62413.6_

8. The operating pressure for undiluted LP-gas systems shall be a maximum of _20psig_ .

a.	5 psig	b.	10 psig
c.	15 psig	(d.)	20 psig

Reference _62413.6.1_

9. Copper tubing to be used with gases shall comply with standard _K or L_ of ASTM B 88 or ASTM B 280.

    a. Type K only
    b. Type L only
    c. Type K or Type L
    d. Type L or Type M

    Reference _6 2412 . 5 . 2_

10. For the field threading of 1-inch metallic pipe, a length of _7/8_ (approximate) shall be provided for the threaded portion.

    a. $^3/_4$ inch
    b. $^7/_8$ inch
    c. 1 inch
    d. $1^1/_8$ inches

    Reference _Table 62414,9.2_

11. In which of the following locations is the installation of gas piping not specifically prohibited?

    a. in a circulating air duct
    b. through a clothes chute
    c. in an elevator shaft
    d. adjacent to an exterior stairway

    Reference _62415 . 1_

12. For gas piping, other than black or galvanized steel, installed in a concealed location, a shield plate need not be provided for protection against physical damage where a bored hole is located a minimum of _1 1/2"_ from the edge of the wood member.

    a. $^5/_8$ inch
    b. 1 inch
    c. $1^1/_4$ inch
    d. $1^1/_2$ inch

    Reference _62415.5_

13. In general, underground gas piping shall be installed a minimum depth of _12"_ below grade.

    a. 6 inches
    b. 12 inches
    c. 18 inches
    d. 24 inches

    Reference _62415.9_

14. The unthreaded portion of gas piping outlets shall extend a minimum of ___1"___ through finished ceilings and walls.

   a. 1 inch   b. 2 inches

   c. 3 inches   d. 4 inches

   Reference _G2415.13_

15. The tracer wire required adjacent to underground nonmetallic gas piping shall be a minimum of __18 AWG__ wire size.

   a. 14 AWG   b. 16 AWG

   c. 18 AWG   d. 20 AWG

   Reference _G2415.143_

16. The inside radius of a bend in a metallic gas pipe shall be a minimum of ___6X___ the outside diameter of the pipe.

   a. six times   b. eight times

   c. twelve times   d. twenty-five times

   Reference _G2416.2 #5_

17. Gas piping shall be tested at a minimum pressure ___1½___ the proposed maximum working pressure, but not less than 3 psig.

   a. equal to   b. of one and one-half times

   c. of twice   d. of two and one-half times

   Reference _G2417.4.1_

18. The minimum duration of a pressure test for gas piping shall be __10__.

   a. 10 minutes   b. 15 minutes

   c. 20 minutes   d. 30 minutes

   Reference _G2417.4.2_

19. Gas piping for other than dry gas conditions shall be sloped a minimum of __¼"–15'__ to prevent traps.

   a. ¹/₈ inch in 10 feet   b. ¹/₈ inch in 15 feet

   c. ¹/₄ inch in 10 feet   d. ¹/₄ inch in 15 feet

   Reference _G2419.1_

20. A gas shutoff valve for an appliance shall be located in the same room as the appliance, with a maximum distance of ___6 FT___ between the valve and the appliance.

    a. 3 feet   b. 4 feet

    (c.) 6 feet   d. 10 feet

    Reference ___62420.5___

21. Except for range and domestic clothes dryer connectors, appliance fuel connectors shall have a maximum overall length of ___3 FT___.

    (a.) 3 feet   b. 5 feet

    c. 6 feet   d. 10 feet

    Reference ___62422.1.2.1___

22. Supports for $^{3}/_{4}$-inch steel gas pipe shall be provided at maximum intervals of ___8 FT___.

    a. 4 feet   b. 6 feet

    (c.) 8 feet   d. 12 feet

    Reference ___Table 62427.1___

23. Gas appliance connectors, where connected to a masonry chimney, shall connect at a point a minimum of ___12"___ above the lowest portion of the interior of the chimney flue.

    (a.) 12 inches   b. 18 inches

    c. 30 inches   d. 36 inches

    Reference ___62425.9___

24. A sauna room provided with a gas-fired sauna heater shall be provided with a minimum ___4"x8"___ ventilation opening located near the top of the door into the sauna room.

    (a.) 4-inch by 8-inch   b. 4-inch by 12-inch

    c. 8-inch by 8-inch   d. 8-inch by 12-inch

    Reference ___62440.7___

25. Unvented gas room heaters shall be limited to a maximum input rating of _40,000_ .

    a. 35,000 Btu/h   (b) 40,000 Btu/h

    c. 50,000 Btu/h   d. 65,000 Btu/h

Reference _G2445.3_

26. Gas-fired equipment and appliances, where suspended above grade level, shall be provided with a minimum clearance of _6"_ inches from adjoining grade.

    a. 2   b. 4

    (c.) 6   d. 8

Reference _G2408..4_

27. Fuel gas piping installed across a roof surface shall be elevated above the roof a minimum of _3½_ inches.

    a. $1^1/_2$ b. 2

    (c) $3^1/_2$ d. 6

Reference _G2415.7_

28. Which of the following test mediums shall not be used in the testing of a fuel gas piping system?

    a. air   b. carbon dioxide

    c. nitrogen   (d.) oxygen

Reference _G2417.2_

29. One-inch O.D. smooth-wall tubing used as fuel gas piping shall be supported horizontally at maximum intervals of _8'_ feet.

    a. 4   b 6

    (c) 8   d. 10

Reference _Table G2424.1_

30. Where a post-mounted illuminating gas appliance is installed on a steel pipe post having a height of 30 inches, what is the minimum size Schedule 40 steel pipe required?

    (a.) $^3/_4$ inch   b. 1 inch

    c. $1^1/_2$ inch   d. $2^1/_2$ inch

Reference _G2450.3_

31. In the sizing of a fuel gas piping system, a barbeque without a manufacturer's input rating shall be rated for ___42,000___ input btu/h.

    a. 25,000   b. 35,000

    c. 40,000   d. 50,000

    Reference _Table G2413.2_

32. The unthreaded portion of a gas piping outlet passing up through an outdoor patio shall extend a minimum of ___2"___ inch(es) above the patio.

    a. 1   b. 2

    c. 4   d. 6

    Reference _G2415.13_

33. Where approved and not susceptible to physical damage, an individual gas line to an outdoor barbeque grill shall be installed a minimum of ___8___ inches below finished grade.

    a. 6   b. 8

    c. 12   d. 18

    Reference _G2415.9.1_

34. An appliance fuel connector for a gas range shall have a maximum overall length of _____ feet.

    a. 2   b. 3

    c. 5   d. 6

    Reference _G2422.1.2.1_

35. Suspended-type gas-fired unit heaters shall have a minimum clearance to combustible materials of ___18"___ inches on the sides.

    a. 6   b. 12

    c. 15   d. 18

    Reference _G2444.4_

2006 IRC Chapters 25, 26 and 27
General Plumbing Requirements and Plumbing Fixtures

OBJECTIVE: To provide an understanding of the general requirements for the installation of plumbing systems, including inspections and tests; piping installation, protection and support; and plumbing fixtures.

REFERENCE: Chapters 25, 26 and 27, 2006 *International Residential Code*

KEY POINTS:
- ~~P2503.4~~ By what method must the building sewer system be tested and inspected? The DWV system? The water-supply system? ~~P2503.6~~ ~~P2503.5~~
- ~~P2603.2.1~~ — Where installed through framing members in concealed locations, how shall piping be protected? How should it be protected from breakage where passing through or under walls? ~~-P2603.3~~
- ~~P2603.5~~ — How shall a soil pipe, waste pipe or building drain be installed where passing under a footing or through a foundation wall?
- ~~P2603.C~~ — What are the installation limitations for piping installed in areas subject to freezing temperatures? To what minimum depth must water service pipe be installed? How shall trenches be located in relationship to footings or bearing walls? ~~-P2604.1~~
- ~~P2604.1~~ — How must piping be installed when located in a trench? How is backfilling to be
- ~~P2604.3~~ — accomplished? How is trenching installed parallel to footings to be located? ~~-P2604.4~~
- ~~P2605.1.1-4~~ — What are the general provisions for the support of piping? What is the maximum horizontal spacing between supports for various types of pipe material? Between vertical supports? ~~Table 2605.1~~
- ~~P2606.1~~ → How are exterior openings for pipes to be waterproofed?
- ~~P2608.1~~ → How must piping be listed and identified? When is third-party testing required? Third-party certification? ~~-2608.4~~
- ~~P2703.1~~ — What is the minimum required diameter for fixture tail pieces serving sinks, dishwashers, laundry tubs, bathtubs and similar fixtures? For bidets, lavatories and similar fixtures? ~~-P2703.2~~
- ~~P2704.1~~ → What type of fixture connection must be provided with an access panel or utility space? What size of access opening must be provided?
- ~~2705.1/2~~ What are the general requirements for the installation of plumbing fixtures regarding support? Watertightness? ~~4 5~~

335

- What horizontal side clearances are mandated for water closets and bidets? What minimum clearance is required at the front of water closets, bidets and lavatories?

- What is the minimum required extension for a standpipe? The maximum allowable extension? Under what conditions may a laundry tray waste connect into a standpipe for an automatic clothes washer drain? 2706.2.1

- What is the minimum permitted size for a shower compartment? What alternative size is permitted? Which direction must hinged shower doors swing? What is the minimum required opening size? How are shower receptors to be constructed and installed? 2709.1

- How are water closets regulated in regard to flushing devices, water supplies and flush valves?

- What is the minimum waste outlet size for a lavatory? A bathtub? A sink? A laundry tub? A food-waste grinder? A floor drain? 2719.1

- How may a sink and dishwasher discharge through a single trap? A sink, dishwasher and food grinder? 2717.3

- What is temperature limit for water supplied to a shower? A bathtub? A whirlpool bathtub? A bidet? 2721.2

Code Text: *New plumbing work and parts of existing systems affected by new work or alterations shall be inspected by the building official to ensure compliance with the requirements of the IRC. DWV systems shall be tested on completion of the rough piping installation by water or air with no evidence of leakage. After the plumbing fixtures have been set and their traps filled with water, their connections shall be tested and proved gas tight and/or water tight.*

Discussion and Commentary: All plumbing and drainage work must be tested in an appropriate manner to verify that it is leak free. The building sewer must be tested with a minimum 10-foot head of water. Both rough plumbing and finished plumbing are tested with a variety of methods available. Water supply piping and any backflow prevention assemblies must also be inspected and tested.

Fill to required height

Cap or plug all vents, fixtures and building drain

30"

Building drain

Building sewer

DWV systems:

Water:
    10-foot head for 15 minutes
Air:
    5 psi for 15 minutes

Building sewer:
    Fill with water for a 10-foot head for 15 minutes

For SI: 1 inch = 25.4 mm, 1 foot = 304.8 mm, 1 psi = 6.895 kPa.

A-433

Once the water supply system is complete, it must be tested under a minimum water pressure equivalent to the working pressure of the system. As an alternative, a minimum 50 psi air test is permitted for the testing of all systems other than plastic.

Code Text: *Pipes passing through or under walls shall be protected from breakage. Pipes passing through concrete or cinder walls and floors, cold-formed steel framing or other corrosive material shall be protected against external corrosion by a protective sheathing or wrapping or other means that will withstand any reaction from lime and acid of concrete, cinder or other corrosive material. Sheathing or wrapping shall allow for expansion and contraction of piping to prevent any rubbing action. Minimum wall thickness of material shall be 0.025 inch (0.64 mm).*

Discussion and Commentary: Pipes made from brass, copper, cast iron and steel are subject to corrosion when exposed to the lime and acid of concrete and cinder or other corrosive material such as soil; therefore, a coating or wrapping is necessary. Typical protective coatings include coal tar wrapper with paper, epoxy or plastic coatings.

Annular spaces between sleeves and pipes must be filled or tightly caulked. Where the penetration of a pipe occurs in a fire-resistant assembly, protection must comply with Section R317.3 for through-penetrations. This may require a listed firestop system.

Code Text: *Any pipe that passes under a footing or through a foundation wall shall be provided with a relieving arch; or there shall be built into the masonry wall a pipe sleeve two pipe sizes greater than the pipe passing through.*

Discussion and Commentary: Piping installed within or under a footing or foundation wall must be structurally protected from any transferred loading from the foundation wall or footing. This protection may be provided through a relieving arch or a pipe sleeve. When a sleeve is used, it should be sized such that it is two pipe sizes larger than the penetration pipe. This space will allow for any differential movement of the pipe. Providing structural protection to the piping system ensures that the piping will not be subjected to the undue stresses that may cause the piping to rupture and leak.

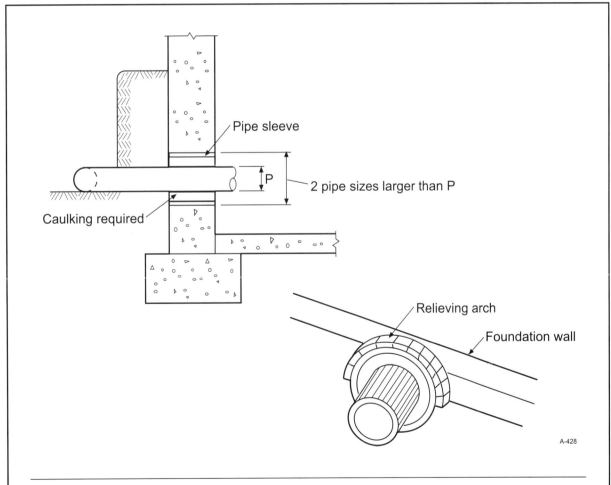

Piping located in concealed spaces within the interior of the building must also be protected from damage if installed through holes or notches in studs, joists and similar framing members. Where the hole or notch is less than $1^1/_2$ inches from the nearest member edge, steel protective shield plates are required.

Topic: Freezing

Category: General Plumbing Requirements

Reference: IRC P2603.6

Subject: Structural and Piping Protection

Code Text: *In localities having a winter design temperature of 32°F (0°C) or lower . . . a water, soil or waste pipe shall not be installed outside of a building, in exterior walls, in attics or crawl spaces, or in any other place subjected to freezing temperature unless adequate provision is made to protect it from freezing by insulation or heat or both.*

Discussion and Commentary: Occupied spaces utilize the building's comfort heating system to assist in protecting pipes from freezing. Where the temperature of the air surrounding the insulation remains low for a significant period of time, such as in a vacant building, insulation alone will not provide adequate protection without the addition of heat. In such conditions, the water in the pipe will freeze regardless of the amount of insulation used, and heat must be provided—often in the form of electric resistance heat tape or cable inside the insulation next to the piping.

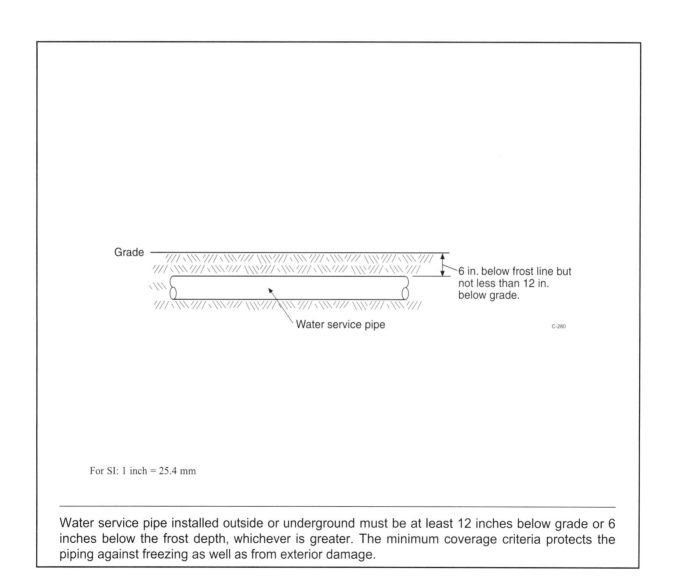

Grade

6 in. below frost line but not less than 12 in. below grade.

Water service pipe

C-280

For SI: 1 inch = 25.4 mm

Water service pipe installed outside or underground must be at least 12 inches below grade or 6 inches below the frost depth, whichever is greater. The minimum coverage criteria protects the piping against freezing as well as from exterior damage.

Code Text: *Where trenches are excavated such that the bottom of the trench forms the bed for the pipe, solid and continuous load-bearing support shall be provided between joints. Where over-excavated, the trench shall be backfilled to the proper grade with compacted earth, sand, fine gravel or similar granular material. Piping shall not be supported on rocks or blocks at any point. Rocky or unstable soil shall be over excavated by two or more pipe diameters and brought to the proper grade with suitable compacted granular material.*

Discussion and Commentary: Pipe is protected from damage by proper backfilling techniques. Backfilling is accomplished in 6-inch layers, with each layer being tamped into place. The material used for backfilling must be evenly distributed to avoid any pipe movement. Large rocks or frozen chunks of dirt must not be located within 12 inches of the pipe.

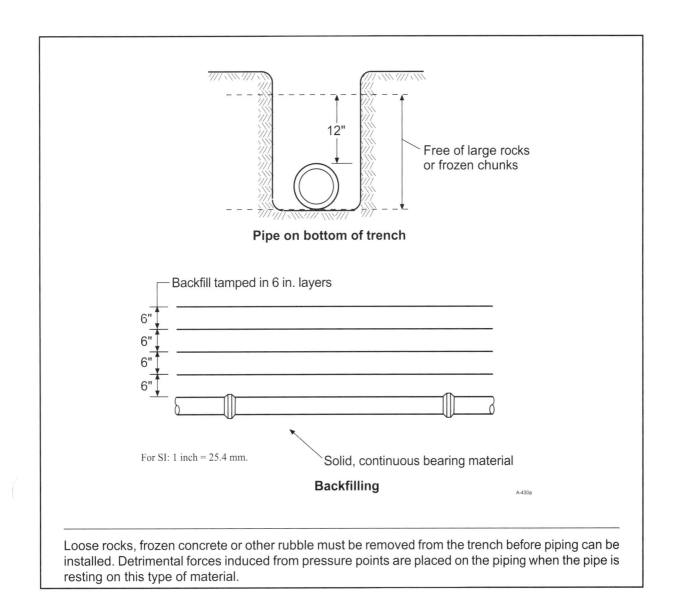

12"

Free of large rocks or frozen chunks

Pipe on bottom of trench

Backfill tamped in 6 in. layers

6"
6"
6"
6"

For SI: 1 inch = 25.4 mm.

Solid, continuous bearing material

Backfilling

A-430a

Loose rocks, frozen concrete or other rubble must be removed from the trench before piping can be installed. Detrimental forces induced from pressure points are placed on the piping when the pipe is resting on this type of material.

Code Text: *Trenching installed parallel to footings shall not extend below the 45-degree (0.79 mm) bearing plane of the bottom edge of a wall or footing.*

Discussion and Commentary: A footing requires a minimum load-bearing area to distribute the weight of the building. That plane extends downward at approximately a 45-degree angle from the base of the footing. The trench containing the piping must not be located below that load-bearing plane, to avoid undermining the building footing. The key to the provision is the term "parallel." Where extending a water or sewer pipe beneath a footing, such installation is usually perpendicular to the foundation and footing. To avoid undermining the load-bearing capacity of the soil, extend the trench beyond the load-bearing plane.

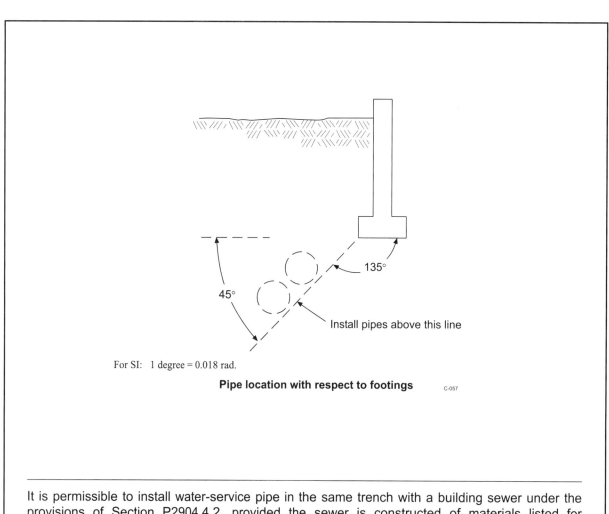

For SI: 1 degree = 0.018 rad.

Pipe location with respect to footings   C-057

It is permissible to install water-service pipe in the same trench with a building sewer under the provisions of Section P2904.4.2, provided the sewer is constructed of materials listed for underground use within a building. Otherwise, various methods of separation are described.

Code Text: *Piping shall be supported so as to ensure alignment and prevent sagging, and allow movement associated with the expansion and contraction of the piping system. Piping in the ground shall be laid on a firm bed for its entire length, except where support is otherwise provided. Hangers and strapping shall be of approved material that will not promote galvanic action. Rigid support sway bracing shall be provided at changes in direction greater than 45 degrees (0.79 rad) for pipe sizes 4 inches (102 mm) and larger. Piping shall be supported at distances not to exceed those indicated in Table P2605.1.*

Discussion and Commentary: For larger pipes, hangers alone may not be sufficient to resist the forces created by water movement within the piping. Therefore, rigid bracing is required to restrict or eliminate lateral movement of both horizontal and vertical piping.

TABLE P2605.1
PIPING SUPPORT

PIPING MATERIAL	MAXIMUM HORIZONTAL SPACING (feet)	MAXIMUM VERTICAL SPACING
ABS pipe	4	10[b]
Aluminum tubing	10	15
Brass pipe	10	10
Cast-iron pipe	5[a]	15
Copper or copper alloy pipe	12	10
Copper or copper alloy tubing ($1^1/_4$ inch diameter and smaller)	6	10
Copper or copper alloy tubing ($1^1/_2$ inch diameter and larger)	10	10
Cross-linked polyethylene (PEX) pipe	2.67 (32 inches)	10[b]
Cross-linked polyethylene/aluminum/cross-linked polyethylene (PEX-AL-PEX) pipe	2.67 (32 inches)	4[b]
CPVC pipe or tubing (1 inch in diameter and smaller)	3	10[b]
CPVC pipe or tubing ($1^1/_4$ inch in diameter and larger)	4	10[b]
Lead pipe	Continuous	4
PB pipe or tubing	2.67 (32 inches)	4
Polyethylene/aluminum/polyethylene (PE-AL-PE) pipe	2.67 (32 inches)	4[b]
Polypropylene (PP) pipe or tubing 1 inch and smaller	2.67 (32 inches)	10[b]
Polypropylene (PP) pipe or tubing, $1^1/_4$ inches and larger	4	10[b]
PVC pipe	4	10[b]
Stainless steel drainage systems	10	10[b]
Steel pipe	12	15

For SI: 1 inch = 25.4 mm, 1 foot = 304.8 mm.
a. The maximum horizontal spacing of cast-iron pipe hangers shall be increased to 10 feet where 10-foot lengths of pipe are installed.
b. Midstory guide for sizes 2 inches and smaller.

Depending on the type of pipe material, the maximum vertical and horizontal spacing between supports can vary. Whereas supports must occur at frequent intervals for horizontal piping due to the potential for sagging, vertical piping typically requires support only at each story height.

Code Text: *Fixture tail pieces shall be not less than $1^1/_2$ inches (38 mm) in diameter for sinks, dishwashers, laundry tubs, bathtubs and similar fixtures, and not less than $1^1/_4$ inches (32 mm) in diameter for bidets, lavatories and similar fixtures.*

Discussion and Commentary: Tail pieces are short lengths of pipe attached directly to a fixture by means of a flange for connection to other piping or traps. Tail pieces, as well as traps, continuous wastes, waste fittings and overflow fittings, must be of No. 20 gage minimum thickness when constructed of seamless drawn brass. Plastic tubular fittings are permitted when in conformance with ASTM F 409. The minimum size of $1^1/_2$ inches for sinks, dishwashers, laundry tubs and bathtubs is consistent with the minimum sizes for fixture drains, traps and trap arms.

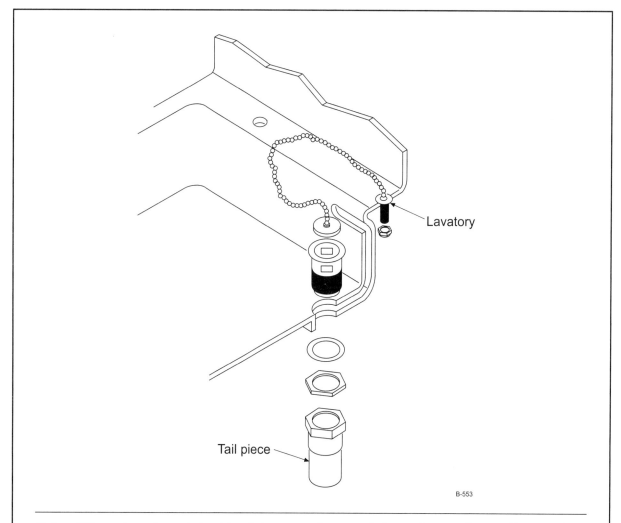

Lavatory

Tail piece

B-553

Fixture fittings, faucets and plumbing fixtures require smooth impervious surfaces with no surface defects that could reduce the efficiency, effectiveness or sanitary requirements of the code. Table P2701.1 lists the appropriate material standards for various plumbing items.

Code Text: *Floor-outlet or floor-mounted fixtures shall be secured to the drainage connection and to the floor, when so designed, by screws, bolts, washers, nuts and similar fasteners of copper, brass or other corrosion-resistant material. Wall-hung fixtures shall be rigidly supported so that strain is not transmitted to the plumbing system. Where fixtures come in contact with walls, the contact area shall be water tight. Plumbing fixtures shall be usable. The location of piping, fixtures or equipment shall not interfere with the operation of windows or doors.*

Discussion and Commentary: The installation of fixtures shall be accomplished based on structural, sanitation and operational considerations.

Apply sealing compound around edge of hole. Drop lavatory into hole. Install clips as shown and tighten firmly. Remove excess sealing.

Water-tight joints surrounding fixtures

B-568

Water closets and bidets must be installed to provide a minimum of 15 inches from the centerline to adjacent walls, creating a minimum compartment size of 30 inches. In front of the water closet, a minimum of 21 inches is mandated. This clearance also applies to lavatories and bidets.

Code Text: *Standpipes shall extend a minimum of 18 inches (457 mm) and a maximum of 42 inches (1067 mm) above the trap weir. Access shall be provided to all standpipe traps and drains for rodding. A laundry tray waste line is permitted to connect into a standpipe for the automatic clothes washer drain. The standpipes shall not be less than 30 inches (762 mm) as measured from the crown weir. The outlet of the laundry tray shall be a maximum horizontal distance of 30 inches (762 mm) from the standpipe trap.*

Discussion and Commentary: A standpipe can serve as an indirect waste receptor. The established minimum receptor height aims to provide a minimal retention capacity and head pressure to prevent overflow. This height limitation is especially necessary when an indirect waste pipe receives a high rate of discharge, such as the pumped discharge from a clothes washer.

Discharge from clothes washer

18" min
42" max

2" min drain

For SI: 1 inch = 25.4 mm.

Standpipes

C-276

The maximum height requirement intends to control the waste flow velocity at the trap inlet. Excessive velocity promotes trap siphonage. Although the standpipe itself can be concealed within construction, the point of indirect connection must be accessible to drain cleaning equipment.

Code Text: *Shower compartments shall have at least 900 square inches (0.6 m²) of interior cross-sectional area. Shower compartments shall not be less than 30 inches (752 mm) in minimum dimension measured from the finished interior dimension of the shower compartment, exclusive of fixture valves, shower heads, soap dishes, and safety grab bars or rails.* See exceptions for showers 1) with fold-down seats, and 2) having compartments at least 25 inches wide and 1,300 square inches in cross-sectional area. *The minimum required area and dimension shall be measured from the finished interior dimension at a height equal to the top of the threshold at a point tangent to its centerline and shall be continued to a height not less than 70 inches (1778 mm) above the shower drain outlet. Hinged shower doors shall open outward.*

Discussion and Commentary: The dimensional criteria facilitate safety and comfort based on typical body movement while showering. Fold-down seats can encroach into the required dimensions, provided the 900 square inches are available when the seats are folded.

Smooth, noncorrosive, nonabsorbent waterproof material

Finished interior surface

Centerline of threshold

30" min.

Plan

30" min. diameter 900 sq. in. min area*

Threshold

*25" min dimension permitted if 1,300 sq. in. min area

For SI: 1 inch = 24.5 mm, 1 square inch = 645 mm².

Shower enclosure

A-384a

Every shower must have a control valve capable of protecting an individual from being scalded. These devices are also required to protect against rapid temperature fluctuation by automatically maintaining the discharge temperature to plus or minus 3°F of the selected temperature. The high stop limit must be set to limit water temperature to a maximum of 120°F.

Code Text: *The shower compartment access and egress opening shall have a minimum clear and unobstructed finished width of 22 inches (559 mm).*

Discussion and Commentary: It is important that an adequate width be provided to and from a shower compartment. The requirement for 22 inches of clear width provides for functional access by the shower user, including the service and maintenance of the compartment. In addition, the established width provides access for emergency response and/or rescue, should the need arise.

For SI: 1 inch = 24.5 mm

The minimum clear width of 22 inches as established by the code is consistent with the criteria utilized throughout the means of egress provisions of the *International Building Code* for limited travel by a single occupant.

Code Text: *Shower receptors shall have a finished curb threshold not less than 1 inch (25 mm) below the sides and back of the receptor. The curb shall be not less than 2 inches (51 mm) and not more than 9 inches (229 mm) deep when measured from the top of the curb to the top of the drain. The finished floor shall slope uniformly toward the drain not less than $^1/_4$ unit vertical in 12 units horizontal (2-percent slope) nor more than $^1/_2$ inch (13 mm), and floor drains shall be flanged to provide a water-tight joint in the floor.*

Discussion and Commentary: When a shower is built in place, a shower pan is required under the shower's finished floor. Various materials are used for shower linings, including sheet lead, sheet copper, PVC sheet, chlorinated polyethylene sheet and hot mopping. The shower pan is designed to collect any water that leaks through the shower floor.

Shower pan installation

The shower pan must connect to the shower drain with a flashing flange with weep holes to allow the leaking water to enter the drainage system. A clamping ring or similar device is used to clamp down the waterproof membrane and make a watertight joint.

Topic: Sink and Dishwasher   **Category:** Plumbing Fixtures

Reference: IRC P2717.2, P2717.3   **Subject:** Dishwashing Machines

Code Text: *A sink and dishwasher are permitted to discharge through a single $1^1/_2$-inch (38 mm) trap. The discharge pipe from the dishwasher shall be increased to a minimum of $^3/_4$ inch (19 mm) in diameter and shall be connected with a wye fitting to the sink tailpiece. The dishwasher waste line shall rise and be securely fastened to the underside of the counter before connecting to the sink tailpiece. The combined discharge from a sink, dishwasher and waste grinder is permitted to discharge through a single $1^1/_2$ -inch (38 mm) trap.*

Discussion and Commentary: Dishwashing machines may connect directly to the drainage system. An indirect connection by air gap or air break is not required. The direct connection is acceptable because the potable water supply is adequately protected against backflow.

For SI: 1 inch = 25.4 mm.

Dishwasher waste connections

A-395a

Because of the type of discharge and the possible contents within the discharge, a domestic dishwasher is permitted to connect directly to a food waste grinder or a kitchen sink tailpiece. The dishwasher line shall rise and be fastened to the underside of the counter before connection.

Code Text: *Access shall be provided to circulation pumps in accordance with the fixture manufacturer's installation instructions. Where the manufacturer's instructions do not specify the location and minimum size of field fabricated access openings, a 12-inch by 12-inch (304 mm by 304 mm) minimum size opening shall be installed to provide access to the circulation pump. A door or panel shall be permitted to close the opening.*

Discussion and Commentary: In order to allow for adequate access to the circulation pump of a whirlpool bathtub, an access opening must be provided. Athough a minimum size of 12 inches by 12 inches is established by the code, the opening must also be large enough to permit the removal and replacement of the pump.

Whirlpool bathtub

Min 12" by 12" access panel

For SI: 1 inch = 24.5 mm

An increase in the size of the access opening is required where the circulation pump is not located in close proximity to the opening. A minimum 18-inch by 18-inch access openings is mandated where the circulation pump is located more than 2 feet from the access opening.

Study Session 14
IRC Chapters 25, 26 and 27

1. Testing of the building sewer shall be accomplished with a minimum _10'_ head of water with the ability to maintain such pressure for a minimum of _15 min_.

    a. 8-foot, 10 minutes   b. 8-foot, 15 minutes

    c. 10-foot, 10 minutes   (d) 10-foot, 15 minutes

    Reference _P2503.4_

2. When tested by air, the DWV system shall be tested at _5 PSI_ or 10 inches of mercury column with the ability to maintain such pressure for a minimum of _15 MIN_.

    a. 5 psi, 10 minutes   (b) 5 psi, 15 minutes

    c. 10 psi, 10 minutes   d. 10 psi, 15 minutes

    Reference _P2503.5.1 & 2_

3. The water-supply system shall be proved tight under a water pressure not less than the working pressure of the system, or by a minimum _50 PSI_ air test for other than plastic piping.

    a. 5 psi   b. 10 psi

    c. 20 psi   (d) 50 psi

    Reference _P2503.6_

4. Where copper piping is to pass through holes bored in wood studs, the pipe shall be protected by shield plates except for those holes that are a minimum of __1½"__ from the nearest edge.

    a. $^5/_8$ inch   b. $^3/_4$ inch

    c. $1^3/_8$ inches   ⓓ. $1^1/_2$ inches

Reference __P2603.2.1__

5. Shield plates required to protect piping installed through notches or bored holes in framing members shall extend a minimum of __2"__ above sole plates and below top plates.

    a. $^1/_2$ inch   b. 1 inch

    ⓒ 2 inches   d. 3 inches

Reference __P2603.2.1__

6. Unless provided with a relieving arch, a building drain passing through a masonry foundation wall shall pass through a pipe sleeve sized __two__ greater than the pipe passing through.

    a. 1 inch   b. 2 inches

    c. one pipe size   ⓓ two pipe sizes

Reference __P2603.5__

7. Where buried, water service pipe shall be installed a minimum of __12"__ deep.

    a. 6 inches   ⓑ. 12 inches

    c. 24 inches   d. 30 inches

Reference __P2603.C__

8. Where a frost line is established at 30 inches, water service pipe shall be installed a minimum of __36__ below grade.

    a. 30 inches   b. 32 inches

    ⓒ 36 inches   d. 42 inches

Reference __P2603.6__

9. Where unstable soil conditions exist in an area to be trenched for a piping installation, the trench shall be overexcavated by a minimum of ___two___.

    a. 4 inches

    b. 6 inches

    c. one pipe diameter

    d. two pipe diameters

    Reference _P2604.1_

10. The backfilling of a trench where piping is installed shall be done in maximum ___6"___ layers of loose earth.

    a. 4-inch

    b. 6-inch

    c. 8-inch

    d. 12-inch

    Reference _P2604.3_

11. One-half-inch copper water pipe installed horizontally shall be supported at maximum intervals of ___12'___.

    a. 3 feet

    b. 4 feet

    c. 6 feet

    d. 12 feet

    Reference _T P2605.1_

12. PB pipe shall be supported at maximum intervals of ___32___ when installed horizontally.

    a. 32 inches

    b. 36 inches

    c. 60 inches

    d. 72 inches

    Reference _T P2605.1_

13. A fixture tail piece for a laundry tub shall be a minimum of ___1½"___ in diameter.

    a. 1 inch

    b. $1^1/_4$ inches

    c. $1^1/_2$ inches

    d. 2 inches

    Reference _P2703.1_

14. Unless provided with an approved arrangement for access purposes, a fixture with a concealed slip-joint connection shall be provided with an access panel or utility space a minimum of _____ in its smallest dimension.

    (a.) 12 inches   b. 18 inches

    c. 30 inches   d. 36 inches

Reference P2704.1

15. Standpipes shall extend a minimum of ___18___ and a maximum of ___42___ above the trap weir.

    a. 18 inches, 30 inches   (b.) 18 inches, 42 inches

    c. 24 inches, 30 inches   d. 24 inches, 42 inches

Reference P2706.2

16. The outlet of a laundry tray shall be a maximum horizontal distance of ___30___ from the standpipe trap.

    a. 18 inches   (b.) 30 inches

    c. 42 inches   d. 60 inches

Reference P2706.2.1

17. In general, shower compartments shall have a minimum interior cross-sectional area of ___900___ square inches.

    (a) 900   b. 1,080

    c. 1,296   d. 1,440

Reference P2708.1

18. The curb for a shower receptor shall have a minimum depth of ___2___ and a maximum depth of ___9___ when measured from the top of the curb to the top of the drain.

    a. 1 inch, 6 inches   b. 1 inch, 9 inches

    c. 2 inches, 6 inches   (d.) 2 inches, 9 inches

Reference P2709.1

19. A lavatory shall have a waste outlet a minimum of __1¼"__ in diameter.

    a. 1 inch

    (b.) 1¼ inches

    c. 1½ inches

    d. 2 inches

    Reference _P2711.3_

20. Unless an approved alternative is provided, flush valve seats in tanks for flushing water closets shall be a minimum of __1"__ above the flood-level rim of the bowl to which it is connected.

    a. ½ inch

    (b.) 1 inch

    c. 1½ inches

    d. 2 inches

    Reference _P2712.1_

21. Bathtubs shall have outlets and overflows a minimum of __1½"__ in diameter.

    a. 1 inch

    b. 1¼ inches

    (c.) 1½ inches

    d. 2 inches

    Reference _P2713.1_

22. Sinks shall have a waste outlet a minimum of __1½"__ in diameter.

    a. 1¼ inches

    (b.) 1½ inches

    c. 2 inches

    d. 3½ inches

    Reference _P2714.1_

23. Food waste grinders shall be connected to a drain having a minimum diameter of __1½__ .

    a. 1¼ inches

    (b.) 1½ inches

    c. 2 inches

    d. 3½ inches

    Reference _P2716.1_

24. Floor drains shall have waste outlets a minimum of __2__ in diameter.

    a. 1¼ inches

    b. 1½ inches

    (c.) 2 inches

    d. 3½ inches

    Reference _P2719.1_

25. What is the minimum size drain required for a macerating toilet system?

    (a.) $^3/_4$ inch   b. $1^1/_4$ inches

    c. $1^1/_2$ inches   d. 2 inches

Reference P2723.2

26. Where a smoke test is utilized to verify gas tightness of the DWV system, a pressure equivalent to a 1-inch water column shall be applied and maintained for a minimum time period of 15 minutes.

    a. 10   (b.) 15

    c. 30   d. 60

Reference P2503.5.2 # 2.1

27. Where an air test of the rough plumbing installation requires a gauge pressure of 5 psi, the testing gauge shall have increments of .10 psi.

    (a.) 0.10   b. 0.50

    c. 1.00   d. 2.00

Reference P2503.8

28. Trenches used for the installation of piping that are excavated parallel to footings shall extend a maximum of $+5°$ below the bearing plane of the bottom edge of the footing.

    a. 30 degrees   (b.) 45 degrees

    c. 12 inches   d. 30 inches

Reference P2604.4

29. Which of the following materials and products shall be certified by an approved third-party certification agency?

    a. plumbing fixtures

    b. pecial waste system components

    c. subsoil drainage system components

    (d.) water distribution system safety devices

Reference TP2609.4

30. There shall be a minimum of _2'_ inches clearance in front of a water closet to any wall, fixture or door.

    a. 18   (b.) 21

    c. 24   d. 27

Reference _P2705.1   #5_

31. The access and egress opening for a shower compartment shall have a minimum and unobstructed clear width of _22_ inches.

    (a.) 22   b. 24

    c. 30   d. 32

Reference _P2708.1.1_

32. A water-temperature-limiting device shall be provided to limit the hot water supplied to a bathtub or whirlpool bathtub to a maximum of _120_ °F.

    a. 110   b. 115

    (c.) 120   d. 125

Reference _P2713.3_

33. Where a sink and a dishwasher discharge through a single $1^1/_2$-inch trap, the discharge pipe from the dishwasher shall be a minimum of _3/4"_ inch in diameter.

    (a.) $^3/_4$

    b. 1

    c. $1^1/_4$

    d. $1^1/_2$

Reference _P2717.2_

34. Where the manufacturer's instructions do not specify the minimum required size of a field-fabricated access opening to the circulation pump of a whirlpool bathtub, a minimum _12x12"_ opening shall be installed.

    a. 10-inch by 10-inch   (b.) 12-inch by 12-inch

    c. 10-inch by 16-inch   d. 12-inch by 18-inch

Reference _P2720.1_

35. The discharge water temperature from a bidet fitting shall be limited to a maximum of
    ___110°F___ by a water-temperature-limiting device.

    a. 98°F   b. 105°F

    c. 110°F   d. 112°F

Reference ___P2721.2___

2006 IRC Chapters 28, 29 and 30
Water Heaters, Water Supply and Distribution, and Sanitary Drainage

OBJECTIVE: To provide an understanding of the general provisions for the installation of water heaters, the supply and distribution of water, and the materials, design, construction and installation of sanitary drainage systems.

REFERENCE: Chapters 28, 29 and 30, 2006 *International Residential Code*

KEY POINTS:
- ~P2801.5~ ~P2801.5~ When is a water heater required to be installed in a pan? What type of pan is mandated? ~P2801.5.1~ How big must it be? How shall it be drained? ~P2801.5.1~
- ~P2801.6~ What special condition applies where a water heater having an ignition source is installed in a garage?
- ~P2802.1~ What special conditions are applicable where a combination water heater-space heating system is utilized?
- ~P2803.1~ What type of relief valve or valves is required for a water heater? How is a pressure-relief valve regulated? A temperature-relief valve? ~P2803.3~ ~P2803.4~
- ~P2803.6.1 1-13~ How is the water heater discharge pipe to be installed? What is the minimum required size of the discharge pipe? Where is the termination point to be located?
- ~P2902.3~ How shall backflow preventers be applied?
- ~P2902.4~ What special conditions exist where both potable and nonpotable water distribution systems are installed? How shall such systems be identified? ~2901.1~
- ~? 2902.4~ How is the potable water supply system to be protected? What methods of backflow protection are acceptable?
- ~P2902.3.1~ What is the minimum permitted air gap for various fixtures? Where shall the critical level of an atmospheric-type vacuum breaker be set? ~P2902.3.3~ ~P2905.2~
- ~P2902S.1~ How must the potable water supply to boilers be protected? To heat exchangers? To lawn ~P2902.5.3~ irrigation systems? To automatic fire sprinkler systems? To solar systems? ~P2902S.4 & 5~
- ~T2903.2~ What are the maximum water consumption flow rates and quantities for various plumbing fixtures and fittings?
- ~P2903.3~ What is the minimum permitted static pressure at the building entrance for water service?
- ~P2903.3.1~ The maximum static pressure? How should thermal expansion be addressed? ~P2903.4~

KEY POINTS:
(Cont'd)

2904-2

- What is the minimum size for a water service pipe? How should the water distribution system be sized? P2903-7 1-4   P290+.d-2

- How should water service pipe be installed? What special requirements apply where the building sewer piping is installed in the same trench? What special concerns are addressed in the installation of water distribution pipe? 2904.5

P303.2 1-6 —
- What types of joints and connections are prohibited in the DWV system? How are approved joints to be made?

P304.1 &
Table P304.1 —
- How is the load on the DWV-system piping determined? What type of fittings are to be used? P3205.1

- Where shall cleanouts be located? What should be done at changes in direction? What clearances are needed in front of cleanouts? What is the minimum size of a cleanout? P301-325

P307.1 —
- What are the conditions of installation for sewage ejectors or sewage pumps?

Topic: Required Pan

Reference: IRC P2801.5

Category: Water Heaters

Subject: General Provisions

Code Text: *Where water heaters or hot water storage tanks are installed in locations where leakage of the tanks or connections will cause damage, the tank or water heater shall be installed in a galvanized steel pan having a minimum thickness of 24 gage (0.016 inch)(0.4 mm) or other pans for such use. The pan shall be not less than $1^{1}/_{2}$ inches (38 mm) deep and shall be of sufficient size and shape to receive all dripping and condensate from the tank or water heater. The pan shall be drained by an indirect waste pipe having a minimum diameter of $^{3}/_{4}$ inch (19 mm).*

Discussion and Commentary: The pan catches water from a leaking tank, a leaking connection to the tank or condensate from the tank, and it may also be used as the indirect waste receptor for the relief valve discharge.

For SI: 1 inch = 25.4 mm.

B-556

The pan must not connect directly to the drainage system. An indirect waste pipe is necessary to prevent backflow from the drainage system into the pan. Because the relief valve discharge piping typically terminates in the pan, the waste pipe must be as large as the discharge pipe.

Topic: Garage Installations	**Category:** Water Heaters
Reference: IRC P2801.6	**Subject:** General Provisions

Code Text: *Water heaters having an ignition source shall be elevated such that the source of ignition is not less than 18 inches (457 mm) above the garage floor.*

Discussion and Commentary: Unless properly elevated, a fuel-burning water heater installed in a garage poses a serious fire hazard. A garage is commonly used as a storage area for gasoline and common household items such as paint, paint thinner and solvents. Fumes from these flammable liquids tend to collect near the floor and produce vapors denser than air. The 18-inch requirement is intended to prevent an energy or heat source from becoming a source of ignition. The ignition source could be a pilot flame, burner, burner igniter or electrical component capable of producing a spark or an arc. The 18-inch elevation must be increased where required by the manufacturer's installation instructions.

For SI: 1 inch = 25.4 mm.

Gas-fired water heater Installation in a garage

A-440

The provision addresses all types of water heaters—not just those that are gas- or oil-fired. It applies equally to gas-fired water heaters of the direct vent type. The requirement would not apply only in the unlikely event that the water heater were a true sealed combustion chamber type.

Topic: Relief Valves

Reference: IRC P2803.1, P2803.3, P2803.4

Category: Water Heaters

Subject: Relief Valves

Code Text: *Appliances and equipment used for heating water or storing hot water shall be protected by 1) a separate pressure-relief valve and a separate temperature-relief valve; or 2) a combination pressure- and temperature-relief valve. Pressure-relief valves . . . shall be set to open at least 25 psi (172 kPa) above the system pressure but not over 150 psi (1034 kPa). The relief-valve setting shall not exceed the tank's rated working pressure. Temperature-relief valves . . . shall be installed such that the temperature-sensing element monitors the water within the top 6 inches (152 mm) of the tank. The valve shall be set to open at a maximum temperature of 210°F (99°C).*

Discussion and Commentary: A combination temperature- and pressure-relief valve or separate temperature-relief and pressure-relief valves protect the water heater against possible explosion.

Pressure-relief mechanism

Temperature-sensing probe

Temperature- and pressure-relief valve

B-557

Relief valves are emergency devices and not intended to function continuously. Any discharge should not go unnoticed, because when a relief valve discharges, it indicates that something is wrong with the system. The discharge point should always be readily observable.

Code Text: *A potable water supply system shall be designed and installed in such a manner as to prevent contamination from nonpotable liquids, solids or gases being introduced into the potable water supply. Connections shall not be made to a potable water supply in a manner that could contaminate the water supply or provide a cross-connection between the supply and source of contamination unless an approved backflow-prevention device is provided. Cross-connections between an individual water supply and a potable public water supply shall be prohibited.*

Discussion and Commentary: Arguably, the most important aspect of the plumbing provisions of the IRC is the protection of potable water systems. History is filled with local and widespread occurrences of sickness and disease caused by not safeguarding the water supply. It is imperative that the potable water supply be maintained in a safe-for-drinking condition at all times and at all outlets.

Potential cause of back siphonage

B-558

As a general condition, all dwelling units must be provided with potable water. If the building contains both a potable and nonpotable water-distribution system, each system needs to be clearly identified by an appropriate method, such as a color marking or metal tags.

Code Text: *The minimum air gap shall be measured vertically from the lowest end of a water supply outlet to the flood level rim of the fixture or receptor into which such potable water outlets discharge. The minimum required air gap shall be twice the diameter of the effective opening of the outlet, but in no case less than the values specified in Table P2902.3.1. An air gap is required at the discharge point of a relief valve or piping.*

Discussion and Commentary: An air gap is not a device and has no moving parts. It is, therefore, the most effective and dependable method of preventing backflow. The potable water opening or outlet is terminated at an elevation above the level of the source of contamination. If the appropriate minimum air gap is provided, the potable water could only be contaminated if the entire area or room flooded to a depth that would submerge the potable water opening or outlet.

For SI: 1 inch = 25.4 mm.

An air gap installation must always be constructed so that the air gap remains permanently fixed. Air gaps may be fabricated from commercially available plumbing components or obtained as separate units and integrated into plumbing and piping systems.

Topic: Design Criteria
Reference: IRC P2903.1, P2903.2

Category: Water Supply and Distribution
Subject: Water-supply System

Code Text: *Attics containing appliances requiring access shall be provided with an opening and a clear and unobstructed passageway large enough to allow removal of the largest appliance, but not less than 30 inches (762 mm) high and 22 inches (559 mm) wide and not more than 20 feet (6096 mm) in length when measured along the centerline of the passageway from the opening to the appliance.*

Discussion and Commentary: The water distribution system must be capable of supplying a sufficient volume of potable water at pressures adequate to enable all plumbing fixtures, devices and appurtenances to function properly and without undue noise. Each fixture has a given flow rate and minimum water supply pressure needed to properly operate the fixture.

TABLE P2903.1
REQUIRED CAPACITIES AT
POINT OF OUTLET DISCHARGE

FIXTURE AT POINT OF OUTLET	FLOW RATE (gpm)	FLOW PRESSURE (psi)
Bathtub	4	8
Bidet	2	4
Dishwasher	2.75	8
Laundry tub	4	8
Lavatory	2	8
Shower	3	8
Shower, temperature controlled	3	20
Sillcock, hose bibb	5	8
Sink	2.5	8
Water closet, flushometer tank	1.6	15
Water closet, tank, close coupled	3	8
Water closet, tank, one-piece	6	20

For SI: 1 gallon per minute = 3.785 L/m,
1 pound per square inch = 6.895 kPa.

TABLE P2903.2
MAXIMUM FLOW RATES AND CONSUMPTION FOR
PLUMBING FIXTURES AND FIXTURE FITTINGS[b]

PLUMBING FIXTURE OR FIXTURE FITTING	PLUMBING FIXTURE OR FIXTURE FITTING
Lavatory faucet	2.2 gpm at 60 psi
Shower head[a]	2.5 gpm at 80 psi
Sink faucet	2.2 gpm at 60 psi
Water closet	1.6 gallons per flushing cycle

For SI: 1 gallon per minute = 3.785 L/m,
1 pound per square inch = 6.895 kPa.
a. A handheld shower spray is also a shower head.
b. Consumption tolerances shall be determined from referenced standards.

The fixtures listed in Table P2903.2 function in a cyclical manner or consume water continuously for the duration of the fixture use. Water is conserved by limiting the flow rate for manually controlled fixtures and by limiting the volume per cycle usage for cyclically operating fixtures.

Code Text: *Supply loads in the building water-distribution system shall be determined by total load on the pipe being sized, in terms of water-supply fixture units (w.s.f.u.), as shown in Table P2903.6, and gallon per minute (gpm) flow rates. For fixtures not listed, choose a w.s.f.u. value of a fixture with similar flow characteristics.*

Discussion and Commentary: A prescriptive method for estimating the water demand for a building has evolved through the years and has been accepted as a standard method of sizing water systems. The method is based upon weighing fixtures in accordance with their water supply load-producing effects on the water distribution system. This approach couples the flow characteristics of the various fixtures with probability curves of simultaneous flushing of the fixtures.

TABLE P2903.6
WATER-SUPPLY FIXTURE-UNIT VALUES FOR VARIOUS PLUMBING FIXTURES AND FIXTURE GROUPS

TYPE OF FIXTURES OR GROUP OF FIXTURES	WATER-SUPPLY FIXTURE-UNIT VALUE (w.s.f.u.)		
	Hot	Cold	Combined
Bathtub (with/without overhead shower head)	1.0	1.0	1.4
Clothes washer	1.0	1.0	1.4
Dishwasher	1.4	—	1.4
Full-bath group with bathtub (with/without shower head) or shower stall	1.5	2.7	3.6
Half-bath group (water closet and lavatory)	0.5	2.5	2.6
Hose bibb (sillcock)[a]	—	2.5	2.5
Kitchen group (dishwasher and sink with/without garbage grinder)	1.9	1.0	2.5
Kitchen sink	1.0	1.0	1.4
Laundry group (clothes washer standpipe and laundry tub)	1.8	1.8	2.5
Laundry tub	1.0	1.0	1.4
Lavatory	0.5	0.5	0.7
Shower stall	1.0	1.0	1.4
Water closet (tank type)	—	2.2	2.2

For SI: 1 gallon per minute = 3.785 L/m.

a. The fixture unit value 2.5 assumes a flow demand of 2.5 gpm, such as for an individual lawn sprinkler device. If a hose bibb/sill cock will be required to furnish a greater flow, the equivalent fixture-unit value may be obtained from this table or Table P2903.6(1).

The water service pipe must be sized in accordance with Section P2903.7, but in no case may it be less in size than $^3/_4$ inch. Using a prescriptive approach, the size of the water distribution system including the service pipe, meter and main distribution pipe can be determined.

Code Text: *Each dwelling unit shall be provided with an accessible main shutoff valve near the entrance of the water service. The valve shall be of a full-open type having nominal restriction to flow, with provision for drainage such as a bleed orifice or installation of a separate drain valve. A readily accessible full-open valve shall be installed in the cold-water supply pipe to each water heater at or near the water heater. Valves serving individual fixtures, appliances, risers and branches shall be provided with access. An individual shutoff valve shall be required on the fixture supply pipe to each plumbing fixture other than bathtubs and showers.*

Discussion and Commentary: All gate valves and "rated" globe valves are fullway. Gate valves are easily identified by observing the stem when opening the valve. All gate valve stems have stationary stems, unlike globe valves, which have rising stems. Nonrated globe valves are not fullway and are referred to as compression stops.

"V" shaped gate

Typical gate valve

Washer compression stop

Typical globe valve

B-560

A hose bibb exposed to freezing conditions must be equipped with a stop and waste valve inside the building so that it can be drained or controlled during freezing periods. A frostproof bibb installed where the stem extends into conditioned space is not required to be valved separately.

Topic: Fittings
Reference: IRC P3002.3

Category: Sanitary Drainage
Subject: Materials

Code Text: *Fittings shall be approved and compatible with the type of piping being used and shall be of a sanitary or DWV design for drainage and venting as shown in Table P3002.3. Waterpipe fittings shall be permitted for use in engineer designed systems where the design indicates compliance with Section P3101.2.1. Drainage fittings shall have a smooth interior waterway of the same diameter as the piping served. Threaded drainage pipe fittings shall be of the recessed drainage type, black or galvanized. Drainage fittings shall be designed to maintain one-fourth unit vertical in 12 units horizontal (2-percent slope) grade.*

Discussion and Commentary: Sanitary, DWV, or water-pipe fittings are acceptable in the dry section of the vent systems. Drainage fittings should have no ledges, shoulders or reductions that can retard or obstruct drainage flow in the piping.

Vent tee

Reducing vent tee, single and double

Long sweep $^1/_4$ bend with low-heel inlet

Long sweep $^1/_4$ bend with high-heel inlet

Sanitary tees, 45° wyes, combination wyes and $^1/_8$ bends

Sanitary tee

45° wye single and double

Combination wye and $^1/_8$ bend single and double

Reducing sanitary tees, 45° wyes, combination wyes and $^1/_8$ bends

45° wye, reducing single and double

Combination wye and $^1/_8$ bend, reducing single and double

60° wye, single and double

60° reducing wye, single and double

For SI: 1 degree = 0.1745 rad.

Only materials approved for the specific type of installation should be used for DWV systems. These materials are available in several levels of quality, which vary with regard to characteristics and properties; thus, each may be used only in accordance with its referenced standard.

Code Text: *Drainage pipe cleanouts shall be installed not more than 100 feet (30 480 mm) apart in horizontal drainage lines measured from the upstream entrance of the cleanout. Cleanouts shall be installed at each fitting with a change of direction more than 45 degrees (0.79 rad) in the building sewer, building drain and horizontal waste or soil lines. Where more than one change of direction occurs in a run of piping, only one cleanout shall be required in each 40 feet (12 192 mm) of developed length of the drainage piping.*

Discussion and Commentary: Because a cleanout is designed to provide access into the drainage system, the cleanout itself must be accessible, regardless if the piping it serves is concealed or in a location not readily accessible. The cleanout is required to extend up to or above the finished grade either inside or outside the building. A cleanout must be accessible for use wherever it is located in the drainage system and must not be covered by any permanent, nonaccessible construction.

For SI: 1 foot = 304.8 mm.

Cleanouts

B-562

For a cleanout to be used, there must be adequate clear space provided to facilitate the cleaning operation. For pipes 3 inches or larger, at least 18 inches must be provided in front of the cleanout, measured perpendicular to the opening. Only 12 inches are needed on smaller pipes.

Code Text: *An offset in a vertical drain, with a change of direction of 45 degrees (0.79 rad) or less from the vertical, shall be sized as a straight vertical drain. A stack with an offset of more than 45 degrees (0.79 rad) from the vertical shall be sized in accordance with three listed conditions. In soil or waste stacks below the lowest horizontal branch, there shall be no change in diameter required if the offset is made at an angle not greater than 45 degrees (0.79 rad) from the vertical. If an offset greater than 45 degrees (0.79 rad) from the vertical is made, the offset and stack below it shall be sized as a building drain.*

Discussion and Commentary: Depending on the conditions involved, there are various methods for sizing offsets in drainage piping. The most complicated of the methods involves horizontal offsets above the lowest branch.

Horizontal offset above lowest branch

Horizontal offset below lowest branch

For SI: 1 degree = 0.1745 rad.

B-563

Horizontal offsets below the lowest branch should not change size if the offset makes an angle less than 45 degrees from the vertical. Where the offset exceeds 45 degrees from the vertical, it should be sized as a building drain in accordance with Table P3005.4.2.

Topic: Sewage Ejectors or Sewage Pumps **Category:** Sanitary Drainage
Reference: IRC P3007.1, P3007.2   **Subject:** Sumps and Ejectors

Code Text: *A sewage ejector, sewage pump, or grinder pump receiving discharge from a water closet shall have minimum discharge velocity of 1.9 feet per second (0.579 m/s) throughout the discharge piping to the point of connection with a gravity building drain, gravity sewer or pressure sewer system. Building drains which cannot be discharged to the sewer by gravity flow shall be discharged into a tightly covered and vented sump from which the contents shall be lifted and discharged into the building gravity drainage system by automatic pumping equipment.*

Discussion and Commentary: A nongrinding ejector pump must be capable of passing $1^1/_2$-inch-diameter solids. A 2-inch backwater valve is required to prohibit sewage from backflowing into the basement. Where in compliance with ASME A112.3.4 or CSA B45.9, macerating toilet systems must be permitted as an alternate to the sewage pump or ejector system.

Backwater or check valve

2 in. min dia. must be able to pass a $1^1/_2$ in. dia. solid ball

Discharge

For SI: 1 inch = 25.4 mm.

Sewage ejector and pump

B-564

The portion of the building drain below the sewer grade should be installed and vented in the same manner as the gravity system. Vents should be independent of the vents serving the gravity system and must be carried up to open air.

Code Text: *For water service systems sizes up to and including 2 inches (51 mm), a device for controlling pressure shall be installed where, because of thermal expansion, the pressure on the downstream side of a pressure-reducing valve exceeds the pressure-reducing valve setting. Where a backflow prevention device, check valve or other device is installed on a water supply system using storage water heating equipment such that thermal expansion causes an increase in pressure, a device for controlling pressure shall be installed.*

Discussion and Commentary: In the connection of a water distribution system to water-heating appliances there is the potential for the migration of heated water into the water distribution piping. In a typical water distribution system, the water will expand into the water service and into the public main if the water is not withdrawn from the system at an outlet or connection. If the expansion of water is not accommodated in the system, dangerously high pressures can develop that can cause damage to piping, components and the water heater.

The possibility of water hammer is addressed through a requirement for the installation of a water-hammer arrestor where quick-closing valves are used.

Code Text: *Water-service pipe is permitted to be located in the same trench with a building sewer provided such sewer is constructed of materials listed for underground use with a building in Section P3002.1. If the building sewer is not constructed of materials listed in Section P3002.1, the water-service pipe shall be separated from the building sewer by a minimum of 5 feet (1524 mm), measured horizontally, of undisturbed or compacted earth or place on a solid ledge at least 12 inches (305 mm) above and to one side of the highest point in the sewer line.* See exception where water service piping crosses sewer piping.

Discussion and Commentary: The separation of water service from the building sewer is intended to reduce the possibility of contaminating the potable water supply. When there is a leak in the building sewer, the soil becomes contaminated around the water pipe. This would put the potable water supply at risk should the water service pipe have a subsequent failure.

For SI: 1 inch = 25.4 mm, 1 foot = 304.8 mm.

Whereas the fundamental means of reducing the possibility of soils contamination is installing the building sewer and water service pipe in two separate trenches located an adequate distance apart, a single trench is permitted where the building sewer is constructed of the appropriate materials.

Quiz

Study Session 15
IRC Chapters 28, 29 and 30

1. A required overflow pan for a water heater shall be a minimum of ___1½___ in depth and of sufficient size and shape to collect all dripping and condensate.

    a.   1 inch   (b) 1¹/₂ inches

    c.   2 inches   d.   3 inches

    Reference _P2801.5.1_

2. Where the termination of a drain pan for a water heater extends to the exterior of the building, it shall terminate a minimum of ___6___ and a maximum of ___24___ above the adjacent ground surface.

    a.   3 inches, 12 inches   b.   3 inches, 24 inches

    c.   6 inches, 12 inches   (d.) 6 inches, 24 inches

    Reference _P2801.5.2_

3. Where installed in a garage, water heaters having an ignition source shall be elevated such that the source of ignition is a minimum of ___18"___ above the garage floor.

    a.   6 inches   b.   12 inches

    (c.) 18 inches   d.   24 inches

    Reference _P2801.6_

4. A water heater pressure-relief valve shall be set to open at least ___25 PSI___ above the system pressure but not over ___150 PSI___.

      a. 15 psi, 125 psi   b. 15 psi, 150 psi

      c. 25 psi, 125 psi   **d.** 25 psi, 150 psi

      Reference ___P2803.3___

5. A water heater temperature-relief valve shall be set to open at a maximum temperature of ___210°F___.

      a. 150°F   b. 165°F

      c. 180°F   **d.** 210°F

      Reference ___P2803.4___

6. Where located close to a wall, what is the minimum required air gap for a lavatory with an effective opening of $^1/_2$ inch?

      a. 1 inch   **b.** $1^1/_2$ inch

      c. 2 inches   d. 3 inches

      Reference T P2902.3.1

7. Approved deck-mounted vacuum breakers shall be installed with the critical level a minimum of ___1"___ above the flood level rim.

      a. 1 inch   b. 2 inches

      c. 4 inches   d. 6 inches

      Reference P2902.4.2

8. In determining the peak demand for the water service and water distribution service, a flow rate of ___4___ and a flow pressure of ___8 PSI___ shall be used for a bathtub at the point of outlet discharge.

      a. 3 gpm, 8 psi   b. 3 gpm, 20 psi

      c. 4 gpm, 8 psi   d. 5 gpm, 8 psi

      Reference T P2903.1

9. A water closet shall have a maximum consumption of __1.6__ gallons per flushing cycle.

   a. 1.0   (b.) 1.6

   c. 2.0   d. 2.3

   Reference __P2903.2__

10. A handheld shower spray shall have a maximum flow rate of __2.5 GPM .70__

    a. 2.2 gpm at 60 psi   b. 2.2 gpm at 80 psi

    c. 2.5 gpm at 60 psi   (d.) 2.5 gpm at 80 psi

    Reference __TP2903.2   A__

11. What are the minimum and maximum static pressures required for a water service?

    (a.) 40 psi, 80 psi   b. 40 psi, 100 psi

    c. 50 psi, 80 psi   d. 50 psi, 100 psi

    Reference __P2903.3 0 P2903.3.1__

12. When sizing a water distribution system, a tank-type water closet shall have a water-supply fixture unit value of __2.2__.

    a. 1.0   b. 1.4

    c. 2.2   d. 2.7

    Reference __T2903.6__

13. Where the minimum available water pressure is 45 psi, distribution piping with a maximum developed length of 200 feet and serving 32 fixture units shall have a minimum diameter of __1 1/4__.

    a. $^{1}/_{2}$ inch   b. $^{3}/_{4}$ inch

    c. 1 inch   (d.) $1^{1}/_{4}$ inches

    Reference __TP2903.7__

14. Pipe and fittings used in the water supply system shall have a maximum of __8 %__ lead content.

    a. 1 percent   b. 2 percent

    c. 5 percent   (d.) 8 percent

    Reference __P2904.2.1__

15. Bends of polyethlene pipe shall have a minimum installed radius of pipe curvature greater than ___30___ pipe diameters.

    a. 15   b. 20

    c. 25   (d.) 30

Reference ___P2904.3___

16. Hot-water-distribution piping within dwelling units shall have a minimum pressure rating of ___100RF___ at 180°F.

    a. 80 psi   (b.) 100 psi

    c. 110 psi   d. 125 psi

Reference ___P2904.5___

17. What is the minimum required radius for bends in copper tubing?

    (a.) four tube diameters   b. six tube diameters

    c. 4 inches   d. 6 inches

Reference ___P2905.1___

18. What is the calculated load on DWV-system piping for two bath groups, a kitchen group, a laundry group and a clothes washer standpipe?

    (a.) 15 d.f.u.   b. 16 d.f.u.

    c. 18 d.f.u.   d. 20 d.f.u.

Reference ___P3004.1___

19. Drainage pipe cleanouts shall be installed a maximum of ___100'___ apart in horizontal drainage lines.

    a. 40 feet   b. 60 feet

    c. 80 feet   (d.) 100 feet

Reference ___P3005.2.2___

20. The minimum size of a cleanout serving a 3-inch pipe shall be ___3"___.

    (a.) 3 inches   b. 2$\frac{1}{2}$ inches

    c. 2 inches   d. 1$\frac{1}{2}$ inches

Reference ___P3005.2.4___

21. Where more than one change in direction occurs in a run of piping in a drainage system, only one cleanout is required for each _____40'_____ of run.

    a. 25 feet   (b) 40 feet

    c. 50 feet   d. 75 feet

    Reference _P3005.2.4_

22. What is the minimum required clearance in front of a cleanout for a 2-inch drainage pipe?

    (a) 12 inches   b. 18 inches

    c. 24 inches   d. 30 inches

    Reference _P3005 2.5_

23. Three-inch horizontal drainage piping shall be installed at a minimum uniform slope of _____1/8'_____ per foot.

    (a) $^1/_8$ inch   b. $^3/_{16}$ inch

    c. $^1/_4$ inch   d. $^5/_{16}$ inch

    Reference _P3005.3_

24. Below grade drain pipes shall be a minimum of _____1 1/2_____ in diameter.

    (a) $1^1/_2$ inches   b. $2^1/_2$ inches

    c. 3 inches   d. 4 inches

    Reference _P3005 4.1_

25. What is the maximum number of fixture units allowed to be connected to a 3-inch-diameter vertical drain stack?

    a. 10   b. 12

    c. 20   (d) 48

    Reference _P3005.4.1_

26. A water heater discharge pipe serving a pressure-relief valve shall terminate atmospherically a maximum of _____6_____ inch(es) above the floor or waste receptor.

    a. 1   b. 2

    c. 4   (d) 6

    Reference _P2803.6.1 #10_

27. A hose connection backflow preventer shall conform to which of the following applicable standards?

   a. ASSE 1012   b. ASSE 1020

   c. ASSE 1052   d. ASSE 1056

   Reference TP 2902 3

28. Water service pipe installed underground and outside of a structure shall have a minimum working pressure rating of ___160___ psi at 73°F.

   a. 100   b. 140

   c. 160   d. 175

   Reference P2904.4

29. Back-to-back water closet connections to double sanitary tee patterns are permitted provided the horizontal developed length between the outlet of the water closet and the connection to the double sanitary tee is a minimum of ___18"___ inches.

   a. 18   b. 24

   c. 30   d. 36

   Reference P3005 11 Exp.

30. The discharge pipe of a macerating toilet system shall be a minimum of _____ inch(es) in diameter.

   a. $^3/_4$ b. $1^1/_4$

   c. $1^1/_2$ d. 2

   Reference P3007 1 Exp.

31. Where chemicals are intended to be introduced into a lawn irrigation system, the potable water supply shall be protected against backflow by a(n) _____.

   a. atmospheric-type vacuum breaker

   b. double check-valve assembly

   c. pressure-type vacuum breaker

   d. reduced pressure principle backflow preventer

   Reference P2902.5.3

32. Where polyethylene plastic pipe is used in water service piping, bends must occur a minimum of ___10___ pipe diameters of any fitting or valve.

    a. 4   b. 6

    c. 10   d. 12

    Reference ___P2904.3___

33. Galvanized steel DWV pipe used within a building shall be maintained a minimum of ___6"___ inches above ground.

    a. 2   b. 3

    c. 4   d. 6

    Reference ___P3002.1___

34. A quarter-bend fitting is permitted for a vertical to horizontal change in direction in drainage piping provided the fixture drain is a maximum of ___2___ inches.

    a. $1^1/_4$

    b. $1^1/_2$

    c. 2

    d. 3

    Reference ___T 3005.1 USA___

35. Where drainage pipe cleanouts are not provided near the base of each vertical waste or soil stack, they shall be located outside the building a maximum of ___3___ feet from the building wall.

    a. 3   b. 5

    c. 10   d. 12

    Reference ___P305.26___

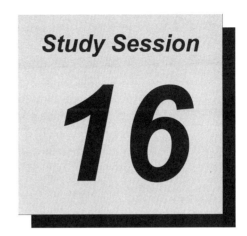

2006 IRC Chapters 31 and 32
Vents and Traps

OBJECTIVE: To provide an understanding of the installation requirements for piping, tubing and fittings for vent systems; vent sizes and connections; and fixture traps.

REFERENCE: Chapters 31 and 32, 2006 *International Residential Code*

KEY POINTS:
- What is the purpose of a vent piping system? ~P2102.2 3102.2
  - P3102.1 — Where must the required vent pipe terminate? How must it be installed? What is its minimum required size? P3102.3
  - P3103.1 — At what minimum height must an open vent pipe terminate above a roof? What if the roof is used for a purpose other than weather protection? — P3103.1
- 3" P3103.2 — What is the minimum required size for a vent extension through a roof or wall in areas subject to cold climates? Where must any increase in size occur? — P3103.2 (if frost occurs)
  - · P3103.5 — How must a vent terminal be located in relationship to a door, openable window or similar air intake? To an adjoining building? — P3103.5
  - TP3105.1 — What is the maximum distance allowed between a fixture trap and its vent? The minimum
    - 3105.3 — distance required? What is the maximum fall due to pipe slope? — P3105.2
  - P3107.1 — How many traps or trapped fixtures may be vented by a common vent? How is the
    - P3107.2 — connection regulated where the drains connect at the same level? At different levels? → P3107.3
  - P3108.1 — Under what conditions is a wet vent permitted? What are the limitations of a wet vent? ← How is a vertical wet vent regulated? P3108.4
  - P3109.4 — How is a waste stack to be sized?   P3110.2
  - P3110.1 — Where is circuit venting acceptable? Where shall the connection be located? What is the maximum slope of the vent section of the horizontal branch drain? ← P3110.3
  - P3111.1 — Where may a combination waste and vent system be used? How shall it be installed? What is the minimum size of the vent pipe? ← TP3111.3
  - P3112.1 — What fixtures are permitted to be vented by an island fixture venting system? Where is the connection to be located? ← P3112.2
  - P3113.1 — What is the minimum required diameter of various vents and venting methods?
  - P3114.1 — How are air admittance valves regulated? Where are such valves permitted? Where should they be located? P3114.4   P3114.3

KEY POINTS: • How shall traps be designed? How shall they be constructed? What types of traps are
(Cont'd) prohibited? *P3201.5 #1 -5*

P3201.2 • What is the minimum required liquid seal for a trap? The maximum allowable trap seal?
What special provisions apply to floor drains?

P3201.6 • Where shall a trap be located in relationship to the fixture outlet? What is the maximum
vertical distance permitted between the fixture outlet and the trap weir? The maximum
horizontal distance?

P3201.7 • What is the minimum trap size for various plumbing fixtures?

Topic: Roof Extension

Category: Vents

Reference: IRC P3103.1

Subject: Vent Terminals

Code Text: *Open vent pipes that extend through a roof shall be terminated at least 6 inches (152 mm) above the roof or 6 inches (152 mm) above the anticipated snow accumulation, whichever is greater, except that where a roof is to be used for any purpose other than weather protection, the vent extensions shall be run at least 7 feet (2134 mm) above the roof.*

Discussion and Commentary: At a very minimum, open vent pipes shall extend at least 6 inches above the roof surface. It is anticipated that the roof will be unoccupied, so the only concern is that gases and odors will discharge well above the roof surface. If it is expected that snow will accumulate on the roof at various times, then the height must be increased so that the vent pipes will maintain a height of at least 6 inches above the snow level. This determination is to be made by the building official based on local information. However, if the roof is used as an occupiable space, such as an entertainment deck or observation platform, the termination point must be above the level of individuals occupying the roof.

For SI: 1 inch = 25.4 mm.

In cold climates, the terminal opening may become closed with frost. This is caused when the flow of warm, moist air rising through the vent stack comes in contact with the frigid outside air, forming frost on the vent's interior. The chance of closure is reduced by increasing the vent size.

Topic: Location

Reference: IRC P3103.5

Category: Vents

Subject: Vent Terminals

Code Text: *An open vent terminal from a drainage system shall not be located less than 4 feet (1219 mm) directly beneath any door, openable window, or other air intake opening of the building or of an adjacent building, nor shall any such vent terminal be within 10 feet (3048 mm) horizontally of such an opening unless it is at least 2 feet (610 mm) above the top of such opening.*

Discussion and Commentary: The vent terminal opening should not be located directly beneath any opening where sewer gas and odors can enter the building. Sewer gases are undesirable, unhealthy and potentially explosive. For the same reasons, the proximity of any vent terminal to openings within a 10-foot horizontal distance, whether in the same building or an adjacent building, are also regulated.

For SI: 1 inch = 25.4 mm.

A-410a

Where a vent terminal extends through an exterior wall, the termination point must be at least 10 feet from any adjoining property line. Vertically, the vent terminal must extend at least 10 feet above the highest grade located within 10 feet horizontally.

Code Text: *Each fixture trap shall have a protecting vent located so that the slope and the developed length in the fixture drain from the trap weir to the vent fitting are within the requirements set forth in Table P3105.1. See exception for self-siphoning fixtures. The total fall in a fixture drain resulting from pipe slope shall not exceed one pipe diameter, nor shall the vent pipe connection to a fixture drain, except for water closets, be below the weir of the trap. A vent shall not be installed within two pipe diameters of the trap weir.*

Discussion and Commentary: The length of each trap arm should not exceed that specified in Table P3105.1, measured from the weir of the trap to the vent along the centerline of the pipe. The measured length must include turns and offsets.

Developed length of fixture branch from trap weir to vent along centerline of pipe

Vent pipe

Waste pipe

Trap arm length

A

B (Trap weir)

Separate points A and B by a min of 2 pipe diameters

Crown vent limitations

B-569

TABLE P3105.1
MAXIMUM DISTANCE OF FIXTURE TRAP FROM VENT

SIZE OF TRAP (inches)	SLOPE (inch per foot)	DISTANCE FROM TRAP (feet)
$1^{1}/_{4}$	$^{1}/_{4}$	5
$1^{1}/_{2}$	$^{1}/_{4}$	6
2	$^{1}/_{4}$	8
3	$^{1}/_{8}$	12
4	$^{1}/_{8}$	16

For SI: 1 inch = 25.4 mm, 1 foot = 304.8 mm, 1 inch per foot = 83.3 mm/m.

A vent must be installed at least two pipe diameters from the trap weir. Where a lesser distance is provided, blockage is possible due to the action of the drainage flowing through the trap. The flow direction and velocity forces waste up into the connection, eventually clogging it fully.

Code Text: *Each trap and trapped fixture is prmitted to be provided with an individual vent. The individual vent shall connect to the fixture draine of the trap or trapped fixture being vented.*

Discussion and Commentary: The simplest form of venting a trap or trapped fixture is an individual vent for each trap. A single vent pipe is connected between the trap of a fixture and the branch connection to the drainage system. With a properly installed individual vent, only the drainage of the fixture served is flowing past the vent.

The sizes of individual vents are to be in accordance with Section P3113. Sizing is based on the minimum required diameter of the drain served and the total developed length. Any vent must be a minimum of one-half the diameter of the drain that it serves, with a minimum diameter of $1^{1}/_{4}$ inches.

Topic: Connections	**Category:** Vents
Reference: IRC P3107.1, P3107.2	**Subject:** Common Vent

Code Text: *An individual vent is permitted to vent two traps or trapped fixtures as a common vent. The traps or trapped fixtures being common vented shall be located on the same floor level. Where the fixture drains being common vented connect at the same level, the vent connection shall be at the interconnection of the fixture drains or downstream of the interconnection.*

Discussion and Commentary: A common vent installed vertically may be used to protect two fixture traps when the drains connect at the same level on a vertical drain. A typical installation is two horizontal drains connecting together to a vertical drain to a double-pattern fitting. The extension of the vertical pipe serves as the vent. The vent may connect at the interconnection of the fixture drains or downstream along the horizontal drain.

For SI: 1 inch = 25.4 mm.

Two fixtures within a single story may connect at different levels to a vertical drain and still be common-vented, provided the drain pipe between the upper and lower fixtures is oversized. Although drainage flows in the pipe between the two fixtures, the piping is not classified as a wet vent.

Code Text: *Any combination of fixtures within two bathroom groups located on the same floor level are permitted to be vented by a horizontal wet vent. The wet vent shall be considered the vent for the fixtures and shall extend from the connection of the dry vent along the direction of the flow in the drain pipe to the most downstream fixture drain connection.*

Discussion and Commentary: A wet vent is a vent pipe that is wet because is also conveys drainage. The wet venting concept is based on employing oversized piping to allow for the flow of air above the waste flow. The low probability of simultaneous fixture discharges and the low-flow velocity that results from fixtures on the same floor level provide the necessary degree of reliability that there will always be adequate volume within the wet vent pipe to permit required airflow.

For SI: 1 inch = 25.4 mm.

Wet venting

A-405

Wet venting is permitted for both vertical and horizontal branch drains and includes what previously was referred to as stack venting. The vertical wet vent must extend from its connection to the dry vent down to the lowest fixture drain connection.

Code Text: *A waste stack shall be considered a vent for all of the fixtures discharging to the stack where installed in accordance with the requirements of Section P3109. The waste stack shall be vertical, and both horizontal and vertical offsets shall be prohibited between the lowest fixture drain connection and the highest fixture drain connection to the stack.. Every fixture drain shall connect separately to the waste stack. The stack shall not receive the discharge of water closets or urinals. A stack vent shall be installed for the waste stack. The size of the stack vent shall not be less than the size of the waste stack.*

Discussion and Commentary: A waste stack vent employs a waste stack as a vent for the fixtures. The principles of use are based on some of the original research that was done in plumbing. The system has been identified by a variety of names, including vertical wet vent and multi-floor stack venting.

For SI: 1 inch = 25.4 mm.

Because the drainage stack serves as the vent, there are certain limitations placed on the design of a waste stack vent. The system is specifically identified as a "waste" stack vent because it prohibits the connection of water closets and urinals. Only "waste" can discharge to the stack.

Code Text: *A maximum of eight fixtures connected to a horizontal branch drain shall be permitted to be circuit vented. Each fixture drain shall connect horizontally to the horizontal branch being circuit vented. The horizontal branch drain shall be classified as a vent from the most downstream fixture drain connection to the most upstream fixture drain connection to the horizontal branch.*

Discussion and Commentary: The principle of circuit venting is that the flow of drainage never exceeds a half-full flow condition. The air for venting the fixtures circulates in the top half of the horizontal branch drainpipe. The flow velocity in the horizontal branch is slow and nonturbulent, thereby preventing pressure differentials from affecting the connecting fixtures.

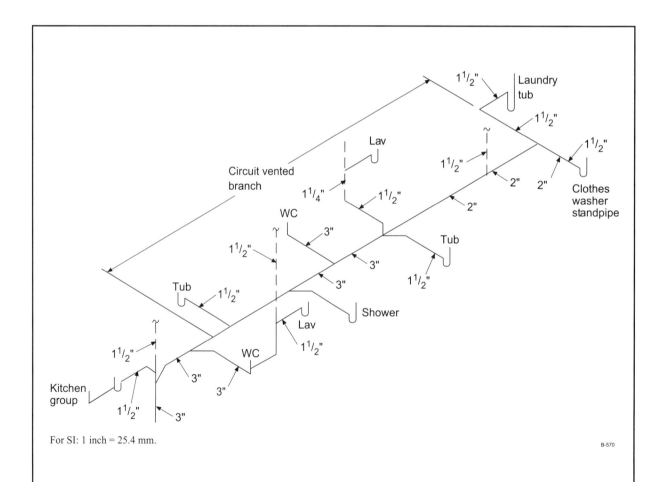

For SI: 1 inch = 25.4 mm.

B-570

The connection of the circuit vent must occur between the two most upstream fixture drains. The circuit vent pipe cannot receive the drainage of any soil or waste. In addition, the slope of the vent section of the horizontal branch drain is limited to 1:12 (8-percent slope).

Code Text: *The only vertical pipe of a combination drain and vent system shall be the connection between the fixture drain of a sink, lavatory, floor drain, standpipe or drinking fountain, and the horizontal combination waste and vent pipe. The maximum vertical distance shall be 8 feet (2438 mm).*

Discussion and Commentary: A combination drain and vent system is a means of extending the distance from a trap to its vent for an unlimited distance because the vent and drain are literally one. The number of fixtures connecting to a combination drain and vent system is also unlimited, provided that the fixtures are floor drains, sinks, standpipes, drinking fountains or lavatories. The system is useful in areas that cannot accommodate vertical vent risers, such as where island sinks are installed or where floor drains are located in large open areas.

For SI: 1 inch = 25.4 mm.

Combination drain and vent

A-408

In a combination drain and vent system, the drain also serves as the vent for the fixture. The system is intended to be a horizontal piping system, with the only vertical piping (limited to 8 feet) being the connection to a sink, lavatory standpipe or drinking fountain located above the level of the vent.

Code Text: *Island fixture venting shall not be permitted for fixtures other than sinks and lavatories. The island fixture vent shall connect to the fixture drain as required for an individual or common vent. The vent shall rise vertically to above the drainage outlet of the fixture being vented before offsetting horizontally or vertically downward. The vent or branch vent for multiple island fixture vents shall extend to a minimum of 6 inches (152 mm) above the highest island fixture being vented before connecting to the outside vent terminal.*

Discussion and Commentary: It is not uncommon for one or more fixtures to be located without the nearby walls or partitions necessary for conventional venting methods to be used. Through the use of loop venting, the loop is typically able to be located below the counter height. The island fixture vent must rise vertically to a height above the fixture drain before it turns down.

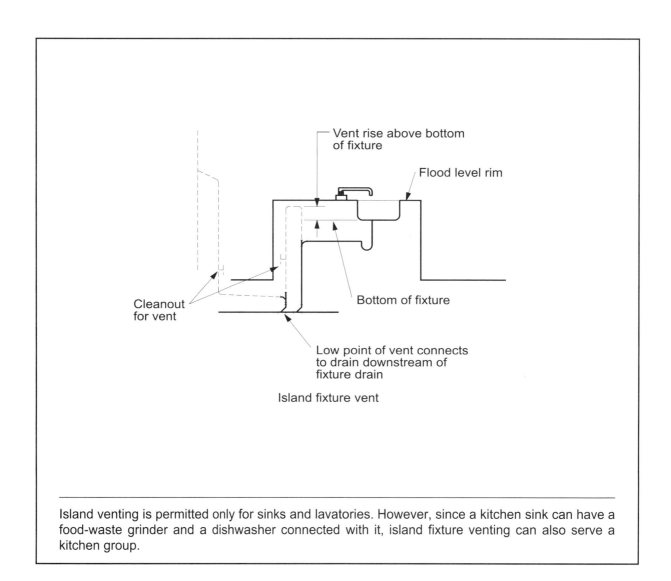

Island venting is permitted only for sinks and lavatories. However, since a kitchen sink can have a food-waste grinder and a dishwasher connected with it, island fixture venting can also serve a kitchen group.

Code Text: *Individual and branch air admittance valves shall be located a minimum of 4 inches (102 mm) above the horizontal branch drain or fixture drain being vented. Stack-type air admittance valves shall be located a minimum of 6 inches (152 mm) above the flood level rim of the highest fixture being vented. The air admittance valve shall be located within the maximum developed length permitted for the vent. The air admittance valve shall be installed a minimum of 6 inches (152 mm) above insulation materials where installed in attics.*

Discussion and Commentary: Individual vents, branch vents, circuit vents and stack vents may all terminate with a connection to an air admittance valve rather than extend to the open air. Air enters the upper plumbing drainage system when negative pressures develop in the piping system. The device closes by gravity and seals the vent terminal when the internal pressure is equal to or greater than atmospheric pressure.

Flood-level rim

Air admittance valve

4" min.

For SI: 1 inch = 25.4 mm.

A-456

Because an air admittance valve is capable of relieving only negative pressures, at least one stack vent or vent stack must extend to the outdoors. The vent to the open air serves as the positive pressure relief for the drainage system.

Code Text: *Traps shall have a liquid seal not less than 2 inches (51 mm) and not more than 4 inches (102 mm). Traps for floor drains shall be fitted with a trap primer or shall be of the deep seal design. Traps shall be set level with respect to their water seals and shall be protected from freezing. Trap seals shall be protected from siphonage, aspiration or back pressure by an approved system of venting.*

Discussion and Commentary: A trap is a simple method used to keep sewer gases from emanating out of the drainage system. The water seal prevents the sewer gases and aerosol-borne bacteria from entering the building space. The configuration of a trap interferes with the flow of the drainage; however, the interference is minimal because of the construction of the trap and the relatively high inlet velocity of the waste flow.

For SI: 1 inch = 25.4 mm.

Installation of trap

A-424

A trap is a simple U-shaped piping arrangement that offers minimal resistance to flow. The only type of fixture trap permitted is the "P" trap. All other types, such as bell traps, drum traps, "S" traps, and traps with moving parts, have undesirable characteristics and are prohibited.

Code Text: *Each plumbing fixture shall be separately trapped by a water seal trap.* See three conditions where separate trapping is not required. *The vertical distance from the fixture outlet to the trap weir shall not exceed 24 inches (610 mm) and the horizontal distance shall not exceed 30 inches (762 mm) measured from the centerline of the fixture outlet to the centerline of the inlet of the trap. Fixtures shall not be double trapped.*

Discussion and Commentary: The limit on trap location is based on multiple criteria. It is desirable to locate the trap as close as possible to the fixture to minimize the amount of drainpipe on the inlet side of the trap. Buildup on the wall of the fixture outlet pipe will breed bacteria and cause odors to develop. The vertical distance is also limited to control the velocity of the drainage flow. If the trap has too great a vertical separation from the fixture, the velocity of flow at the trap inlet may create self-siphoning of the water in the trap. The horizontal dimension is believed to be the maximum length through which the flow energy is able to move waste.

For SI: 1 inch = 25.4 mm.

Double-trapping a fixture is prohibited because of the unnecessary additional obstruction of flow and potential for stoppages. Double-trapping will cause a volume of air to be trapped between two trap seals, and the "air-bound" drain will impede the flow.

Topic: Size of Fixture Traps

Reference: IRC P3201.7

Category: Traps

Subject: Fixture Traps

Code Text: *Fixture trap size shall be sufficient to drain the fixture rapidly and not less than the size indicated in Table P3201.7. A trap shall not be larger than the drainage pipe into which the trap discharges.*

Discussion and Commentary: Where a fixture trap is properly sized, it allows the fixture to drain rapidly and function as intended. Although the code does not regulate for maximum size, an oversized trap will not scour itself and will be prone to clogging. If the trap is larger than the drainage pipe into which it discharges, there is a greater potential for clogging because the reduction does not allow a water velocity that is necessary to scour the trap.

TABLE P3201.7
SIZE OF TRAPS AND TRAP ARMS FOR PLUMBING FIXTURES

PLUMBING FIXTURE	TRAP SIZE MINIMUM (inches)
Bathtub (with or without shower head and/or whirlpool attachments)	$1^{1}/_{2}$
Bidet	$1^{1}/_{4}$
Clothes washer standpipe	2
Dishwasher (on separate trap)	$1^{1}/_{2}$
Floor drain	2
Kitchen sink (one or two traps, with or without dishwasher and garbage grinder)	$1^{1}/_{2}$
Laundry tub (one or more compartments)	$1^{1}/_{2}$
Lavatory	$1^{1}/_{4}$
Shower (based on the total flow rate through showerheads and bodysprays) Flow rate: 5.7 gpm and less / More than 5.7 gpm up to 12.3 gpm / More than 12.3 gpm up to 25.8 gpm / More than 25.8 gpm up to 55.6 gpm	$1^{1}/_{2}$ / 2 / 3 / 4
Water closet	Note a

For SI: 1 inch = 25.4 mm.
a. Consult fixture standards for trap dimensions of specific bowls.

In Section P3201.6, the provisions of Exception 2 allow a set of not more than three similar fixtures (lavatories, kitchen sinks or laundry tubs) to be drained by one trap. The trap size need not be increased because, in most instances, side-by-side fixtures are either served by a single faucet or do not experience heavy use simultaneously.

Quiz

Study Session 16
IRC Chapters 31 and 32

1. Where the accumulation of snow is not a concern, open vent pipes shall extend a minimum of ___6___ inches above a roof used solely for weather protection.

   (a.) 6   b. 12

   c. 18   d. 24

   Reference _P3103. 1_

2. Where a vent extension through a roof is increased in size due to cold climatic conditions, the size increase shall occur inside the structure a minimum of ___12___ inches below the roof.

   a. 6   (b) 12

   c. 24   d. 36

   Reference _P3103.2_

3. Horizontal vent pipes forming branch vents shall be a minimum of ___6___ inches above the flood level rim of the highest fixture served.

   a. 3   b. 4

   (c.) 6   d. 12

   Reference _P3104.5_

4. Where a roof is to be used for any purpose other than weather protection, the vent extension shall run a minimum of _____ above the roof.

    a. 2 feet   b. 5 feet

    (c.) 7 feet   d. 10 feet

Reference P3103.1

5. Unless located a minimum of ___24___ above the top of the opening, an open vent terminal from a drainage system shall be located at least 10 feet horizontally from a door, openable window or other air intake opening of the building.

    a. 12 inches   (b.) 24 inches

    c. 36 inches   d. 60 inches

Reference P3103.5

6. Every dry vent shall rise vertically a minimum of ___6"___ above the flood level rim of the highest trap or trapped fixture being vented.

    a. 1 inch   b. 2 inches

    c. 4 inches   (d) 6 inches

Reference P3104.4

7. A connection between a vent pipe and a vent stack or stack vent shall be made a minimum of ___6"___ above the flood level rim of the highest fixture served by the vent.

    a. 1 inch   b. 2 inches

    c. 4 inches   (d.) 6 inches

Reference P3104.5

8. A maximum distance of ___6"___ is permitted between a 1$^{1}/_{2}$-inch fixture trap and the vent fitting.

    a. 4 feet   b. 5 feet

    (c.) 6 feet   d. 8 feet

Reference T3105.1

9. The total fall in a fixture drain due to pipe slope is limited to a maximum of ___1 P.D.___ .

    a. 1 inch   b. one pipe diameter

    c. 2 inches   d. two pipe diameters

    Reference ___P3105.2___

10. A vent shall be installed a minimum of ___Two___ from the trap weir.

    a. 1 inch   b. one pipe diameter

    c. 2 inches   d. two pipe diameters

    Reference ___P3105.3___

11. An individual vent may vent a maximum of ___Two___ trap(s) or trap fixture(s) as a common vent.

    a. one   b. two

    c. three   d. four

    Reference ___P3107.1___

12. Where fixture drains connect at different levels and the vent connects as a vertical extension of the vertical drain, the vertical drain pipe connecting the two fixture drains shall be considered the vent for the lower fixture drain. If the vent pipe size is 2 inches, what is the maximum permitted discharge from the upper fixture drain?

    a. 1 d.f.u.   b. 4 d.f.u.

    c. 6 d.f.u.   d. 8 d.f.u.

    Reference ___3107.3___

13. A vertical wet vent shall have a minimum pipe size of ___3"___ where serving a total fixture unit load of 9 d.f.u.

    a. 2 inches   b. 2$\frac{1}{2}$ inches

    c. 3 inches   d. 4 inches

    Reference ___3107.3___

14. A 3-inch waste stack vent may be used for a maximum total discharge to the stack of __24 d.f.u.__.

    a. 8 d.f.u.   b. 12 d.f.u.

    (c.) 24 d.f.u.   d. 32 d.f.u.

    Reference _P3109.4_

15. What is the maximum number of fixtures connected to a horizontal branch drain that are permitted to be circuit vented?

    a. two   b. four

    c. six   (d.) eight

    Reference _P3110.1_

16. Which one of the following fixtures shall not be served by a combination waste and vent system?

    (a) laundry tub   b. floor drain

    c. standpipe   d. lavatory

    Reference _P3111.1_

17. The maximum permitted slope of a horizontal combination waste and vent pipe shall be __1/2__ unit vertical in 12 units horizontal.

    a. one-eighth   b. one-fourth

    (c.) one-half   d. one

    Reference _P3111.2.1_

18. A $2\frac{1}{2}$-inch diameter pipe used as a combination waste and vent shall serve a maximum of __6__ fixture units where connected to a horizontal branch.

    a. 3   (b.) 6

    c. 12   d. 26

    Reference _T3111.3_

19. Vents having a minimum developed length exceeding __40__ shall be increased by one nominal pipe size for the entire developed length of the vent pipe.

    a. 25

    b. 40

    c. 50

    d. 60

Reference __P3113.1__

20. Where used in a vent system, an individual air admittance valve shall be located a minimum of __4"__ above the horizontal branch drain or fixture drain being vented.

    a. 1 inch

    b. 2 inches

    c. 4 inches

    d. 6 inches

Reference __P3114.4__

21. Traps shall have a minimum liquid seal of __2__ inches and a maximum seal of __4__ inches.

    a. 2, 4

    b. 2, 6

    c. 3, 4

    d. 3, 6

Reference __P3201.2__

22. Where a fixture is separately trapped, what is the maximum vertical distance permitted from the fixture outlet to the trap weir?

    a. 6 inches

    b. 12 inches

    c. 18 inches

    d. 24 inches

Reference __P3201.6__

23. Where common trapped fixture outlets are permitted, they shall be located a maximum of __30"__ apart.

    a. 18 inches

    b. 24 inches

    c. 30 inches

    d. 36 inches

Reference __P3201.6 Exc. #2__

24. What is the minimum required trap size for a floor drain?

    a. 1¹/₄ inches

    b. 1¹/₂ inches

    c. 2 inches

    d. 2¹/₂ inches

Reference __TP3201.7__

25. The minimum required size for a trap arm serving a clothes washer standpipe shall be
_____2"_____.

    a. $1^1/_4$ inches   b. $1^1/_2$ inches

    (c.) 2 inches   d. $2^1/_2$ inches

Reference _P3201.7_

26. An open vent terminal extending through an exterior wall shall terminate a minimum of
_____10_____ feet from the lot line.

    a. 4   b. 5

    c. 6   (d.) 10

Reference _P3103.6_

27. In circuit venting, the maximum slope of the vent section of the horizontal branch drain shall be _____1_____ unit(s) vertical in 12 units horizontal.

    a. $^1/_4$ b. $^1/_2$

    (c.) 1   d. 2

Reference _P3110.3_

28. What is the maximum vertical distance between the fixture drain of a sink and a horizontal combination waste and vent pipe?

    a. 30 inches   b. 4 feet

    c 6 feet   (d.) 8 feet

Reference _P3111.2_

29. The vent or branch vent for multiple island fixture vents shall extend a minimum of _____6_____ inch(es) above the highest island fixture being vented before connecting to the outside vent terminal.

    a. 1   b. 2

    c. 4   (d.) 6

Reference _P3112.2_

30. Stack-type air admittance valves shall be located a minimum of ___6___ inches above the flood level rim of the highest fixture being vented.

    a. 4   (b.) 6

    c. 12   d. 15

    Reference _P3114.4_

31. Offsets in a stack vent installed for the waste stack shall be located a minimum of ___6___ inches above the flood level of the highest fixture.

    a. 2   b. 4

    (c) 6   d. 8

    Reference _P3109.3_

32. Vent pipes shall be a minimum of _1¼_ inches in diameter.

    (a) 1¼   b. 1½

    c. 2   d. 2½

    Reference _P3113.1_

33. The air pressure relief pipe from a pneumatic sewage ejector shall be a minimum of _1¼_ inch(es) in size.

    a. ³/₄   b. 1

    (c) 1¼   d. 1½

    Reference _P3113.4.2_

34. Where the developed length of a sump vent is 110 feet, what is the minimum required vent size for a sump with a sewage pump having a discharge capacity of 60 gallons per minute?

    a. 1¼ inches   b. 1½ inches

    (c.) 2 inches   d. 2½ inches

    Reference _T3113.4.1_

35. The horizontal distance from a fixture outlet to the trap weir shall be a maximum of _____30"_____ inches measured from the centerline of the fixture outlet to the centerline of the trap inlet.

a. 24

(b.) 30

c. 36

d. 48

Reference P3201.6

2006 IRC Chapters 33, 34, 35 and 36

General Electrical Requirements, Definitions, Services, and Branch Circuits and Feeder Requirements

OBJECTIVE: To obtain an understanding of the general requirements for electrical systems, equipment and components, including provisions addressing service conductors, branch circuits and feeders.

REFERENCE: Chapters 33, 34, 35 and 36, 2006 *International Residential Code*

KEY POINTS:
• What minimum working clearances are required at energized equipment and panelboards? How is clearance above a panelboard regulated? What is the minimum required headroom?

• What materials are permitted for use as conductors? What is the minimum permitted size?

• How are stranded conductors regulated? Conductors in parallel? How are conductors to be identified?

• What is the difference between "accessible" and "readily accessible"? How do "damp," "dry," and "wet" locations differ? How is "rainproof" different from "rain tight"?

• What are the requirements for "labeled" and "listed" equipment or materials? How do these terms related to that of "identified"?

• Where must the service disconnecting means be located? What is the maximum number of disconnects?

• How is the minimum load for ungrounded service conductors to be calculated?

• What minimum clearances are mandated for overhead service-drop conductors from doors, porches, decks, stairs and balconies? Above roofs? Above pedestrian areas and sidewalks? Over residential property and driveways? Over public streets and alleys or parking areas subject to truck traffic?

• How are connections at service heads to be installed?

• What is the basic method for creating a grounding electrode system? When is the use of a metal underground water pipe permitted for such a such system? A concrete-encased electrode? Ground rings? Plate electrodes?

• What is a "made" electrode? When are such electrodes permitted? What minimum sizes are required for rod and pipe electrodes? How are they to be installed? What is the minimum required resistance to ground? How may this be achieved?

409

KEY POINTS:
(Cont'd)

- Where is bonding required to ensure electrical continuity? What methods of bonding are permitted?
- What are the rating requirements for branch circuits serving lighting units and general utilization equipment? Fixed utilization equipment? Motors? Ranges and cooking appliances? Heating loads? Air-conditioning and heat pumps?
- How many branch circuits are required to serve the kitchen and dining area? The laundry area? Bathrooms?
- How is the size of feeders to be determined? Conductors?

Code Text: *Wood-framed structural members shall not be drilled, notched or altered in any manner except as provided for in* the IRC. *Electrical installations in hollow spaces, vertical shafts, and ventilation or air-handling ducts shall be made so that the possible spread of fire or products of combustion will not be substantially increased. Electrical penetrations through fire-resistance-rated walls, partitions, floors or ceilings shall be protected by approved methods to maintain the fire-resistance rating of the element penetrated. Penetrations through fireblocking and draftstopping shall be protected in an approved manner to maintain the integrity of the element penetrated.*

Discussion and Through penetrations and membrane penetrations of fire-resistive elements must be
Commentary: protected in accordance with the requirements of Section R317.3. Where fireblocking and draftstopping is mandated by Section R602.8 and Section R502.12, respectively, the penetration by electrical installations must also be done in an appropriate manner.

Wood stud

Through penetration
Sec. R317.3.1

Bored hole
Sec. R602.6, #2

Membrane penetrations
Sec. R317.3.2

Notch
Sec. R602.6, #1

A-379a

For the notching and boring of structural wood framing members, Sections R502.8, R602.6 and R802.7 must be followed. These requirements are in addition to those mandated in Table E3702.1 for the protection of the electrical cables and conductors installed in such framing.

Code Text: *Except as otherwise specified in Chapters 33 through 42, the dimension of the working space in the direction of access to panelboards and live parts likely to require examination, adjustment, servicing or maintenance while energized shall be not less than 36 inches (914 mm) in depth. In addition to the 36-inch dimension (914 mm), the work space shall not be less than 30 inches (762 mm) wide in front of the electrical equipment and not less than the width of such equipment. The work space shall be clear and shall extend from the floor or platform to a height of 6.5 feet (1981 mm).*

Discussion and Commentary: A minimum depth of 3 feet in front of the access point to an electrical panelboard is needed to allow for adequate clearance for a person from any live parts. It provides enough working depth to both maintain a safe distance and allow free movement when a person is servicing or maintaining the equipment.

6' 6" min

36" min   30" min

B-572

For SI: 1 inch = 25.4 mm, 1 foot = 304.8 mm.

Regardless of the size of the panelboard, a minimum clear width of 30 inches is required for working space. It need not be centered in front of the equipment but should be located for safe access. If the panelboard exceeds 30 inches in width, the clear space must extend the full width.

Code Text: *A dedicated space directly over a panelboard that extends from the panelboard to the structural ceiling or to a height of 6 feet (1829 mm) above the panelboard, whichever is lower, and has a width and depth equal to the equipment shall be dedicated and kept clear of equipment unrelated to the electrical equipment. Piping, ducts or equipment unrelated to the electrical equipment shall not be installed in such dedicated space. See exception for suspended ceilings with removable panels.*

Discussion and Commentary: In addition to the working space that must extend vertically in front of a panelboard, the area directly above the panelboard itself is regulated. This area must be free of any equipment or building features that would interfere with the direct run of cables and/or conduits to the equipment. The only exception allows for a suspended ceiling to encroach into the dedicated panelboard space, provided it has ceiling panels that are removable.

Dedicated space clear of unrelated equipment

Min of 6 ft or to structural ceiling

B-572a

For SI: 1 foot = 304.8 mm.

This provision reserves the area above the panelboard for the exclusive use of the electrical system and its components. Plumbing and HVAC piping, ducts and equipment must be located outside the designated area.

Code Text: *Where conductors are to be spliced, terminated or connected to fixtures or devices, a minimum length of 6 inches (150 mm) of free conductor shall be provided at each outlet, junction or switch point. The required length shall be measured from the point in the box where the conductor emerges from its raceway or cable sheath. Where the opening to an outlet, junction, or switch point is less than 8 inches (200 mm) in any dimension, each conductor shall be long enough to extend at least 3 inches (75 mm) outside of such opening.*

Discussion and Commentary: To provide adequate conductor length for connection purposes, at least 6 inches of free conductor must be available. The conductor extension also ensures that undue stress will not be placed on the conductor during or after the splice or connection.

Where the opening of the outlet, switch or junction box is limited in size to the point where the connection work must be done outside of the box, a minimum conductor length of 3 inches is required beyond the front edge of the box.

Code Text: *The minimum load for ungrounded service conductors and service devices that serve 100 percent of the dwelling unit load shall be computed in accordance with Table E3502.2. Ungrounded service conductors and service devices that serve less than 100 percent of the dwelling unit load shall be computed as required for feeders in acordance with Chapter 36.*

Discussion and Commentary: Table E3502.2 sets forth a procedure for the calculation of the minimum service load. It is based on the square footage of the dwelling unit, the number of small appliance circuits and the rating of any fastened-in-place, permanently connected or dedicated circuit-supplied appliances such as ranges, ovens, clothes dryers and water heaters. A demand factor is applied to the calculated sum, to which is added a load for the air-conditioning and/or heating equipment.

TABLE E3502.2
MINIMUM SERVICE LOAD CALCULATION

LOADS AND PROCEDURE
3 volt-amperes per square foot of floor area for general lighting and general use receptacle outlets.
Plus
1,500 volt-amperes total number of 20-ampere-rated small appliance and laundry circuits.
Plus
The nameplate volt-ampere rating of all fastened-in-place, permanently connected or dedicated circuit-supplied appliances such as ranges, ovens, cooking units, clothes dryers and water heaters.
Apply the following demand factors to the above subtotal:
The minimum subtotal for the loads above shall be 100 percent of the first 10,000 volt-amperes of the sum of the above loads plus 40 percent of any portion of the sum that is in excess of 10,000 volt-amperes.
Plus the largest of the following:
Nameplate rating(s) of the air-conditioning and cooling equipment.
Nameplate rating(s) of the heating where a heat pump is used without any supplemental electric heating.
Nameplate rating of the electric thermal storage and other heating systems where the usual load is expected to be continuous at the full nameplate value. Systems qualifying under this selection shall not be figured under any other category in this table.
One-hundred percent of nameplate rating of the heat pump compressor and sixty-five percent of the supplemental electric heating load for central electric space-heating systems. If the heat pump compressor is prevented from operating at the same time as the supplementary heat, the compressor load does not need to be added to the supplementary heat load for the total central electric space-heating load.
Sixty-five percent of nameplate rating(s) of electric space-heating units if less than four separately controlled units.
Forty percent of nameplate rating(s) of electric space-heating units of four or more separately controlled units.
The minimum total load in amperes shall be the volt-ampere sum calculated above divided by 240 volts.

Undergrounded service conductors must have an ampacity of no less than the load to be served. The minimum rating of the undergrounded service conductors for a single-family dwelling is 100 amperes, with an allowance of 60 amperes for other installations.

Code Text: *Conductors used as ungrounded service entrance conductors, service lateral conductors, and feeder conductors that serve as the main power feeder to a dwelling unit shall be those listed in Table E3503.1. Ungrounded service conductors shall have a minimum size in accordance with Table E3503.1. The grounded conductor size shall not be less than the maximum unbalance of the load and its size shall not be smaller than the required minimum grounding electrode conductor size specified in Table E3503.1.*

Discussion and Commentary: The minimum conductor size listed in Table E3503.1 is based on the maximum allowable load. Conductor size differs based on the allowable ampacity and the type of conductor material, either copper or aluminum. The same criteria are used for sizing the grounding electrode conductor.

TABLE E3503.1
SERVICE CONDUCTOR AND GROUNDING ELECTRODE CONDUCTOR SIZING

CONDUCTOR TYPES AND SIZES—THHN, THHW, THW, THWN, USE, XHHW, THW-2, THWN-2, XHHW-2, SE, USE-2 (Parallel sets of 1/0 and larger conductors are permitted in either a single raceway or in separate raceways)		ALLOWABLE AMPACITY	MINIMUM GROUNDING ELECTRODE CONDUCTOR SIZE[a]	
Copper (AWG)	Aluminum and copper–clad aluminum (AWG)	Maximum load (amps)	Copper (AWG)	Aluminum (AWG)
4	2	100	8[b]	6[c]
3	1	110	8[b]	6[c]
2	1/0	125	8[b]	6[c]
1	2/0	150	6[c]	4
1/0	3/0	175	6[c]	4
2/0	4/0 or two sets of 1/0	200	4[d]	2[d]
3/0	250 kcmil or two sets of 2/0	225	4[d]	2[d]
4/0 or two sets of 1/0	300 kcmil or two sets of 3/0	250	2[d]	1/0[d]
250 kcmil or two sets of 2/0	350 kcmil or two sets of 4/0	300	2[d]	1/0[d]
350 kcmil or two sets of 3/0	500 kcmil or two sets of 250 kcmil	350	2[d]	1/0[d]
400 kcmil or two sets of 4/0	600 kcmil or two sets of 300 kcmil	400	1/0[d]	3/0[d]

For SI: 1 inch = 25.4 mm.

a. Where protected by a ferrous metal raceway, grounding electrode conductors shall be electrically bonded to the ferrous metal raceway at both ends.

b. Eight AWG grounding electrode conductors shall be protected with metal conduit or nonmetallic conduit.

c. Where not protected, 6 AWG grounding electrode conductors shall closely follow a structural surface for physical protection. The supports shall be spaced not more than 24 inches on center and shall be within 12 inches of any enclosure or termination.

d. Where the sole grounding electrode system is a ground rod or pipe as covered in Section E3508.2, the grounding electrode conductor shall not be required to be larger than 6 AWG copper or 4 AWG aluminum. Where the sole grounding electrode system is the footing steel as covered in Section E3508.1.2, the grounding electrode conductor shall not be required to be larger than 4 AWG copper conductor.

An underground service conductor that serves a workshop, detached garage or similar accessory structure need only have a minimum rating of 60 amperes. The sizing of such conductors shall be in accordance with Chapter 36 for feeders.

Code Text: *Open conductors and multiconductor cables without an overall outer jacket shall have a clearance of not less than 3 feet (914 mm) from the sides of doors, porches, decks, stairs, ladders, fire escapes, and balconies, and from the sides and bottom of windows that open.*

Discussion and Commentary: The clearances for overhead service drops are limited to conductors and cables that are not installed in a raceway or provided with an approved outer jacket. The requirement reduces the potential for damage to the service conductors and limits the risk of accidental contact. The clearance requirement from openable windows applies only to the sides and below the window opening. Service conductors and drip loops located just above an opening are deemed to be out of reach.

3 ft clearance

Clearance not required from nonopenable window

Clearance not required above openings

Clearance not required between window and SE cable or conductors in raceway

For SI: 1 foot = 304.8 mm.

B-575

The need for a minimum 3-foot separation is based on the potential proximity of people to the service conductors. It is expected that contact is possible from open windows, porches, stairs, decks and similar occupiable areas. No clearance is required from nonopenable windows.

Code Text: *Conductors shall have a vertical clearance of not less than 8 feet (2438 mm) above the roof surface. The vertical clearance above the roof level shall be maintained for a distance of not less than 3 feet (914 mm) in all directions from the edge of the roof. See four exceptions allowing for reduced clearances.*

Discussion and Commentary: The general provisions mandate at least an 8-foot clearance between a service-drop conductor and a roof. This clearance must not only occur directly above the roof surface but also extend horizontally at least 3 feet, except for the point where the service drop is attached to the side of the building. However, the separation may be reduced to 3 feet where the roof slope is 4:12 or greater. It is anticipated that travel across such a steeply-sloping roof will be minimal.

For SI: 1 inch = 25.4 mm, 1 foot = 304.8 mm.

B-576

It is possible that the clearance will need to be increased to more than 8 feet where the roof is expected to have pedestrian access. This could occur where there is a rooftop court or sundeck or where the roof is also the upper deck of a parking garage.

Code Text: *Service-drop conductors shall have the following minimum clearances from final grade: 1) for service-drop cables supported on and cabled together with a grounded bare messenger wire, the minimum vertical clearance shall be 10 feet (3048 mm) at the electric service entrance to buildings, at the lowest point of the drip loop of the building electric entrance, and above areas or sidewalks accessed by pedestrians only; 2) twelve feet (3658 mm)—over residential property and driveways; and 3) eighteen feet (5486 mm)—over public streets, alleys, roads or parking areas subject to truck traffic.*

Discussion and Commentary: The minimum clearances for service-drop conductors are based on the expected height of potential hazards. Where there is no anticipated vehicle traffic, at least 10 feet of vertical clearance is mandated. Where vehicles are present, higher clearances are necessary.

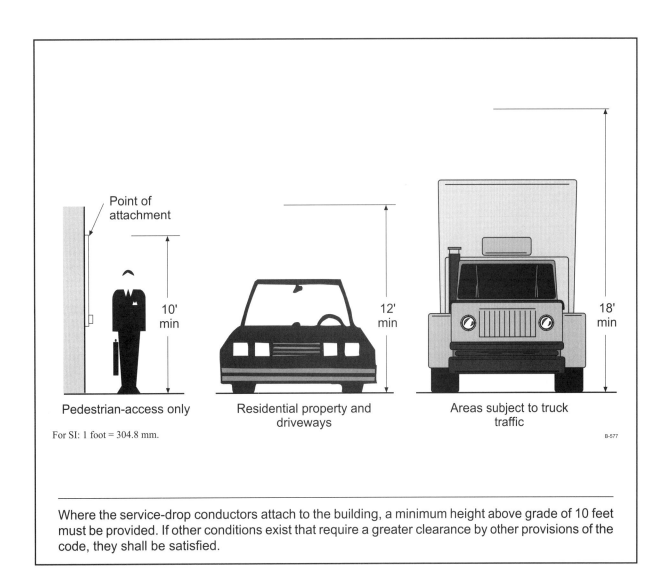

For SI: 1 foot = 304.8 mm.

B-577

Where the service-drop conductors attach to the building, a minimum height above grade of 10 feet must be provided. If other conditions exist that require a greater clearance by other provisions of the code, they shall be satisfied.

Code Text: *All electrodes specified in Sections E3508.1.1* (metal underground water pipe), *E3508.1.2* (concrete-encased electrode), *E3508.1.3* (ground rings), *E3508.1.4* (rod and pipe electrodes) *and E3508.1.5* (plate electrodes) *that are present at each building or structure served shall be bonded together to form the grounding electrode system. Where none of these electrodes are available, on or more of the electrodes specified in Sections E3508.1.3, E3508.1.4 and E3508.1.5 shall be installed and used.* See exception for concrete-encased electrodes of existing buildings.

Discussion and Commentary: One method of providing a grounding electrode is the use of a metal underground water pipe that is in direct contact with the earth for a minimum distance of 10 feet. Where this method is used, it is necessary to also provide a supplementary electrode, such as electrode of pipe, conduit or rod, a ground ring, or a plate electrode.

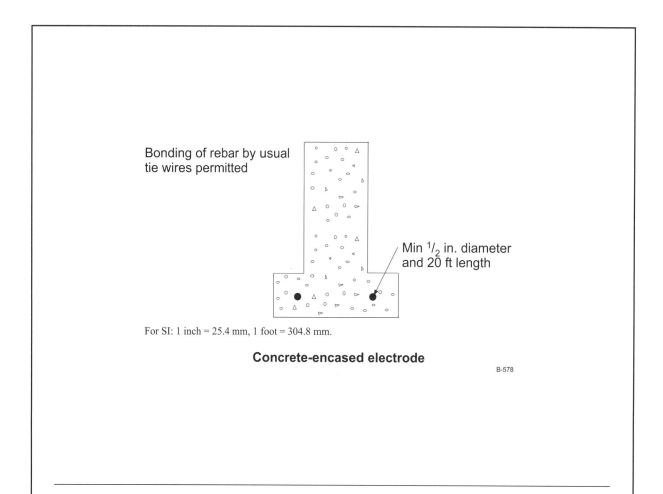

Bonding of rebar by usual tie wires permitted

Min ¹/₂ in. diameter and 20 ft length

For SI: 1 inch = 25.4 mm, 1 foot = 304.8 mm.

Concrete-encased electrode

B-578

One or more complying reinforcing bars can be used as a grounding electrode when encased in concrete. The bar must be at least ¹/₂ inch in diameter, be at least 20 feet in length, and be located within and near the bottom of the concrete foundation or footing.

Code Text: *Where more than one rod, pipe or plate electrode is used, each electrode of one grounding system shall be not less than 6 feet (1829 mm) from any other electrode of another grounding system. Two or more grounding electrodes that are effectively bonded together shall be considered as a single grounding electrode system.*

Discussion and Commentary: Electrodes of rods or pipes can be utilized in a grounding electrode system. The minimum length of such made electrodes shall be 8 feet, embedded below the permanent moisture level if possible. Pipes or conduit must be a minimum of $^3/_4$-inch diameter in trade size, while iron or steel rods are to be at least $^5/_8$-inch diameter.

For SI: 1 inch = 25.4 mm, 1 foot = 304.8 mm, 1 degree = 0.1745 rad.

Rod and pipe electrodes   B-579

At least 8 feet of the electrode must be in contact with the soil. The electrode should be driven vertically unless soil conditions are prohibitive. In such cases, the electrode can be driven at an angle of up to 45 degrees or placed in a minimum 30-inch deep trench.

Code Text: *The requirements for circuits having two or more outlets, or receptacles, other than the receptacle circuits of Section E3603.2* (kitchen and dining area receptacles), *are summarized in Table E3602.13. Branch circuits in dwelling units shall supply only loads within that dwelling unit or loads associated only with that dwelling unit. Branch circuits required for the purpose of lighting, central alarm, signal, communications or other needs for public or common areas of a two-family dwelling shall not be supplied from equipment that supplies an individual dwelling unit.*

Discussion and Commentary: Whereas a 20-ampere rated receptacle is not permitted to be installed on a 15-ampere rated circuit, it is acceptable for a 15-ampere rated receptacle to be located on a 20-ampere circuit. However, Section E3902.1.1 requires that a single receptacle on an individual branch be rated equal to the rating of the branch circuit.

TABLE E3602.13
BRANCH-CIRCUIT REQUIREMENTS—SUMMARY[a,b]

	CIRCUIT RATING		
	15 amp	20 amp	30 amp
Conductors: Minimum size (AWG) circuit conductors	14	12	10
Maximum overcurrent-protection device rating Ampere rating	15	20	30
Outlet devices: Lampholders permitted Receptacle rating (amperes)	Any type 15 maximum	Any type 15 or 20	N/A 30
Maximum load (amperes)	15	20	30

a. These gages are for copper conductors.
b. N/A means not allowed.

A two-family dwelling may have areas that are common to both residences, such as landscape lighting and exterior accent lighting. Under such conditions, the power for the common area loads must be supplied from a separate service and panelboard.

Code Text: *Central heating equipment other than fixed electric space heating shall be supplied by an individual branch circuit. A minimum of two 20-ampere-rated branch circuits shall be provided to serve all wall and floor receptacle outlets located in the kitchen, pantry, breakfast area, dining area or similar area of a dwelling. A minimum of one 20-ampere-rated branch circuit shall be provided for receptacles located in the laundry area and shall serve only receptacle outlets located in the laundry area. A minimum of one 20-ampere branch circuit shall be provided to supply the bathroom receptacle outlet(s). The minimum number of branch circuits shall be determined from the total calculated load and the size or rating of the circuits used.*

Discussion and Commentary: In addition to general lighting and convenience outlet circuits, a number of specific areas of a dwelling unit must be provided with their own dedicated branch circuits.

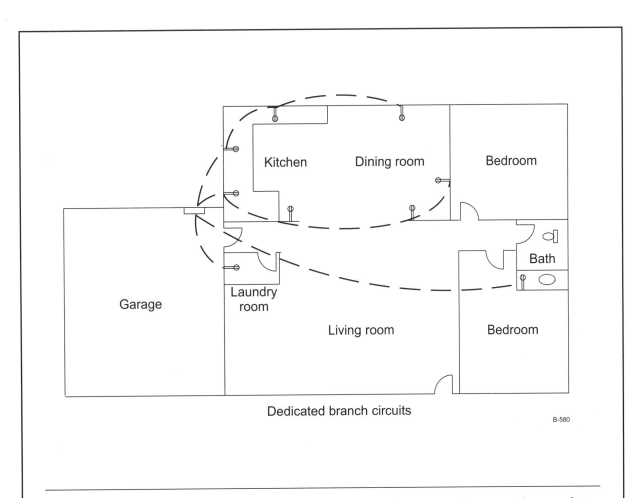

Dedicated branch circuits

B-580

The receptacles located at the kitchen countertop must be served by a minimum of two 20-ampere-rated branch circuits. One or both of the circuits may also be used to satisfy the requirements for the other outlets located in the kitchen and dining areas.

Topic: Circuit Identification

Reference: IRC E3606.2

Category: Branch Circuits and Feeders

Subject: Panelboards

Code Text: *All circuits and circuit modifications shall be legibly identified as to their clear, evident, and specific purpose or use. The identification shall include sufficient detail to allow each circuit to be distinguished from all others. The identification shall be included in a circuit directory located on the face of the panelboard enclosure or inside the panel door.*

Discussion and Commentary: The importance of adequately identifying the various circuits cannot be overstated. The code mandates that all circuits be noted in a legible fashion, indicating their purpose in a clear, evident and specific manner. It is also important to adequately differentiate each circuit from all of the others.

Circuits to be clearly and specifically identified

Panelboard

The language of the code requires some degree of interpretation by the building official, as the provision is primarily performance in nature. However, it is obvious that generic terminology such as "lighting" or "kitchen" is not acceptable.

Quiz

Study Session 17
IRC Chapters 33, 34, 35 and 36

1. Energized parts operating at a minimum of ___50 v___ shall be guarded against accidental contact by people through the use of approved enclosures.

    (a.) 50 volts   b. 60 volts

    c. 90 volts   d. 110 volts

    Reference E 3304.8

2. An electrical panelboard requiring access while energized shall be provided with a minimum of ___36___ in depth measured in the direction of access.

    a. 30 inches   (b.) 36 inches

    c. 42 inches   d. 48 inches

    Reference E3305.2

3. Work space in front of an electrical panelboard shall be a minimum of ___30"___ in width, but no less than the width of the panelboard.

    (a.) 30 inches   b. 36 inches

    c. 42 inches   d. 48 inches

    Reference E3305.2

4. The dedicated space above an electrical panelboard shall be a minimum of _6 FT_ high or to the structural ceiling, whichever is lower.

    a. 4 feet   b. 5 feet

    (c.) 6 feet   d. 6.5 feet

Reference _E 3305.2_

5. The minimum size of electrical conductors for feeders and branch circuits, when of copper, shall be ____14____ AWG.

    a. No. 16   (b.) No. 14

    c. No. 12   d. No. 10

Reference _E3306.2_

6. Where electrical conductors are to be spliced, terminated or connected to fixtures or devices, a minimum length of __6"__ of free conductor shall be provided at each outlet, junction or switch point.

    a. 4 inches   (b.) 6 inches

    c. 8 inches   d. 12 inches

Reference _E3306.10.3_

7. Insulated grounded conductors of sizes 6 AWG and smaller may be identified by all but which one of the following methods?

    a. continuous white outer finish

    b. continuous natural gray outer finish

    (c.) continuous black outer finish

    d. three continuous white stripes (on other than green insulation)

Reference _E3307.1_

8. Equipment grounding conductors of sizes 6 AWG and smaller may be identified by all but which one of the following methods?

    (a.) continuous white color

    b. continuous green color

    c. bare

    d. continuous green color with one or more yellow stripes

Reference _E3307.2_

9. In general, which of the following continuous colors is permitted for the insulation on ungrounded conductors?

    a. gray   b. white

    c. green   (d.) red

Reference _E3307.3_

10. What is the minimum permitted size of a grounded service conductor serving a maximum load of 150 amps?

    a. No. 1 aluminum   (b.) No. 1 copper

    c. No. 1/0 aluminum   d. No. 1/0 copper

Reference _TE3503.1_

11. An open service conductor without an overall outer jacket shall have a minimum clearance of ___3"___ from the sides of doors, porches and openable windows.

    (a.) 3 feet   b. 4 feet

    c. 6 feet   d. 8 feet

Reference _E3504.1_

12. An overhead service conductor shall have a minimum clearance above a 3:12 roof of ___8'___.

    a. 3 feet   b. 6 feet

    c. 7 feet   (d.) 8 feet

Reference _E3504.2.1_

13. Service-drop conductors shall have a minimum clearance of ___12'___ over residential property and driveways.

    a. 8 feet   b. 10 feet

    (c.) 12 feet   d. 15 feet

Reference _E3504.2.2.II 2_

14. In no case shall the point of attachment of service-drop conductors to a building be less than ___10f___ above finished grade.

    a. 8 feet   (b.) 10 feet

    c. 12 feet   d. 14 feet

    Reference __E3504.3__

15. A metal underground water pipe used as part of the grounding electrode system shall be in direct contact with the earth for a minimum of ___Nf___.

    a. 8 feet   (b) 10 feet

    c. 15 feet   d. 20 feet

    Reference __E3500.1.1__

16. Where more than one rod, pipe or plate electrode is used in one grounding electrode system, each electrode shall be located a minimum of ___6 ft___ from any other electrode of another grounding system.

    a. 5 feet   (b.) 6 feet

    c. 10 feet   d. 20 feet

    Reference __E3508.3__

17. Where a grounding electrode is made of an iron or steel rod, the rod shall be a minimum of ___5/8___ in diameter and a minimum of ___8___ in length.

    a. $^1/_2$ inch, 8 feet   b. $^1/_2$ inch, 10 feet

    (c.) $^5/_8$ inch, 8 feet   d. $^5/_8$ inch, 10 feet

    Reference __E3508.1 & #2__

18. Where used outdoors, an aluminum or copper-clad aluminum grounding conductor shall be installed a minimum of ___18___ from the earth.

    a. 6 inches   b. 12 inches

    (c.) 18 inches   d. 30 inches

    Reference __E3510.2__

19. The rating of any one cord- and plug-connected utilization equipment not fastened in place shall be a maximum of _____ of the branch-circuit ampere rating.

   (a.) 80 percent   b. 100 percent

   c. 110 percent   d. 125 percent

   Reference _E3602. 3_

20. A minimum of _Two_ branch circuit(s) shall be provided to serve receptacles located in the kitchen, pantry, breakfast area and dining area.

   a. one 15-ampere-rated   b. one 20-ampere-rated

   c. two 15-ampere-rated   (d.) two 20-ampere-rated

   Reference _E3603.2_

21. A minimum of _1_ branch circuit(s) shall be provided for receptacles located in the laundry room and a minimum of _1_ branch circuit(s) shall be provided to supply the bathroom receptacle outlet(s).

   a. one 15-ampere-rated, one 15-ampere-rated

   (b.) one 20-ampere-rated, one 20-ampere-rated

   c. one 15-ampere-rated, two 15-ampere-rated

   d. one 20-ampere-rated, two 20 ampere-rated

   Reference _E3603.3 d3603 4_

22. In the calculation of electrical feeder loads, an electric clothes dryer load shall be considered _5,000_ VA for each dryer circuit or the nameplate rating load of each dryer, whichever is greater.

   a. 4,000   (b.) 5,000

   c. 5,500   d. 6,000

   Reference _T3604,2 (1)_

23. When sizing feeder conductors, a minimum unit load of _3 VA_ shall constitute the minimum lighting and convenience receptacle load for each square foot of floor area.

   a. two volt-amperes   (b.) three volt-amperes

   c. five volt-amperes   d. six volt-amperes

   Reference _E3604.4_

24. What is the maximum overcurrent-protection-device rating permitted for No. 12 copper conductors?

    a. 15 amps   (b.) 20 amps

    c. 25 amps   d. 30 amps

    Reference _3605.53_

25. Which one of the following ratings is not a standard ampere rating for fuses and inverse time circuit breakers?

    a. 25 amperes   b. 50 amperes

    (c.) 75 amperes   d. 100 amperes

    Reference _3605.6_

26. Where metallic plugs or plates are used with nonmetallic electrical enclosures, they shall be recessed a minimum of _1/4_ inch from the outer surface of the enclosure.

    a. $^1/_8$ (b.) $^1/_4$

    c. $^3/_8$ d. $^1/_2$

    Reference _E3304.5_

27. Electrical panelboards and overcurrent protection devices shall not be located in _a_.

    (a.) clothes closets   b. sleeping rooms

    c. garages   d. storage rooms

    Reference _E3305.4_

28. Where the opening to an outlet, junction, or switch point is less than 8 inches in any dimension, each electrical conductor shall be long enough to extend a minimum of _3_ inches outside of such opening.

    a. 2   b. (3)

    c. 4   d. 6

    Reference _E3306.10.3_

29. Electrical service disconnecting means shall not be installed in __d__.

   a. storage closets   b. laundry rooms

   c. sleeping rooms   (d.) bathrooms

   Reference E3501 6.2

30. In order to be used as a part of the grounding electrode system, interior metal water pipe shall be located a maximum of __5__ feet from the entrance to the building.

   a.. 3   (b) 5

   c. 6   d. 10

   Reference E3508.1.1

31. Electrical conductors located above a roof shall be installed with a minimum vertical clearance of __A__ feet where the roof has a slope of 8:12.

   a.) 3   b. 4

   c. 6   d. 7

   Reference E3504.2.1 Exp. 2

32. Service-entrance cables shall be supported by straps or other approved means within __12__ inches of every electrical service head and at maximum intervals of __30__ inches.

   a. 8, 30   b. 8, 36

   c.) 12, 30   d. 12, 36

   Reference E3505.7

33. A plate electrode installed as a part of a grounding electrode system shall be located a minimum of __30__ inches below the surface of the earth.

   a. 12   b. 18

   c. 24   (d.) 30

   Reference E3508.1.5

34. Where an electrical branch circuit has multiple outlets, what is the minimum required size of circuit conductors for a circuit rated at 30 amps?

    a. 14 AWG   b. 12 AWG

    (c.) 10 AWG   d. 8 AWG

Reference _E3602.13_

35. An overcurrent protection device shall be installed so that the center of the switch or circuit breaker, when in its highest position, is located a maximum of ___C___ above the floor or working platform.

    a. 5 feet, 0 inches   b. 6 feet, 4 inches

    (c.) 6 feet, 7 inches   d. 7 feet, 0 inches

Reference _E3605.7 #5_

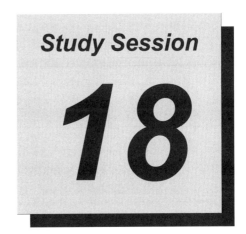

Study Session

18

2006 IRC Chapters 37, 38, 39, 40 and 41

Lighting Distribution, Devices and Lighting Fixtures, Appliance Installation and Swimming Pools

OBJECTIVE: To gain an understanding of the provisions regulating wiring methods, power and lighting distribution, devices and lighting fixtures, appliance installation and swimming pools.

REFERENCE: Chapters 37, 38, 39, 40 and 41, 2006 *International Residential Code*

KEY POINTS:

- How are electrical cables to be installed where they are run in attics across the top of structural members? Where they are run parallel to framing members? How must they be protected from physical damage?

- What are the minimum cover requirements for direct buried cable or raceways installed underground? How are direct buried cables to be protected from damage where emerging from the ground?

- How must backfill be placed in an excavation containing electrical cables or raceways?

- What method is used to determine the minimum spacing requirements for convenience receptacles? What wall spaces are to be considered? How are floor receptacles regulated?

- Where are small appliance receptacles regulated? How many receptacles must be provided? How must they be distributed?

- In what manner are counter receptacles to be located? How do the provisions differ for island counter spaces? For peninsular counter spaces?

- How many wall receptacles are required in a bathroom? Outdoors? In laundry areas? In basements and garages? In hallways?

- Where is ground-fault circuit-interrupter protection mandated?

- In what locations are arc-fault circuit interrupter protection mandated?

- In what rooms and spaces must wall-switch-controlled lighting outlets be installed? Under what conditions are wall-switch-controlled receptacles permitted as an alternate to lighting outlets?

- How is nonmetallic-sheathed cable to be installed in nonmetallic boxes? What is the minimum required depth of such boxes? How is box volume and fill calculated? How are boxes to be installed?

- How are luminaires to be installed in wet or damp locations? In bathtub and shower areas? In clothes closets?

KEY POINTS: *(Cont'd)*

- What are the limits for supporting luminaires by the screw shell of a lampholder? By an outlet box? What clearances are required between a recessed luminaire and thermal insulation? Other combustible material?

- How is track lighting to be installed? Where is track lighting prohibited?

- Where shall receptacle outlets be located in relationship to a pool, spa or hot tub? Which receptacles require ground-fault circuit-interrupter protection? How are luminaires, lighting outlets and ceiling-suspended paddle fans regulated?

- What parts of a swimming pool, spa or hot tub must be bonded? What are the acceptable bonding methods? Which equipment must be grounded? How is pool equipment regulated?

Code Text: *The allowable wiring methods for electrical installations shall be those listed in Table E3701.2. Single conductors shall be used only where part of one of the recognized wiring methods listed in Table E3701.2. As used in the IRC, abbreviations of the wiring-method types shall be as indicated in Table E3701.2.*

Discussion and Commentary: Single insulated conductors are not permitted as a wiring method without being part of a cable assembly or installed in tubing or conduit. In many older houses, a wiring method known as "knob-and-tube" was used, consisting of single insulated conductors supported in free air on porcelain insulators and, where run through a wood framing member, installed through insulating tubes. Where remodeling, retrofitting or repair is being done in such a house, the transition to new wiring should be to cable.

TABLE E3701.2
ALLOWABLE WIRING METHODS

ALLOWABLE WIRING METHOD	DESIGNATED ABBREVIATION
Armored cable	AC
Electrical metallic tubing	EMT
Electrical nonmetallic tubing	ENT
Flexible metal conduit	FMC
Intermediate metal conduit	IMC
Liquidtight flexible conduit	LFC
Metal-clad cable	MC
Nonmetallic sheathed cable	NM
Rigid nonmetallic conduit	RNC
Rigid metallic conduit	RMC
Service entrance cable	SE
Surface raceways	SR
Underground feeder cable	UF
Underground service cable	USE

In many older houses, the wiring is still sound and in good shape, but when additions or extensions are made, the possibility for arcing, bad connections, heating and overloading come into play. Thus, one of the allowable wiring methods must be utilized.

Code Text: *Wiring methods shall be installed and supported in accordance with Table E3702.1.*

Discussion and Commentary: A number of different installation conditions are addressed in an effort to protect and support various wiring methods permitted by the code. For NM cable and other materials subject to damage, the installation procedures address the common situation of cable within stud cavities. If the cable is run parallel along the side of the stud or furring strip, it must be located at least $1\frac{1}{4}$ inches from the stud edge or otherwise physically protected. The same minimum dimension is necessary where cable is run perpendicular to the vertical framing members through bored holes. Where the mandated depth is not provided, it is common to install a minimum $\frac{1}{16}$-inch-thick steel plate across the stud edge.

TABLE E3702.1
GENERAL INSTALLATION AND SUPPORT REQUIREMENTS FOR WIRING METHODS[a, b, c, d, e, f, g, h, i, j, k]

INSTALLATION REQUIREMENTS (Requirement applicable only to wiring methods marked "A")	AC MC	EMT IMC RMC	ENT	FMC LFC	NM UF	RNC	SE	SR[a]	USE
Where run parallel with the framing member or furring strip, the wiring shall be not less than $1\frac{1}{4}$ inches from the edge of a furring strip or a framing member such as a joist, rafter or stud or shall be physically protected.	A	—	A	A	A	—	A	—	—
Bored holes in framing members for wiring shall be located not less than $1\frac{1}{4}$ inches from the edge of the framing member or shall be protected with a minimum 0.0625-inch steel plate or sleeve, a listed steel plate or other physical protection.	A[k]	—	A[k]	A[k]	A[k]	—	A[k]	—	—
Where installed in grooves, to be covered by wallboard, siding, paneling, carpeting, or similar finish, wiring methods shall be protected by 0.0625-inch-thick steel plate, sleeve, or equivalent, a listed steel plate or by not less than $1\frac{1}{4}$-inch free space for the full length of the groove in which the cable or raceway is installed.	A	—	A	A	A	—	A	A	A
Securely fastened bushings or grommets shall be provided to protect wiring run through openings in metal framing members.	—	—	A[j]	—	A[j]	—	A[j]	—	—
The maximum number of 90-degree bends shall not exceed four between junction boxes.	—	A	A	A	—	A	—	—	—
Bushings shall be provided where entering a box, fitting or enclosure unless the box or fitting is designed to afford equivalent protection.	A	A	A	A	—	A	—	A	—
Ends of raceways shall be reamed to remove rough edges.	—	A	A	A	—	A	—	A	—
Maximum allowable on center support spacing for the wiring method in feet.	4.5[b, c]	10	3[b]	4.5b	4.5[i]	3[d]	2.5[e]	—	2.5[e]
Maximum support distance in inches from box or other terminations.	12[b, f]	36	36	12[b,g]	12[h,i]	36	12	—	12

For SI: 1 inch = 25.4 mm, 1 foot = 304.8 mm, 1 degree = 0.009 rad.
a. Installed in accordance with listing requirements.
b. Supports not required in accessible ceiling spaces between light fixtures where lengths do not exceed 6 feet.
c. Six feet for MC cable.
d. Five feet for trade sizes greater than 1 inch.
e. Two and one-half feet where used for service or outdoor feeder and 4.5 feet where used for branch circuit or indoor feeder.
f. Twenty-four inches where flexibility is necessary.
g. Thirty-six inches where flexibility is necessary.
h. Within 8 inches of boxes without cable clamps.
i. Flat cables shall not be stapled on edge.
j. Bushings and grommets shall remain in place and shall be listed for the purpose of cable protection.
k. See Sections R502.8 and R802.7 for additional limitations on the location of bored holes in horizontal framing members.

A concern in accessible attics is the presence of cables that may be subject to contact and potential physical damage. The code requires protection for these cables, particularly where they are located within 6 feet of the attic access opening.

Code Text: *Direct buried cable or raceways shall be installed in accordance with the minimum cover requirements of Table E3703.1. Direct buried conductors and cables emerging from the ground shall be protected by enclosures or raceways extending from the minimum cover distance below grade required by Section 3703.1 to a point at least 8 feet (2438 mm) above finished grade. In no case shall the protection be required to exceed 18 inches (457 mm) below finished grade.*

Discussion and Commentary: To protect cables or raceways buried below ground level from damage, the code establishes a minimum depth. The burial depth is dependent upon two factors: the location or method of burial, and the type of wiring method. Where direct buried conductors and cables are permitted, they must be protected from below ground level to a point high enough that physical damage is improbable.

TABLE E3703.1
MINIMUM COVER REQUIREMENTS, BURIAL IN INCHES[a, b, c, d, e]

LOCATION OF WIRING METHOD OR CIRCUIT	TYPE OF WIRING METHOD OR CIRCUIT				
	1 Direct burial cables or conductors	2 Rigid metal conduit or intermediate metal conduit	3 Nonmetallic raceways listed for direct burial without concrete encasement or other approved raceways	4 Residential branch circuits rated 120 volts or less with GFCI protection and maximum overcurrent protection of 20 amperes	5 Circuits for control of irrigation and landscape lighting limited to not more than 30 volts and installed with type UF or in other identified cable or raceway
All locations not specified below	24	6	18	12	6
In trench below 2-inch-thick concrete or equivalent	18	6	12	6	6
Under a building	0 (In raceway only)	0	0	0 (In raceway only)	0 (In raceway only)
Under minimum of 4-inch-thick concrete exterior slab with no vehicular traffic and the slab extending not less than 6 inches beyond the underground installation	18	4	4	6 (Direct burial) 4 (In raceway)	6 (Direct burial) 4 (In raceway)
Under streets, highways, roads, alleys, driveways and parking lots	24	24	24	24	24
One- and two-family dwelling driveways and outdoor parking areas, and used only for dwelling-related purposes	18	18	18	12	18
In solid rock where covered by minimum of 2 inches concrete extending down to rock	2 (In raceway only)	2	2	2 (In raceway only)	2 (In raceway only)

For SI: 1 inch = 25.4 mm.
a. Raceways approved for burial only where encased concrete shall require concrete envelope not less than 2 inches thick.
b. Lesser depths shall be permitted where cables and conductors rise for terminations or splices or where access is otherwise required.
c. Where one of the wiring method types listed in columns 1 to 3 is combined with one of the circuit types in columns 4 and 5, the shallower depth of burial shall be permitted.
d. Where solid rock prevents compliance with the cover depths specified in this table, the wiring shall be installed in metal or nonmetallic raceway permitted for direct burial. The raceways shall be covered by a minimum of 2 inches of concrete extending down to the rock.
e. Cover is defined as the shortest distance in inches (millimeters) measured between a point on the top surface of any direct-buried conductor, cable, conduit or other raceway and the top surface of finished grade, concrete, or similar cover.

The backfill used to cover cables or raceways must be placed in such a manner as to avoid damage to the wiring method. It may be necessary to use boards, sleeves, granular material or other suitable methods to protect the raceway or cable from physical damage.

Code Text: *In every kitchen, family room, dining room, living room, parlor, library, den, sun room, bedroom, recreation room, or similar room or area of dwelling units, receptacle outlets shall be installed . . . so that no point measured horizontally along the floor line in any wall space is more than 6 feet (1829 mm), from a receptacle outlet. A wall space shall include: 1) any space that is 2 feet (610 mm) or more in width, (including space measured around corners), and that is unbroken along the floor line by doorways, fireplaces, and similar openings; 2) the space occupied by fixed panels in exterior wall, excluding sliding panels; and 3) the space created by fixed room dividers such as railings and free standing bar-type counters. Receptacle outlets in floors shall not be counted as part of the required number of receptacle outlets except where located within 18 inches (457 mm) of the wall.*

Discussion and Commentary: The maximum spacing between convenience receptacles is basically 12 feet, allowing flexibility in the placement of electric appliances without the need of any extension cord. Those spaces that are subject to the requirement tend to be habitable spaces, those areas of a dwelling used for living, sleeping, dining or cooking.

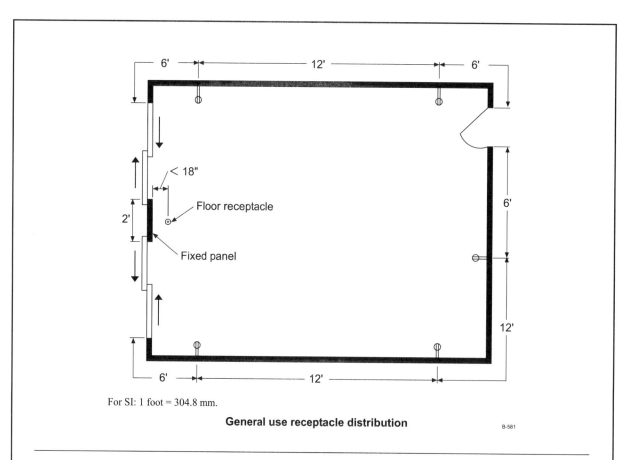

For SI: 1 foot = 304.8 mm.

General use receptacle distribution

B-581

If a wall space is at least 2 feet wide, it must be provided with a receptacle outlet. It is not acceptable to measure across a doorway or similar opening for compliance with the 6-foot rule. An extension cord across a door opening is a considerable hazard.

Code Text: *In kitchens and dining rooms of dwelling units, receptacle outlets for counter spaces shall be installed . . . at each wall counter space 12 inches (305 mm) or wider. Receptacle outlets shall be installed so that no point along the wall line is more than 24 inches (610 mm), measured horizontally from a receptacle outlet in that space. See exception for outlets on a wall directly behind a range or sink. Receptacle outlets shall be located not more than 20 inches (508 mm) above the countertop. Receptacle outlets shall not be installed in a face-up position in the work surfaces or countertops.*

Discussion and Commentary: Receptacle outlets are mandated in a repetitive fashion along a kitchen counter for both safety and convenience. With a number of appliances potentially in use at the same time, and because they are likely to be moved to a number of different positions, it is important that access to multiple receptacles be provided.

For SI: 1 inch = 25.4 mm, 1 foot = 304.8 mm.

Countertop receptacles

B-582

If an island counter space exceeds a specified size, it too must be provided with at least one receptacle outlet. The same holds true for peninsular counter spaces of similar dimensions. Appliance cords pose an obvious problem when extended through work areas or traffic paths.

Code Text: *At least one wall receptacle outlet shall be installed in bathrooms and such outlet shall be located within 36 inches (914 mm) of the outside edge of each lavatory basin. At least one receptacle outlet accessible at grade level and not more than 6 feet 6 inches (1981 mm) above grade, shall be installed outdoors at the front and back of each dwelling unit having direct access to grade. At least one receptacle outlet shall be installed to serve laundry appliances. At least one receptacle outlet, in addition to any provided for laundry equipment, shall be installed in each basement and in each attached garage, and in each detached garage that is provided with electrical power. Hallways of 10 feet (3048 mm) or more in length shall have at least one receptacle outlet.*

Discussion and Commentary: In addition to habitable spaces, those spaces typically defined as not habitable also require a limited number of receptacle outlets. Although the number of outlets mandated may be fewer than required for habitable rooms, it is important that outlets be provided to serve any equipment or activity that is anticipated for the space.

For SI: 1 inch = 25.4 mm, 1 foot = 304.8 mm.

Bathrooms
Sect. 3801.6

Hallways
Sect. 3801.10

B-583

Heating, air-conditioning and refrigeration equipment, other than evaporative coolers, shall be provided with a convenience outlet installed within 25 feet of the equipment. To be used during maintenance or repair activities, the outlet must be located on the same level as the equipment.

Topic: Required GFCI Protection   **Category:** Power and Lighting Distribution

Reference: IRC E3802.1   E3802.7   **Subject:** Ground-fault Protection

Code Text: *(Summary) All 125-volt, single-phase, 15- and 20-ampere receptacles installed in bathrooms; garages; grade-level portions of unfinished accessory buildings used for storage or work areas; outdoors; in a crawl space at or below grade; and unfinished basements shall have ground-fault circuit-interrupter protection.* See exceptions. *All 125-volt, single-phase, 15- and 20-ampere receptacles that serve kitchen countertop surfaces, and all 125-volt, single-phase, 15- and 20-ampere receptacles that are located within 6 feet of a laundry, utility or wet bar sink, shall have ground-fault circuit-interrupter protection.*

Discussion and Commentary: Where there is an unbalance between the ungrounded "hot" conductor and the grounded "neutral" conductor, the purpose of a ground-fault circuit interrupter (GFCI) is to trip the circuit off. Where the unbalance is caused by human contact, this trip prevents serious injury or possibly death.

For SI: 1 foot = 304.8 mm.

B-962

The purpose of the GFCI is to recognize an unbalanced load and shut off the circuit, but it does not reduce the magnitude of the ground-fault current. A very strong, but brief, shock is still encountered by a person during the time period the GFCI device needs to recognize the unbalance and trip.

Code Text: *Nonmetallic boxes shall be used only with nonmetallic-sheathed cable, cabled wiring methods, flexible cords and nonmetallic raceways. See exceptions for special bonding conditions. Where nonmetallic-sheathed cable is used, the cable assembly, including the sheath, shall extend into the box not less than $^1/_4$ inch (6.4 mm) through a nonmetallic-sheathed cable knockout opening. Where nonmetallic-sheathed cable is used with boxes not larger than a nominal size of $2^1/_4$ inches by 4 inches (57 mm by 102 mm) mounted in walls or ceilings, and where the cable is fastened within 8 inches (203 mm) of the box measured along the sheath, and where the sheath extends through a cable knockout not less than $^1/_4$ inch (6.4 mm), securing the cable to the box shall not be required.*

Discussion and Commentary: To provide adequate insulation for the conductors up and into the box, the sheathing must extend into the box a limited distance.

Secured within 8 in. of box

Nonmetallic sheathed cable

Cable to extend at least $^1/_4$ in. into box

Cable need not be secured to box

For SI: 1 inch = 25.4 mm.

Max $2^1/_4$ in. x 4 in. nonmetallic box

B-574b

It is important that boxes located in potentially wet or damp locations be installed in such a manner as to prevent the entry and accumulation of moisture. Where installed in such areas, the boxes must be listed for use in wet locations.

Code Text: *Cord-connected luminaires, chain-, cable-, or cord-suspended luminaires, lighting track, pendants, and ceiling-suspended (paddle) fans shall not have any parts located within a zone measured 3 feet (914 mm) horizontally and 8 feet (2438 mm) vertically from the top of a bathtub rim or shower stall threshold. This zone is all encompassing and includes the zone directly over the tub or shower. Luminaires located in this zone shall be listed for damp locations and where subject to shower spray, shall be listed for wet locations.*

Discussion and Commentary: In most homes, a bathroom ceiling is not high enough to permit any fixture other than a surface-mounted or recessed-type fixture above a bathtub or shower. Fixtures such as chain-hung or pendant types where a cord is exposed above a bathtub or shower could be a safety hazard, and there is more of a danger of electrocution where water is present.

Luminaries in bathtub and shower

For SI: 1 foot = 304.8 mm

Although specific luminaires are prohibited in the tub and shower zone, other types, such as those that are surface-mounted, are acceptable when listed for damp locations. A listing recognizing the allowed use of the luminaire in a wet location is mandated where the luminaire will be subjected to shower spray.

Code Text: *The types of luminaires installed in clothes closets shall limited to surface-mounted or recessed incandescent luminaires with completely enclosed lamps, and surface-mounted or recessed fluorescent luminaires. Surface-mounted incandescent luminaires shall be installed on the wall above the door or on the ceiling, provided there is a minimum clearance of 12 inches (305 mm) (6 inches for surface-mounted fluorescent luminaires) between the fixture and the nearest point of a storage space. Recessed incandescent luminaires with a completely enclosed lamp (and recessed fluorescent luminaires) shall be installed in the wall or on the ceiling provided there is a minimum clearance of 6 inches (152 mm) between the luminaire and the nearest point of a storage space.*

Discussion and Commentary: It is quite probable that a clothes closet is the most common place in a house where a light source will be located adjacent to combustible materials. Therefore, various clearances are set forth based on the type of luminaire installed.

For SI: 1 inch = 25.4 mm.

B-586

It is impossible to fully regulate the use of a space once the dwelling unit is occupied; therefore, the code defines for its purposes the extent of a "storage space." It is bounded by the sides and back of the closet walls and extends to a height of 6 feet or the highest clothes-hanging rod.

Code Text: *Luminaires and lampholders shall be securely supported. A luminaire that weighs more than 6 pounds (2.72 kg) or exceeds 16 inches (406 mm) in any dimension shall not be supported by the screw shell of a lampholder. Per Section 3805.7, Outlet boxes or fittings . . . shall be permitted to support luminaires weighing 50 pounds (23 kg) or less. A luminaire that weighs more than 50 pounds (23 kg) shall be supported independently of an outlet box except where the outlet box is listed for the weight to be supported. A recessed luminaire that is not identified for contact with insulation shall have all recessed parts spaced at least 0.5 inch (12.7 mm) from combustible materials.*

Discussion and Commentary: Lampholders and luminaires should only be carried by the lampholder screw shell when they are of limited weight. The screw shell is not designed as a supporting element beyond supporting a bulb and simple shade or diffuser.

B-587

Where a luminaire exceeds 50 pounds, such as a large chandelier or similar decorative element, it cannot be adequately supported by an outlet box. Therefore, it is necessary to provide an independent means of support, typically addressed in the luminaire's installation instructions.

Topic: Support of Ceiling-suspended Paddle Fans **Category:** Appliance Installation
Reference: IRC E4001.6, E3805.9 **Subject:** General

Code Text: *Ceiling-suspended fans (paddle) shall be supported independently of an outlet box or by a listed outlet box or outlet box system identified for the use and installed in accordance with Section E3805.9. Outlet boxes and outlet box systems used as the sole support of ceiling-suspended fans (paddle) shall be marked by their manufacturer as suitable for this purpose and shall not support ceiling-suspended fans (paddle) that weigh more than 70 pounds (32 kg). For outlet boxes and outlet box systems designed to support ceiling-suspended fans (paddle) that weigh more than 35 pounds (16 kg), the required marking shall include the maximum weight to be supported.*

Discussion and Commentary: Ceiling fans can produce a potentially hazardous condition if not installed correctly. The combination of fan weight and movement of the fan creates a condition that is specifically addressed by the code. It is important to consider both the weight of the fan and any light fixture or other accessory when determining the requirements.

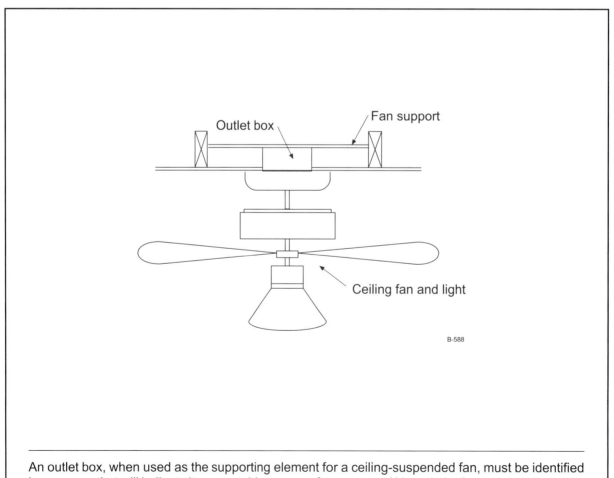

An outlet box, when used as the supporting element for a ceiling-suspended fan, must be identified in a manner that will indicate its acceptable use as a fan support. Although the fan weight is limited to 70 pounds, any fan that weighs over 35 pounds must utilize an outlet box that is specifically marked to indicate the maximum weight that it can support.

Topic: Receptacle Outlets

Category: Swimming Pools

Reference: IRC E4103.1

Subject: Equipment Location and Clearances

Code Text: *Receptacles that provide power for water-pump motors or other loads directly related to the circulation and sanitation system shall be permitted to be located between 5 feet and 10 feet (1524 mm and 3048 mm) from the inside walls of pools and outdoor spas and hot tubs. Other receptacles on the property shall be located not less than 10 feet (3048 mm) from the inside walls of pools and outdoor spas and hot tubs except where permitted by Section E4103.1.3 (restricted space). At least one 125-volt 15- or 20-ampere receptacle supplied by a general-purpose branch circuit shall be located a minimum of 10 feet (3048 mm) from and not more than 20 feet (6096 mm) from the inside wall of pools and outdoor spas and hot tubs except as permitted by Section E4103.1.3.*

Discussion and Commentary: Convenience receptacles must be located at least 10 feet from both indoor and outdoor swimming pools, as well as outdoor spas and hot tubs. At least one outlet shall be provided, but it cannot be located more than 20 feet from the pool edge. In those cases where the pool is within 10 feet of a dwelling and the lot dimensions do not permit the required 10-foot separation, a single receptacle outlet is allowed if it is not located within 5 feet of the pool's inside wall.

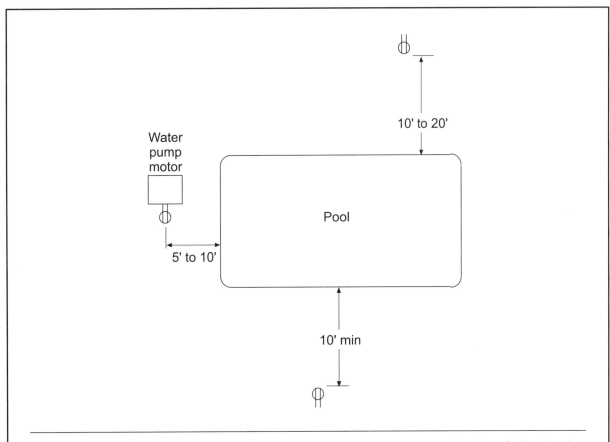

An accesible disconnecting means is required to be readily accessible and in sight of all swimming pools, spas and hot tubs.

Code Text: *Receptacles shall be located not less than 5 feet (1524 mm) from the inside walls of indoor spas and hot tubs. A minimum of one 125-volt receptacle shall be located between 5 feet (1524 mm) and 10 feet (3048 mm) from the inside walls of indoor spas or hot tubs. All 125 volt receptacles rated 30 amperes or less and located within 10 feet (3048 mm) of the inside walls of spas and hot tubs installed indoors shall be protected by ground-fault circuit-interrupters.*

Discussion and Commentary: The requirements applicable to receptacles adjacent to indoor hot tubs and spas are not as restrictive as those for both indoor and outdoor swimming pools, as well as outdoor hot tubs and spas. The limited space available in most indoor spa and hot tub areas would make it impractical to apply the provisions mandated for outdoor facilities.

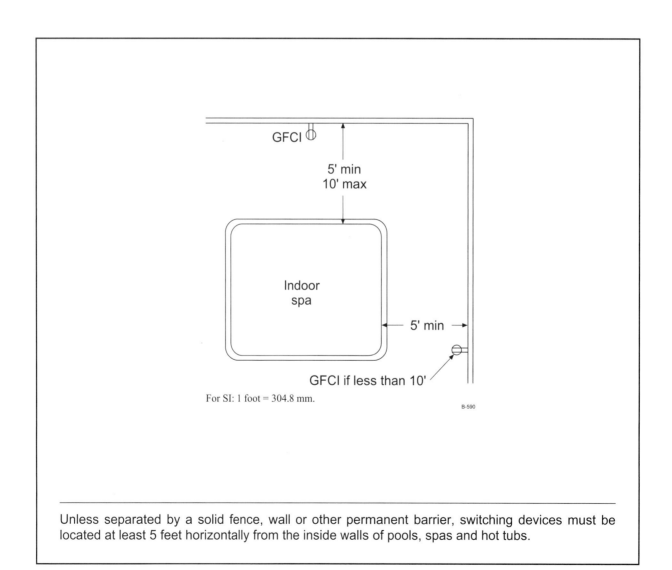

For SI: 1 foot = 304.8 mm.

B-590

Unless separated by a solid fence, wall or other permanent barrier, switching devices must be located at least 5 feet horizontally from the inside walls of pools, spas and hot tubs.

Topic: Luminaires and Ceiling Fans
Reference: IRC E4103.4

Category: Swimming Pools
Subject: Equipment Location and Clearances

Code Text: *In outdoor pool, outdoor spas and outdoor hot tubs areas, luminaires, lighting outlets and ceiling-suspended paddle fans shall not be installed over the pool or over the area extending 5 feet (1524 mm) horizontally from the inside walls of a pool except where no part of the luminaire or ceiling-suspended paddle fan is less than 12 feet (3658 mm) above the maximum water level.*

Discussion and Commentary: The code mandates clearances for luminaires in connection with both outdoor and indoor swimming pools. Typically the provisions for outdoor pools are consistent with those located indoors; however, a reduction in the luminaire or paddle fan height requirements to 7 feet, 6 inches is permitted for indoor installations. To provide this reduced clearance, the luminaires must be of the totally-enclosed type, and a GFCI must be installed in the branch circuit serving the fan or luminaire.

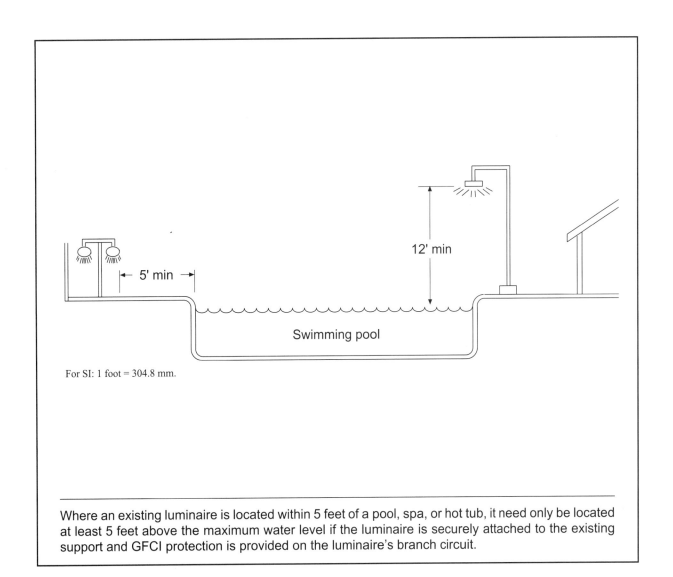

For SI: 1 foot = 304.8 mm.

Where an existing luminaire is located within 5 feet of a pool, spa, or hot tub, it need only be located at least 5 feet above the maximum water level if the luminaire is securely attached to the existing support and GFCI protection is provided on the luminaire's branch circuit.

Code Text: *Except where installed with the clearances specified in Table E4103.5, the following parts of pools and outdoor spas and hot tubs shall not be placed under existing service-drop conductors or any other open overhead wiring; nor shall such wiring be installed above the following: 1) pools and the areas extending 10 feet (3048 mm) horizontally from the inside of the walls of the pool, 2) diving structures, or 3) observation stands, towers, and platforms.*

Discussion and Commentary: When installing a new swimming pool, spa or hot tub, it is important to take into account the position of any overhead service-drop conductors or wiring. If the location of the pool, spa or hot tub brings it within the clearances required by the code, the conductors or wiring must be rerouted to bring the installation into compliance.

TABLE E4103.5
OVERHEAD CONDUCTOR CLEARANCES

	INSULATED SUPPLY OR SERVICE DROP CABLES, 0-750 VOLTS TO GROUND, SUPPORTED ON AND CABLED TOGETHER WITH AN EFFECTIVELY GROUNDED BARE MESSENGER OR EFFECTIVELY GROUNDED NEUTRAL CONDUCTOR (feet)	ALL OTHER SUPPLY OR SERVICE DROP CONDUCTORS (feet)	
		Voltage to ground	
		0-15 kV	Greater than 15 to 50 kV
A. Clearance in any direction to the water level, edge of water surface, base of diving platform, or permanently-anchored raft	22.5	25	27
B. Clearance in any direction to the diving platform	14.5	17	18

For SI: 1 foot = 304.8 mm.

A minimum 10-foot (3048 mm) clearance is required for overhead communications conductors, coaxial cables and their supports that are owned, operated and maintained by a utility. The separation must be maintained to pools, diving boards, observation stands and similar structures.

Study Session 18
IRC Chapters 37, 38, 39, 40 and 41

1. Electrical nonmetallic tubing is an allowable wiring method in all but which one of the following applications?

    a. feeders   b. branch circuits

    c. embedded in masonry   (d.) direct burial

Reference T E3701.4

2. Unless installed parallel to framing members, NM wiring run in an attic accessed by a portable ladder shall be protected from damage where located within _____ of the nearest edge of the attic entrance.

    (a.) 6 feet   b. 7 feet

    c. 8 feet   d. 10 feet

Reference E3702.2.1

3. Type NM wiring shall be supported a maximum of __4.5__ on center.

    a. 3 feet   (b.) 4.5 feet

    c. 6 feet   d. 12 feet

Reference T E3702.1

4. In general, direct burial cable installed below a driveway serving a dwelling unit shall have a minimum burial depth of _____.

   a. 6 inches   b. 12 inches

   c. 18 inches   d. 24 inches

   Reference _E3703.1_

5. Direct buried conductors and cables emerging from the ground shall be protected by enclosures or raceways extending to a minimum height of __8__ above finished grade or the point of entrance to the building.

   a. 7 feet   b. 8 feet

   c. 10 feet   d. 12 feet

   Reference _E3703.3_

6. Receptacle outlets shall be installed so that all points along the floor line in any wall space are a maximum of __6''__ from an outlet in that space when measured horizontally.

   a. 4 feet   b. 6 feet

   c. 8 feet   d. 12 feet

   Reference _E3801.2.1_

7. An unbroken wall space having a minimum width of __24''__ shall be considered a wall space for the purpose of determining receptacle outlet distribution.

   a. 12 inches   b. 18 inches

   c. 24 inches   d. 48 inches

   Reference _E3801.2.2 #1_

8. Floor receptacle outlets cannot be counted as part of the required number of receptacle outlets unless located within __18__ from the wall.

   a. 4 inches   b. 6 inches

   c. 12 inches   d. 18 inches

   Reference _E3801.3_

9. At a kitchen counter, countertop receptacle outlets shall be installed with a maximum horizontal distance of ___24___ from any point along the wall line to an outlet.

   a. 12 inches   b. 18 inches

   (c.) 24 inches   d. 36 inches

   Reference _E3501.4.1_

10. In a bathroom, at least one receptacle outlet shall be located a maximum of ___30"___ from the outside edge of each lavatory.

    a. 18 inches   b. 24 inches

    (c.) 30 inches   d. 36 inches

    Reference _E3501.6_

11. A hallway a minimum of ___10ft___ in length shall be provided with at least one receptacle outlet.

    a. 6 feet   b. 8 feet

    (c.) 10 feet   d. 12 feet

    Reference _E3501.10_

12. All 125-volt, single-phase, 20-ampere receptacles shall be provided with ground-fault circuit-interrupter protection where located within ___6"___ of the outside edge of a laundry sink.

    a. 2 feet   b. 3 feet

    c. 4 feet   (d.) 6 feet

    Reference _E3502.7_

13. Where nonmetallic-sheathed cable is used, the cable assembly including the sheath shall extend into the box a minimum of ___1/4"___ through a nonmetallic-sheathed cable knockout opening.

    (a.) $^1/_4$ inch   b. $^3/_8$ inch

    c. $^1/_2$ inch   d. $^3/_4$ inch

    Reference _E3805.8.1_

14. Electrical outlet boxes shall have a minimum internal depth of _//2'_ .

    (a) ¹/₂ inch   b ⁵/₈ inch

    c. ³/₄ inch   d. 1 inch

    Reference _E3805.1_

15. What is the maximum number of No. 12 conductors permitted for a 4-inch by 1¹/₂-inch standard round box not marked with a cubic-inch capacity?

    a. 4   (b.) 6

    c. 9   d. 10

    Reference _T E3805.12.1_

16. In a wall constructed of wood or other combustible material, outlet boxes shall be installed so that the front edge of the box is set back a maximum of _1/4'_ from the finished surface.

    a. 0 inches; (the box must be flush or project outward)

    b. ¹/₈ inch

    (c.) ¹/₄ inch

    d. /₈ inch

    Reference _E3806.5_

17. All switches shall be installed so that the center of the grip of the operating handle, in its highest position, shall be located a maximum of _6'7''_ above the floor or working platform.

    a. 6 feet, 0 inches   b. 6 feet, 6 inches

    (c.) 6 feet, 7 inches   d. 6 feet, 9 inches

    Reference _E3901.6_

18. What is the required receptacle rating for a 20-ampere branch circuit supplying two or more receptacles or outlets?

    a. 15 amperes only   b. 20 amperes only

    c. 30 amperes only   (d.) 15 amperes or 20 amperes

    Reference _Table E3902.1.2_

19. Cord-connected luminaires, lighting track and pendants shall be located a minimum of _____3_____ horizontally and _____5_____ vertically from the top of a bathtub rim or shower stall threshold.

    a. 2 feet, 7 feet

    b. 2 feet, 8 feet

    c. 3 feet, 7 feet

    d. 3 feet, 8 feet

    Reference _E3903.10_

20. Where installed in a storage closet, a recessed fluorescent luminaire shall be located so as to provide a minimum clearance of _____6"_____ between the fixture and the nearest point of the storage space.

    a. 3 inches

    b. 6 inches

    c. 8 inches

    d. 12 inches

    Reference _E3903.1.1.#4_

21. A luminaire that weighs more than _____6_____ or exceeds _____16"_____ in any dimension shall not be supported by the screw shell of a lampholder.

    a. 4 pounds, 14 inches

    b. 4 pounds, 16 inches

    c. 6 pounds, 14 inches

    d. 6 pounds, 16 inches

    Reference _E3904.4_

22. Ceiling-suspended paddle fans weighing a maximum of _____70 lbs_____ are permitted to be supported by outlet boxes identified for such use.

    a. 25 pounds

    b. 35 pounds

    c. 42 pounds

    d. 70 pounds

    Reference _E3905.9_

23. 125-volt receptacles located a minimum of _____ from the inside walls of a swimming pool need not be protected by a ground-fault circuit-interrupter.

    a. 10 feet

    b. 15 feet

    c. 20 feet

    d. 30 feet

    Reference _E4103.1.4_

24. Receptacles shall be located a minimum of ___5'___ from the inside walls of indoor spas and hot tubs.

    (a.) 5 feet   b.   6 feet

    c.   10 feet   d.   12 feet

    Reference _E4103.4.5_

25. Where underground wiring is installed within 5 feet of the inside walls of a swimming pool, a minimum burial depth of ___6'___ is required for intermediate metal conduit.

    a.   (6) inches   b.   12 inches

    c.   18 inches   d.   30 inches

    Reference _T E4103.6_

26. Where three or more conductors of minimum size ___T___ AWG are run at angles with joists in unfinished basements, additional protection is not required if the cable assembly is attached directly to the bottom of the joists.

    a.   6   (b.) 8

    c.   10   d.   12

    Reference _E3702.4_

27. Underground service conductors not encased in concrete and buried 18 inches or more below grade shall have their location identified by a warning ribbon placed a minimum of _12'_ above the underground installation.

    a.   4   b   .6

    c.   8   (d.) 12

    Reference _E3703.2_

28. All branch circuits that supply 125-volt, single-phase, 15- and 20-ampere outlets installed in dwelling unit _B_ shall be protected by an arc-fault circuit interrupter.

    a   bathrooms   (b.) bedrooms

    c.   kitchens   d.   garages

    Reference _E3802.12_

29. A wall-mounted luminaire weighing a maximum of _6#_ pounds is permitted to be supported on boxes other than those specifically designed for luminaires, provided the luminaire is secured to the box with at least two No. 6 or larger screws.

a. 6   b. 10

c. 14   d. 18

Reference _E3805.6 EXCO._

30. Lighting track shall be located a minimum of _8'_ above the finished floor unless protected from physical damage or the track operates at less than 30 volts rms open-circuit voltage.

a. 5 feet, 0 inches   b. 6 feet, 0 inches

c. 6 feet, 4 inches   d. 6 feet, 8 inches

Reference _E3905.4 #7_

31. The required bathroom receptacle outlet is not required to be mounted on a wall if it is installed on the side or face of the lavatory basin cabinet a maximum of _12"_ inches below the countertop.

a. 6   b. 8

c. 12   d. 18

Reference _E3801.6 EXC_

32. For other than evaporative coolers, a receptacle outlet shall be installed for the servicing of HVAC equipment a maximum of _3_ feet from such equipment.

a. 20   b. 25

c. 40   d. 50

Reference _E3801.11_

33. Unless the outlet box is listed for the weight to be supported, an outlet box may support a luminaire having a maximum weight of _50#_ pounds.

a. 28   b. 35

c. 50   d. 70

Reference _E3805.7_

34. Drywall surfaces that are broken or incomplete shall be repaired so that gaps or open spaces at the edge of a panelboard employing a flush-type cover are a maximum of _____ 1/8" inch in size.

    a. 0 (no gaps permitted)   b. $^1/_{16}$

    c. $^1/_8$ d. $^1/_4$

Reference _E 3607.4_

35. A flexible cord for a cord-and-plug-connected range hood shall have a minimum length of _____ 18" inches and a maximum length of _____ 36" inches.

    a. 18, 36   b. 18, 48

    c. 30, 36   d. 36, 48

Reference _T4001.3_

Answer Keys

Study Session 1
2006 *International Residential Code*

1.	c	Sec. R101.2
2.	b	Sec. R102.1
3.	c	Sec. R102.5
4.	a	Sec. R103.1
5.	d	Sec. R104.1
6.	b	Sec. R104.9.1
7.	b	Sec. R104.10
8.	a	Sec. R104.10.1
9.	c	Sec. R104.11.1
10.	b	Sec. R105.2, #B1
11.	b	Sec. R105.2, #B7
12.	b	Sec. R105.5
13.	a	Sec. R105.7
14.	a	Sec. R106.3.1
15.	b	Sec. R106.5
16.	d	Sec. R109.1
17.	d	Sec. R109.3
18.	d	Sec. R110.3
19.	d	Sec. R110.4
20.	b	Sec. R112.2
21.	c	Sec. R112.3
22.	b	Sec. R113.2
23.	c	Sec. R114.1
24.	d	Chapter 43
25.	a	Chapter 43
26.	d	Sec. R101.3
27.	d	Sec. R105.2, #B9
28.	d	Sec. R105.2, #M7
29.	a	Sec. R109.3
30.	b	Sec. R112.1

31.	c	Sec. R102.4, Exception
32.	c	Sec. R104.7
33.	b	Sec. R105.6
34.	d	Sec. R107.1
35.	d	Sec. R110.1, Exception 2

Study Session 2
2006 *International Residential Code*

1.	a	Figure R301.2(3)	31.	c	Table R301.2.1.2, Note c
2.	d	Figure R301.2(4)	32.	d	Table R301.5, Note i
3.	d	Figure R301.2(5), Note	33.	c	Table R301.5, Note h
4.	c	Figure R301.2(6)	34.	b	Table R302.1
5.	c	Figure R301.1(7)	35.	b	Table R302.1
6.	a	Table R301.2.1.2			
7.	a	Table R301.2.1.3			
8.	b	Sec. R301.2.1.4, #2			
9.	b	Table R301.2.2.1.1			
10.	b	Sec. R301.2.2.3.1, R703.7			
11.	c	Sec. R301.2.2.2.1, #5			
12.	d	Sec. R301.2.2.2.2, #4			
13.	c	Sec. R301.2.3			
14.	c	Table R301.5			
15.	c	Table R301.5, Note b			
16.	c	Table R301.5			
17.	c	Table R301.6			
18.	d	Table R301.7			
19.	b	Table R301.7			
20.	c	Table R302.1			
21.	c	Sec. R302.1, Table R302.1			
22.	b	Sec. R302.1, Exception 1, R105.2, #B1			
23.	d	Table R302.2			
24.	d	Table R302.1, R317.3.1.2			
25.	a	Sec. R302.1, Exception 2			
26.	c	Sec. R301.2.1.4, #3			
27.	b	Sec. R301.2.2, Exception			
28.	b	Sec. R301.2.2.2.1			
29.	a	Sec. R301.3, #2			
30.	a	Table R301.5, Note f			

Study Session 3
2006 *International Residential Code*

1.	c	Sec. R303.1	31.	a	Sec. R303.3
2.	a	Sec. R303.1	32.	d	Sec. R303.6.1
3.	a	Sec. R303.1, Exception 1	33.	a	Table R308.3
4.	d	Sec. R303.3, Exception	34.	c	Sec. R310.1, Exception
5.	b	Sec. R304.1	35.	b	Sec. R310.5
6.	d	Sec. R304.4			
7.	b	Sec. R304.3			
8.	b	Sec. R305.1			
9.	a	Sec. R305.1, Exception 2			
10.	a	Figure R307.1			
11.	a	Figure R307.1			
12.	b	Sec. R307.2			
13.	a	Sec. R308.4, #6			
14.	c	Sec. R308.4, #9			
15.	a	Sec. R308.6.1			
16.	b	Sec. R308.6.8			
17.	b	Sec. R309.1			
18.	c	Sec. R309.1.1			
19.	a	Sec. R309.2			
20.	c	Sec. R310.1			
21.	c	Sec. R310.1.1, Exception			
22.	a	Sec. R310.1.2, R310.1.3			
23.	d	Sec. R310.2			
24.	d	Sec. R309.2			
25.	c	Sec. R308.6.2, #1			
26.	c	Sec. R303.4.1			
27	a	Sec. R303.6			
28.	c	Sec. R305.1, #4			
29.	c	Sec. R308.4, Exception 4			
30.	a	Sec. R308.4, Exception 1			

Study Session 4

2006 *International Residential Code*

1.	b	Sec. R311.3	31.	c	Sec. R311.4.3, Exception 2
2.	c	Sec. R311.4.3	32.	b	Sec. R311.5.3.3
3.	d	Sec. R311.6.1	33.	d	Sec. R311.5.4
4.	c	Sec. R311.6.3	34.	c	Sec. R312.2, Exception 2
5.	b	Sec. R311.6.2, #2	35.	a	Sec. R317.1, Exception 2
6.	c	Sec. R311.5.1			
7.	c	Sec. R311.5.1			
8.	c	Sec. R311.5.3.1, R311.5.3.2			
9.	b	Sec. R311.5.3.1			
10.	c	Sec. R311.5.3.3, Exception 1			
11.	b	Sec. R311.5.3.3			
12.	b	Sec. R311.5.2			
13.	d	Sec. R311.5.8.1			
14.	c	Sec. R311.5.6.1			
15.	b	Sec. R311.5.6.2			
16.	a	Sec. R311.5.6.3, #1			
17.	c	Sec. R312.1			
18.	b	Sec. R312.2			
19.	b	Sec. R313.2			
20.	b	Sec. R314.4			
21.	c	Sec. R315.1			
22.	c	Sec. R317.1			
23.	b	Sec. R317.2, Exception			
24.	b	Sec. R319.1, #2			
25.	d	Sec. R324.1.3			
26.	a	Sec. R311.2.2			
27.	c	Sec. R314.5.11, #2			
28.	d	Sec. R311.5.6.3, #2			
29.	c	Sec. R311.5.6			
30.	d	Sec. R311.4.2			

Study Session 5
2006 *International Residential Code*

1.	b	Sec. R401.3	31.	a.	Sec. R403.1.6, Exception 2
2.	b	Table R401.4.1	32.	d	Sec. R404.4.5
3.	b	Table R402.2	33.	a	Sec. R406.1, #3
4.	c	Table R402.2, Note d	34.	a	Sec. R407.3
5.	c	Table R403.1	35.	c	Sec. R408.3, #1
6.	b	Sec. R403.1.1			
7.	b	Sec. R403.1.1			
8.	b	Figure R403.1(2)			
9.	a	Sec. R403.1.3.1			
10.	d	Sec. R403.1.4			
11.	b	Sec. R403.1.5			
12.	c	Sec. R403.1.6			
13.	a	Sec. R403.1.6			
14.	d	Sec. R403.1.7.3			
15.	b	Sec. R403.3			
16.	c	Table R403.3			
17.	d	Table R404.1.1(1)			
18.	d	Table R404.1.1(5)			
19.	a	Sec. R404.1.6			
20.	b	Sec. R404.1.7, Exception			
21.	d	Sec. R404.2.2			
22.	b	Sec. R404.2.3			
23.	d	Sec. R405.1			
24.	c	Sec. R408.1			
25.	b	Sec. R408.4			
26.	d	Sec. R403.1.4.1, Exception 1			
27.	c	Sec. R403.2			
28.	a	Sec. R403.3.2			
29.	d	Sec. R406.2, #5			
30.	b	Sec. R408.2			

Study Session 6
2006 *International Residential Code*

1.	c	Sec. R502.3.1	31.	c	Sec. R502.6
2.	b	Table R502.3.1(1)	32.	d	Sec. R502.8.1
3.	a	Table R502.3.1(2)	33.	b	Sec. R502.8.1
4.	d	Table R502.5(1)	34.	c	Table R503.2.1.1(1)
5.	b	Table R502.5(1)	35.	c	Sec. R506.2.4
6.	b	Table R502.5(2)			
7.	c	Sec. R502.4			
8.	b	Sec. R502.6			
9.	a	Sec. R502.6.1			
10.	d	Sec. R502.7.1			
11.	c	Sec. R502.8.1			
12.	d	Sec. R502.8.1			
13.	d	Sec. R502.10			
14.	c	Sec. R502.12			
15.	b	Sec. R502.12.1			
16.	d	Table R503.2.1.1(1), Note h			
17.	b	Table R503.2.1.1(2)			
18.	a	Sec. R505.1.1, R301.3, #2			
19.	c	Sec. R505.2.4			
20.	b	Table R505.3.1(2)			
21.	c	Table R505.3.2(1)			
22.	b	Sec. R506.1			
23.	c	Sec. R506.2.1			
24.	a	Sec. R506.2.2, Exception, Table R405.1			
25.	c	Sec. R506.2.3			
26.	a	Table R502.3.3(1)			
27.	c	Table R502.3.3(2)			
28.	d	Sec. R502.4			
29.	b	Sec. R502.10			
30.	b	Sec. R506.2.2			

Study Session 7
2006 *International Residential Code*

1.	c	Table R602.3(5)	31.	d	Sec. R602.6.1
2.	b	Table R602.3.1	32.	c	Sec. R602.11.1
3.	a	Sec. R602.3.2	33.	d	Table R611.2
4.	b	Sec. R602.3.2, Exception	34.	a	Sec. R613.2
5.	b	Table R602.3(3)	35.	d	Sec. R703.6.3
6.	c	Sec. R602.6, #1			
7.	c	Sec. R602.6, #2			
8.	d	Sec. R602.6, #2			
9.	d	Table R602.7.2			
10.	c	Sec. R602.8.1			
11.	a	Sec. R602.8.1.1			
12.	d	Sec. R602.9			
13.	a	Table R602.10.1			
14.	d	Sec. R602.10.1			
15.	b	Sec. R603.1.2			
16.	d	Sec. R603.2, #6			
17.	b	Sec. R606.8			
18.	a	Figure R606.11(1)			
19.	d	Table R702.3.5			
20.	b	Sec. R702.3.6			
21.	a	Sec. R702.3.8			
22.	d	Table R703.4			
23.	a	Sec. R703.7.4.1, Exception			
24.	c	Sec. R703.9			
25.	c	Table R703.7.3			
26.	a	Sec. R602.3.1, Exception 1			
27.	d	Table R602.3(1)			
28.	b	Table R602.3(2)			
29.	c	Sec. R602.6.1			
30.	b	Sec. R703.6.2.1			

Study Session 8

2006 *International Residential Code*

1.	b	Sec. R801.3		31.	a	Sec. R802.3
2.	a	Sec. R802.3		32.	a	Sec. R806.2
3.	b	Sec. R802.3.1		33.	a	Sec. R806.4, #3.2
4.	b	Sec. R802.3.2		34.	d	Sec. R905.2.7.1
5.	c	Table R802.4(1)		35.	c	Sec. R907.3 #4, R903.5.1
6.	a	Table R802.5.1(4)				
7.	a	Sec. R802.5.1				
8.	a	Sec. R802.6				
9.	c	Sec. R802.7.1				
10.	a	Sec. R802.9				
11.	c	Sec. R803.2.1, Table R503.2.1.1(1)				
12.	b	Tables R804.3.3(2), R804.3.3(1)				
13.	c	Sec. R804.3.3.2				
14.	a	Sec. R806.1				
15.	c	Sec. R807.1				
16.	c	Sec. R808.1				
17.	c	Sec. R902.1				
18.	d	Sec. R905.2.6				
19.	d	Sec. R905.2.7				
20.	a	Sec. R905.2.8.3				
21.	c	Sec. R905.3.6				
22.	d	Sec. R905.7.5				
23.	b	Table R905.7.5				
24.	c	Sec. R905.8.6				
25.	c	Sec. R905.8.8				
26.	d	Sec. R802.7.1				
27.	b	Sec. R802.9				
28.	b	Sec. R806.3				
29.	a	Sec. R807.1				
30.	c	Sec. R903.4.1				

Study Session 9
2006 *International Residential Code*

1.	d	Sec. R1003.2	31.	c	Sec. R1001.5.1
2.	a	Sec. R1003.6	32.	b	Sec. R1001.6, Exception
3.	d	Sec. R1003.9	33.	b	Sec. R1001.9.2
4.	c	Sec. R1003.12	34.	a	Sec. R1006.3
5.	b	Sec. R1003.13	35.	c	Sec. N1101.4.1
6.	a	Sec. R1003.17			
7.	c	Sec. R1003.18			
8.	b	Sec. R1001.2			
9.	d	Sec. R1001.3.1			
10.	c	Sec. R1001.5			
11.	b	Sec. R1001.5			
12.	c	Sec. R1001.7			
13.	d	Sec. R1001.9.1			
14.	d	Sec. R1001.10			
15.	b	Sec. R1001.9.2, Exception			
16.	d	Sec. R1001.11			
17.	d	Sec. R1001.11, Exception 4			
18.	c	Sec. R1001.11, Exception 4			
19.	d	Sec. R1004.1			
20.	b	Sec. R1006.4			
21.	a	Sec. N1101.4			
22.	a	Sec. N1101.6.1			
23.	b	Table N1102.1			
24.	c	Sec. N1102.4.2			
25.	b	Sec. N1103.2.1			
26.	b	Sec. R1003.2			
27.	d	Sec. R1003.9.1, #1			
28.	c	Table R1003.20			
29.	c	Sec. R1001.6			
30.	d	Sec. R1002.5			

Study Session 10
2006 *International Residential Code*

1.	b	Sec. M1305.1	31.	c	Sec. M1304.1.3, Exception 2
2.	b	Sec. M1305.1.1	32.	a	Sec. M1307.4.1.1
3.	a	Sec. M1305.1.1	33.	b	Sec. M1308.3
4.	a	Sec. M1305.1.2	34.	d	Sec. M1408.3, #4
5.	a	Sec. M1305.1.3	35.	b	Sec. M1411.3.1
6.	c	Sec. M1305.1.4.1			
7.	c	Sec. M1305.1.4.2			
8.	b	Sec. M1306.2			
9.	b	Table M1306.2			
10.	d	Sec. M1306.2.1, Table M1306.2			
11.	b	Sec. M1307.3			
12.	b	Sec. M1402.3			
13.	a	Sec. M1403.1			
14.	b	Sec. M1403.2			
15.	c	Sec. M1406.5			
16.	d	Sec. M1407.3			
17.	c	Sec. M1408.3, #3			
18.	a	Sec. M1408.4			
19.	a	Sec. M1408.5, #3			
20.	b	Sec. M1409.2, #2			
21.	c	Sec. M1411.2			
22.	b	Sec. M1411.3.1			
23.	b	Sec. M1411.3.1, #1			
24.	a	Sec. M1413.1			
25.	c	Sec. M1414.2			
26.	a	Sec. M1305.1.4			
27.	b	Sec. M1307.2			
28.	c	Sec. M1307.4.1			
29.	d	Sec. M1308.2			
30.	b	Sec. M1410.2			

Study Session 11
2006 *International Residential Code*

1.	b	Sec. M1502.5
2.	c	Sec. M1502.6
3.	d	Sec. M1502.6
4.	d	Sec. M1503.1
5.	a	Sec. M1503.2, Exception, #4
6.	b	Sec. M1505.1
7.	a	Sec. M1505.1
8.	d	Sec. M1601.1.1, #1
9.	c	Sec. M1601.1.1, #2
10.	b	Table M1601.1.1(2)
11.	a	Sec. M1601.1.1, #5
12.	c	Sec. M1601.1.1, #6
13.	b	Sec. M1601.1.2
14.	b	Sec. M1601.1.2
15.	a	Sec. M1601.2.1, #1
16.	a	Sec. M1601.2.1, #3
17.	b	Sec. M1601.2.2
18.	c	Sec. M1601.3.2
19.	d	Sec. M1601.4.2
20.	c	Sec. M1601.4.3
21.	d	Sec. M1602.2, #1
22.	a	Sec. M1602.2, #4
23.	a	Sec. M1602.3
24.	b	Sec. M1901.1
25.	d	Sec. M1902.4
26.	b	Table M1507.3
27.	c	Table M1507.3
28.	a	Sec. M1601.4.4
29.	b	Sec. M1601.4.5
30.	c	Sec. M1902.4

31.	b	Sec. M1502.2
32.	d	Sec. M1502.4
33.	a	Table M1601.1.1(1)
34.	c	Sec. M1601.3.1
35.	b	Sec. M1601.3.6

Study Session 12
2006 *International Residential Code*

1.	d	Sec. M1701.1.1
2.	c	Sec. M1701.4
3.	d	Sec. M1701.5
4.	a	Sec. M1701.5
5.	d	Sec. M1702.2
6.	c	Sec. M1702.2
7.	a	Sec. M1702.2
8.	d	Sec. M1703.2.1
9.	b	Sec. M1703.2.1
10.	a	Sec. M1703.2.1
11.	b	Sec. M1703.5
12.	c	Sec. M1803.3.1
13.	a	Sec. M1804.2.6, #5
14.	c	Sec. M1703.3, #2
15.	c	Sec. M1703.4
16.	b	Table M1803.2
17.	b	Sec. M1803.3
18.	c	Sec. M1803.3.2
19.	c	Table M1803.3.4
20.	d	Sec. M1804.2.3
21.	a	Sec. M1805.3
22.	c	Sec. M2003.1.1
23.	d	Sec. M2103.3
24.	b	Sec. M2203.3
25.	c	Sec. M2301.2.9
26.	d	Sec. M2005.2
27.	a	Table M2101.9
28.	d	Sec. M2101.10
29.	b	Sec. M2201.2
30.	a	Sec. M2204.3

31.	d	Sec. M1703.3, #3
32.	b	Sec. M1804.2.3
33.	d	Sec. M2002.4
34.	a	Sec. M2201.2.2
35.	c	Sec. M2203.4

Study Session 13
2006 *International Residential Code*

1.	d	Sec. G2406.2	31.	c	Table G2413.2
2.	c	Sec. G2408.2	32.	b	Sec. G2415.13
3.	b	Sec. G2408.3	33.	b	Sec. G2415.9.1
4.	d	Table G2409.2	34.	d	Sec. G2422.1.2.1
5.	b	Table G2409.2	35.	d	Sec. G2444.4
6.	a	Sec. G2412.5			
7.	a	Sec. G2413.6			
8.	d	Sec. G2413.6.1			
9.	c	Sec. G2414.5.2			
10.	b	Table G2414.9.2			
11.	d	Sec. G2415.1			
12.	d	Sec. G2415.5			
13.	b	Sec. G2415.9			
14.	a	Sec. G2415.13			
15.	c	Sec. G2415.14.3			
16.	a	Sec. G2416.2, #5			
17.	b	Sec. G2417.4.1			
18.	a	Sec. G2417.4.2			
19.	d	Sec. G2419.1			
20.	c	Sec. G2420.5			
21.	a	Sec. G2422.1.2.1			
22.	c	Table G2424.1			
23.	a	Sec. G2425.9			
24.	a	Sec. G2440.7			
25.	b	Sec. G2445.3			
26.	c	Sec. G2408.4			
27.	c	Sec. G2415.7			
28.	d	Sec. G2417.2			
29.	c	Table G2424.1			
30.	a	Sec. G2450.3			

Study Session 14
2006 *International Residential Code*

1.	d	Sec. P2503.4
2.	b	Sec. P2503.5.1, #2
3.	d	Sec. P2503.6
4.	d	Sec. P2603.2.1
5.	c	Sec. P2603.2.1
6.	d	Sec. P2603.5
7.	b	Sec. P2603.6
8.	c	Sec. P2603.6
9.	d	Sec. P2604.1
10.	b	Sec. P2604.3
11.	d	Table P2605.1
12.	a	Table P2605.1
13.	c	Sec. P2703.1
14.	a	Sec. P2704.1
15.	b	Sec. P2706.2
16.	b	Sec. P2706.2.1
17.	a	Sec. P2708.1
18.	d	Sec. P2709.1
19.	b	Sec. P2711.3
20.	b	Sec. P2712.4
21.	c	Sec. P2713.1
22.	b	Sec. P2714.1
23.	b	Sec. P2716.1
24.	c	Sec. P2719.1
25.	a	Sec. P2723.2
26.	b	Sec. P2503.5.2, #2.1
27.	a	Sec. P2503.8
28.	b	Sec. P2604.4
29.	d	Table P2608.4
30.	b	Sec. P2705.1, #5

31.	a	Sec. P2708.1.1
32.	c	Sec. P2713.3
33.	a	Sec. P2717.2
34.	b	Sec. P2720.1
35.	c	Sec. P2721.2

Study Session 15
2006 *International Residential Code*

1.	b	Sec. P2801.5.1
2.	d	Sec. P2801.5.2
3.	c	Sec. P2801.6
4.	d	Sec. P2803.3
5.	d	Sec. P2803.4
6.	b	Table P2902.3.1
7.	a	Sec. P2902.4.2
8.	c	Table P2903.1
9.	b	Table P2903.2
10.	d	Table P2903.2, Note a
11.	a	Sec. P2903.3, P2903.3.1
12.	c	Table P2903.6
13.	d	Table P2903.7
14.	d	Sec. P2904.2
15.	d	Sec. P2904.3
16.	b	Sec. P2904.5
17.	a	Sec. P2905.1
18.	a	Table P3004.1
19.	d	Sec. P3005.2.2
20.	a	Sec. P3005.2.9
21.	b	Sec. P3005.2.4
22.	a	Sec. P3005.2.5
23.	a	Sec. P3005.3
24.	a	Sec. P3005.4.1, #1
25.	d	Table P3005.4.1
26.	d	Sec. P2803.6.1, #10
27.	c	Table P2902.3
28.	c	Sec. P2904.4
29.	a	Sec. P3005.1.1, Exception
30.	a	Sec. P3007.1, Exception

31.	d	Sec. P2902.5.3
32.	c	Sec. P2904.3
33.	d	Sec. P3002.1
34.	c	Table P3005.1, Note a
35.	a	Sec. P3005.2.6

Study Session 16
2006 *International Residential Code*

1.	a	Sec. P3103.1
2.	b	Sec. P3103.2
3.	c	Sec. P3104.5
4.	c	Sec. P3103.1
5.	b	Sec. P3103.5
6.	d	Sec. P3104.4
7.	d	Sec. P3104.5
8.	c	Table P3105.1
9.	b	Sec. P3105.2
10.	d	Sec. P3105.3
11.	b	Sec. P3107.1
12.	b	Table P3107.3
13.	c	Table P3108.3
14.	c	Table P3109.4
15.	d	Sec. P3110.1
16.	a	Sec. P3111.1
17.	c	Sec. P3111.2.1
18.	b	Table P3111.3
19.	b	Sec. P3113.1
20.	c	Sec. P3114.4
21.	a	Sec. P3201.2
22.	d	Sec. P3201.6
23.	c	Sec. P3201.6, #2
24.	c	Table P3201.7
25.	c	Table P3201.7
26.	d	Sec. P3103.6
27.	c	Sec. P3110.3
28.	d	Sec. P3111.2
29.	d	Sec. P3112.2
30.	b	Sec. P3114.4

31.	c	Sec. P3109.3
32.	a	Sec. P3113.1
33.	c	Sec. P3113.4.2
34.	c	Table P3113.4.1
35.	b	Sec. P3201.6

Study Session 17
2006 *International Residential Code*

1.	a	Sec. E3304.8	31.	a	Sec. E3504.2.1, Exception 2
2.	b	Sec. E3305.2	32.	c	Sec. E3505.7
3.	a	Sec. E3305.2	33.	d	Sec. E3508.1.5
4.	c	Sec. E3305.3	34.	c	Table E3602.13
5.	b	Sec. E3306.3	35.	c	Sec. E3605.7, #5
6.	b	Sec. E3306.10.3			
7.	c	Sec. E3307.1			
8.	a	Sec. E3307.2			
9.	d	Sec. E3307.3			
10.	b	Table E3503.1			
11.	a	Sec. E3504.1			
12.	d	Sec. E3504.2.1			
13.	c	Sec. E3504.2.2, #2			
14.	b	Sec. E3504.3			
15.	b	Sec. E3508.1.1			
16.	b	Sec. E3508.3			
17.	c	Sec. E3508.1.4, #2			
18.	c	Sec. E3510.2			
19.	a	Sec. E3602.3			
20.	d	Sec. E3603.2			
21.	b	Sec. E3603.3, E3603.4			
22.	b	Table E3604.2(1)			
23.	b	Sec. E3604.4			
24.	b	Table E3605.5.3			
25.	c	Sec. E3605.6			
26.	b	Sec. E3304.5			
27.	a	Sec. E3305.4			
28.	b	Sec. E3306.10.3			
29.	d	Sec. E3501.6.2			
30.	b	Sec. E3508.1.1			

Study Session 18

2006 *International Residential Code*

1.	d	Tables E3701.2, E3701.4
2.	a	Sec. E3702.2.1
3.	b	Table E3702.1
4.	c	Table E3703.1
5.	b	Sec. E3703.3
6.	b	Sec. E3801.2.1
7.	c	Sec. E3801.2.2, #1
8.	d	Sec. E3801.2.3
9.	c	Sec. E3801.4.1
10.	d	Sec. E3801.6
11.	c	Sec. E3801.10
12.	d	Sec. E3802.7
13.	a	Sec. E3805.3.1
14.	a	Sec. E3805.4
15.	b	Table E3805.12.1
16.	a	Sec. E3806.5
17.	c	Sec. E3901.6
18.	d	Table E3902.1.2
19.	d	Sec. E3903.10
20.	b	Sec. E3903.11, #4
21.	d	Sec. E3904.4
22.	d	Sec. E4001.6, E3805.9
23.	c	Sec. E4103.1.4
24.	a	Sec. E4103.1.5
25.	a	Table E4103.6
26.	b	Sec. E3702.4
27.	d	Sec. E3703.2
28.	b	Sec. E3802.12
29.	a	Sec. E3805.6, Exception
30.	a	Sec. E3905.4, #8

31.	c	Sec. E3801.6, Exception
32.	b	Sec. E3801.11
33.	c	Sec. E3805.7
34.	c	Sec. E3807.4
35.	a	Table E4001.3